AA002880

Joint Varenna-Lausanne International Workshop 2014

Journal of Physics: Conference Series Volume 561

Varenna, Italy
1-5 September 2014

Editors:

Xavier Garbet **Olivier Sauter**

ISBN: 978-1-63439-670-7
ISSN: 1742-6588

Printed from e-media with permission by:

Curran Associates, Inc.
57 Morehouse Lane
Red Hook, NY 12571

Some format issues inherent in the e-media version may also appear in this print version.

Printed by Curran Associates, Inc. (2015)

For permission requests, please contact the Institute of Physics
at the address below.

Institute of Physics
Dirac House, Temple Back
Bristol BS1 6BE UK

Phone: 44 1 17 929 7481
Fax: 44 1 17 920 0979

techtracking@iop.org

Additional copies of this publication are available from:

Curran Associates, Inc.
57 Morehouse Lane
Red Hook, NY 12571 USA
Phone: 845-758-0400
Fax: 845-758-2634
Email: curran@proceedings.com
Web: www.proceedings.com

Table of contents

Volume 561

2014

Joint Varenna-Lausanne International Workshop 2014
1–5 September 2014, Varenna, Italy

Accepted papers received: 5 November 2014
Published online: 27 November 2014

Preface

Papers

Journal of Physics: Conference Series **561** (2014) 011001 doi:10.1088/1742-6596/561/1/011001

Preface

The 2014 joint Varenna-Lausanne international workshop on the theory of fusion plasmas was once more a great meeting. The programme covers a wide variety of topics, namely turbulence, MHD, edge physics and RF wave heating. The broad spectrum of skills involved in this meeting, from fundamental to applied physics, is striking. The works published in this special issue combine mathematics, numerics and physics at various levels – confirming the increasing integration of expertise in our community.

As an incentive to read this cluster, let us mention a few outstanding results. Several papers address fundamental issues in turbulent transport, in particular the dynamics of structures. It is quite remarkable that this subject is now mature enough to propose signatures that can be tested by measurements. Linear and non linear MHD was also at the forefront. Several works illustrate the increasing level of realistic description of a fusion device, in particular by implementing complicated wall geometries. Moreover some noticeable progress has been made in the understanding of reconnection processes in collisionless regimes. The activity on radio-frequency heating and current drive is well represented, driven by the future operation of W7-X, ITER, and DEMO on a longer time scale. Finally the development of innovative numerical techniques, an old tradition of the conference, has driven several nice articles.

The programme committee is traditionally keen in promoting young scientists. A number of senior scientists also attend the meeting on a regular basis, so that the attendance was nicely balanced. We believe that these efforts have been particularly fruitful this year. The number of young (and less young) faces was particularly impressive and this special issue illustrates this feature.

The success of the 2014 edition brings evidence that the joint Varenna-Lausanne is the right place for presenting the most recent results, exchanging ideas and preparing for the future.

The quality and size of the scientific production is illustrated by the 22 papers which appear in the present volume of Journal of Physics Conference Series – all peer reviewed. Let us mention another set of 19 papers to appear in Plasma Physics and Controlled Fusion. We hope the reader will enjoy this special issue and will find ideas for new bright achievements.

Xavier Garbet, Olivier Sauter
October 23, 2014

Peer review statement

All papers published in this volume of *Journal of Physics: Conference Series* have been peer reviewed through processes administered by the proceedings Editors. Reviews were conducted by expert referees to the professional and scientific standards expected of a proceedings journal published by IOP Publishing.

Simulations of fast-wave current drive in pulsed and steady-state DEMO designs

R. Bilato, M. Brambilla, E. Fable

Max-Planck Institut für Plasmaphysik, EURATOM Association, Garching, Germany

E-mail: roberto(dot)bilato(at)ipp(dot)mpg(dot)de

Abstract. Electromagnetic waves in the ion-cyclotron (IC) range of frequencies are presently investigated as possible current drive (CD) systems in fusion reactors. Among many physical and technical issues, an accurate description of radio-frequency (RF) power absorption by fusion-born alpha particles is of special importance, since RF heating of these particles is not only detrimental for the CD efficiency, but might worsen the operative conditions by increasing their prompt losses.

The capability of the full-wave `TORIC` code has been recently augmented to account for RF absorption by fusion-born alpha particles, calculated to all-orders in finite Larmor radius and with a realistic distribution function. Here, we present simulation with `TORIC` addressing the sensitivity of current drive efficiency on the design of a future reactor, in particular density and temperature profiles, magnetic field intensity, and plasma dimensions. For this purpose, we have investigated possible frequency windows for CD for two proposed versions of the DEMO reactor, namely its pulsed and its more ambitious steady-state design. The important role of the antenna for a realistic estimate of the CD efficiency is pointed out.

1. Introduction

The European Power Plant Physics and Technology (PPPT) task-force, established to identify the main physics and technological challenges for a fusion reactor [1], has recently focused its analysis on two designs of the demonstration reactor DEMO [2]. The "Early DEMO" (DEMO1), based on a "conservative extrapolation" of present knowledge and available technology, is meant to be a short- to medium-term target. As a consequence, DEMO1 is pulsed, with a duty cycle of about 2 hours, less than what is expected in a commercial reactor (about 8 hours). The "advanced DEMO" (DEMO2), by contrast, ambitiously aims to be compact and to operate in steady-state, two features which are economically appealing.

Both approaches require external systems for current drive (CD): in DEMO1 CD is foreseen to extend the pulse length to reduce the material fatigue and to increase the power-plant efficiency, whereas in DEMO2 CD should complement the bootstrap current during the recovery of the flux in the central solenoid. The CD systems used in past and present devices are based either on propagation and absorption of radio-frequency (RF) waves in the lower-hybrid (LH), electron-cyclotron (EC), and ion-cyclotron (IC) ranges of frequencies, or on the injection of energetic neutrals which ionize in the plasma (neutral-beam injection (NBI)) [3, 4]. Numerical simulations have been recently done to quantitatively address the capabilities of an ICRF system in driving current in DEMO [5, 6, 7]. A key issue in optimizing ICRF-CD is to enhance as much as possible the fraction of RF power directly deposited in the electrons by reducing competition by ion absorption. In particular, in an ignited plasma the fraction of RF power absorbed by fusion-born alpha particles can be substantial, and detrimental for the operation of the reactor by increasing their prompt losses and thus the wall thermal loading. Since this fraction depends sensitively on the distribution function of these alpha particles, it is important to model it accurately in the simulations, namely when evaluating the coefficients of the wave equation. In the full-wave TORIC code [8] the slowing-down distribution function [9] has been recently implemented for this purpose [16]. It has been shown that approximating it with an equivalent Maxwellian having the same density and energy content as the true distribution appreciably overestimate the RF power absorbed by energetic particles [16].

2. Constraints on fast wave current drive in ignited plasmas

In contrast with other CD schemes, CD by the fast wave (FW) in the IC range of frequencies in reactor plasmas does not require major technical developments for the sources. The most difficult element for a successful FWCD system design is, however, the launching structure. Even disregarding for the present all technical aspects, the physics of FWCD poses contradictory demands on the geometry of the antenna, obliging to compromises in its design. The sensitivity of the CD efficiency on the plasma parameters, moreover, will require a degree of flexibility not easily realizable.

As well-known, the CD efficiency depends in the first place on the ratio of the parallel phase velocity to the electron thermal speed ($x_e = \omega/k_\parallel v_{\rm the}$ with $v_{\rm the} = \sqrt{2T_e/m_e}$). If toroidicity is ignored, the CD efficiency is largest when either $x_e \gg 1$, which minimizes collisional scattering of resonant electrons, or $x_e \ll 1$, which minimize the power expenditure per momentum transferred. In a tokamak, however, the latter option is not viable: to avoid degradation by toroidal trapping of the resonant electrons x_e cannot be much lower than unity, and the degree of degradation depends on the radial and poloidal localization of the RF power directly absorbed by electrons. The condition $x_e \gtrsim 1$ translates into a constraint on the toroidal wave number n_φ,

$$n_\varphi \lesssim \frac{f_{\rm [MHz]}\, R_{\rm ant[m]}}{3\sqrt{T_{\rm e[keV]}}} \tag{1}$$

where we have approximated $k_\parallel$ with its average value $n_\varphi/R_{\rm ant}$, and $R_{\rm ant}$ is the antenna distance from the torus axis. Finally, it is important to take into account that $Z_{\rm eff}$ has an important influence on the CD efficiency. In TORIC it is implemented the Ehst-Karney parametric formula for the CD efficiency [10].

Since the final goal is to maximize the driven current per unit of total coupled power, in addition to optimize the CD efficiency by an appropriate choice of the wave parallel phase velocity, it is essential also to maximize the fraction of the launched power deposited in the electrons. This is not trivial, because in the IC frequency range electron damping of the fast wave is a much weaker absorption mechanism than IC absorption by ions. At the high temperatures of an ignited plasma this is true even for relatively high IC harmonic resonances, and, in particular, for energetic alpha particles during slowing down, in spite of their great dilution. Windows of predominant electron absorption, therefore, can be found only when all IC resonances are either outside the plasma, or, at least, close to the inner plasma edge, well screened from the antenna by a large plasma layer in which only electron absorption can occur. As the position of these frequency windows depends sensitively on the value of the confining magnetic field, and to a lesser extent on the poloidal location of the launcher, it will be important to match accurately the antenna working range to the final DEMO design.

To maximize the fraction of power deposited in the electrons, it is also clearly desirable to increase as much as possible the strength of electron absorption in any plasma layer. Since the number of resonant electrons decreases exponentially as x_e decreases much below unity, this

	B_0 [T]	R [m]	a [m]	I_p [MA]	q_0 / q_{95}	n_{e0}/ $\langle n_{e0} \rangle$ [10^{19}m^{-3}]	T_{e0} [keV]	Z_{eff}	P_{fus} [GW]
DEMO1 (pulsed)	6.3	9.3	2.6	16.8	1./3.	11.2/9.7	30.7	2.78	1.8
DEMO2 (steady-state)	4.3	8.2	3	19.85	1.5/3.8	9.2/7.6	37.1	3.26	2.1

Table 1. A few of the main parameters of DEMO1 and DEMO2 used in this analysis [2].

poses a demand opposite to equation (1) on the antenna periodicity. Thus, the optimization of the total driven current will be a compromise between good electron absorption and high CD efficiency, which have opposite dependencies on n_φ. This compromise is easier for power absorption by electrons peaked on axis, for which the trapping effects leading to condition (1) are less severe.

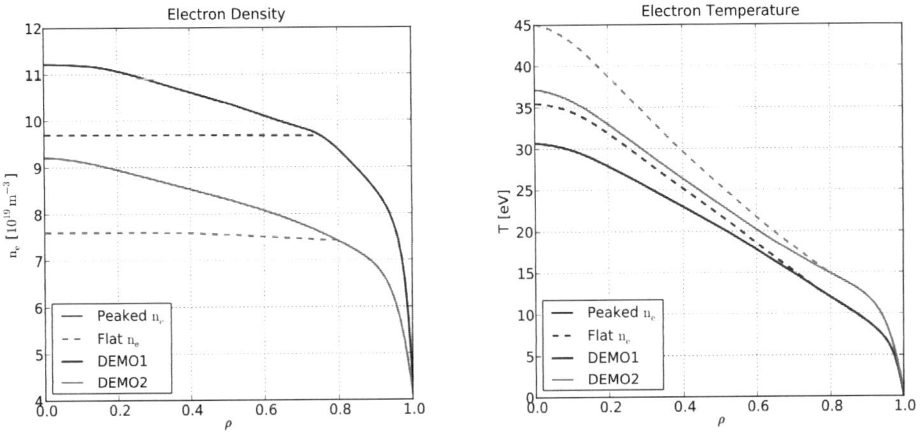

Figure 1. Temperature and electron density profiles for DEMO1 and DEMO2 and for both peaked and flat profiles. The background species are assumed to be in thermal equilibrium.

3. ICRF scenarii and simulations with the full-wave TORIC code

The DEMO conceptual designs are obtained with an optimization procedure of some selected physics, engineering and economical quantities. Zero-dimensional systems codes that perform this optimization analysis, such as PROCESS [11], give the optimal values of a number of global scalar parameters. From these parameters we have calculated the magnetic equilibrium configuration using the Grad-Shafranov solver SPIDER [12] interfaced with the ASTRA transport code [13] for the density and temperature profiles. For this purpose SPIDER was executed in the "prescribed boundary" mode, with elongation and triangularity equal to 1.56 and 0.33 respectively, and with a parabolic q (safety factor) profile. Using the transport model of ASTRA, the kinetic profiles have been iteratively optimized to match the values of fusion power and Greenwald fraction selected by the PROCESS. Since a key issue for DEMO is to mitigate the thermal load of the divertor plates, the present European DEMO designs have for this purpose a relatively high Z_{eff}. The values of Z_{eff} in table 1 have been simulated by diluting the plasma with traces of iron in addition to 5% of He^4 ashes. To investigate the impact of density profiles in the plasma core while using the same equilibria, we have flattened the core density profiles while keeping unchanged the plasma pressure. The latter is achieved by rigidly rescaling the temperatures by a common local factor. The nominal (peaked) and flat profiles are shown in figure (1) together with the corresponding temperature profiles. It is not pretended that the flat-density profiles are realistic.

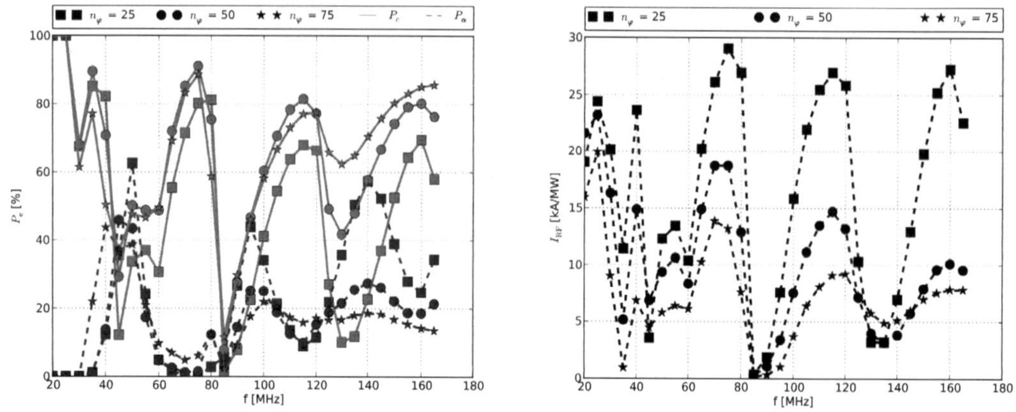

Figure 2. (left) Red and blue lines are the fractions of RF power absorbed by electrons and α particles, respectively, as function of the frequency of the injected waves, in the case of DEMO1 with peaked density profiles. Symbols refer to three different values of n_φ which are linearly scaled with frequency in order to keep constant x_e. (right) Total driven currents.

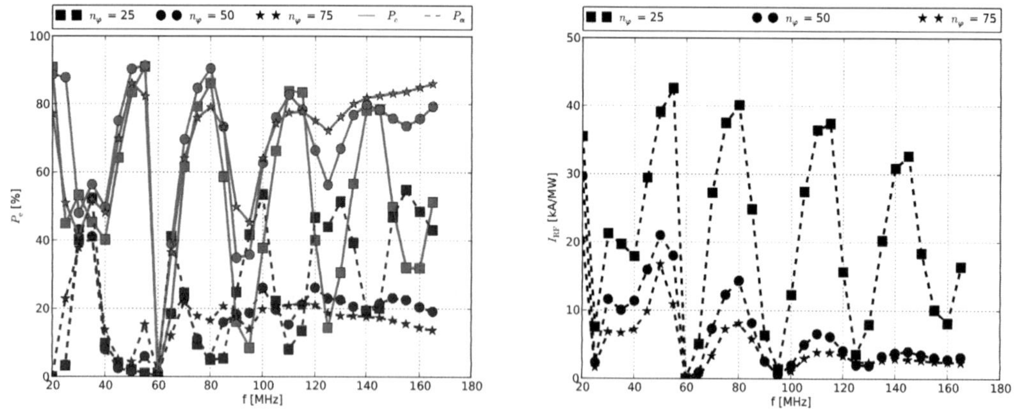

Figure 3. Same of figure 2 for DEMO2 with peaked density profile.

To identify the possible frequency windows [5, 6, 7] for these scenarii, preliminary frequency scans for three values of n_φ are shown in figures 2 and 3. Since the critical parameter is x_e both for electron absorption and for CD efficiency, the frequency scan is performed by keeping constant x_e. This implies that n_φ is varied linearly with frequency. To show the dependence on x_e, we have chosen three values of n_φ, chosen so that their values at 50 MHz were 25, 50, and 75: they are representative of the current spectrum of traditional multi-strap antennas. The symbols in figures 2 and 3 refer to these three values of n_φ. As shown in figure 2(right), when the frequency is lower than about 25 MHz, no ion cyclotron resonances are present inside the plasma, and electrons can absorb RF power without competing with ions. At these low frequencies it is very difficult to obtain the required n_φ values with traditional antennas. The next operative window is around 70 MHz, where only the first harmonic of the tritium, $\omega = 2\,\Omega_{cT}$, falls inside the plasma far on the high field side. Other operative windows are present at higher frequencies starting from 110 MHz. At higher frequencies, however, the RF power absorbed by α particles is not negligible. In DEMO1 at 110 MHz, for example, with $\omega = 2\Omega_{c\alpha}$ close to the magnetic axis, absorption by α is reduced by placing the launcher in the upper or lower part of the outer vessel, far from the equatorial plane, to avoid that RF waves impinge the α IC resonance [6, 7]. Figures 2(right) and 3(right) show the corresponding driven currents. As anticipated, the power P_e absorbed by the electrons increases with increasing n_φ. At the same time the CD efficiency decreases with n_φ, mainly because of toroidal trapping, and as a consequence the total driven current actually decreases with increasing n_φ. Similar considerations hold for DEMO2 when the frequencies are rescaled by about the ratio of the magnetic fields, as shown in figure 3.

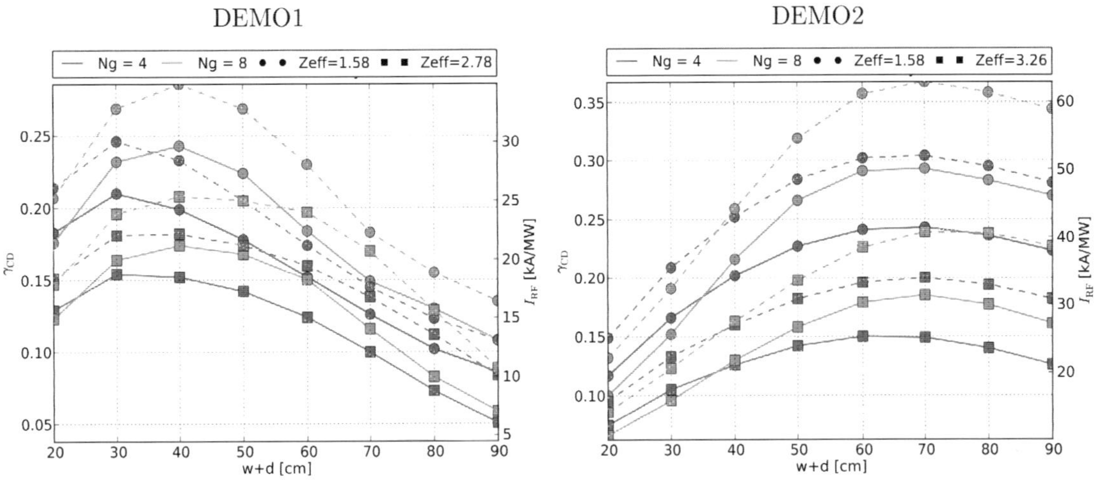

Figure 4. Total driven current as function of the toroidal periodicity of the antenna with two (circle) and four straps (square) (left: DEMO1 at 70 MHz - right: DEMO2 at 55 MHz). The solid and dashed lines are respectively for peaked and flat density profiles (figure 1). For comparison, the same quantities are shown for a lower value of Z_{eff}, namely 1.58. The left axis shows the same values in terms of γ_{CD}. In all these runs, the antenna phasing $\Delta\phi$ is set to $\pi/2$..

4. The role of the antenna

The total amount of FWCD is the result of an interplay among several effects, which in turn depend on plasma and antenna parameters. Therefore, an estimate of the total driven current requires a model of the ICRF launchers. If we rely on simple models of traditional multi-strap antennas, the relevant antenna parameters are the toroidal periodicity, $w + d$ with w the width and d the distance of the straps, the number of straps, N_g, and the relative phase of the currents flowing in neighboring straps, $\Delta\phi$. The Fourier spectrum of the antenna currents has the main peak centered around $\hat{n}_\varphi$ and of width $\Delta\hat{n}_\varphi$ given approximatively by

$$\hat{n}_\varphi \approx \frac{R_{\text{ant}}}{w+d}\Delta\phi, \qquad \Delta\hat{n}_\varphi \approx \frac{2\pi}{N_g(w+d)}\Delta\phi \tag{2}$$

where we have assumed that $\Delta\phi$ is equal for all neighboring straps. The power spectrum of the wave fields inside the plasma is affected by the presence of an evanescence layer between the antenna and the low-density cutoff layer [14]. This evanescence layer works as low-pass n_φ filter, which tailors the antenna spectrum by cutting the high n_φ part of the spectrum. As a consequence, the main peak of the field spectrum is shifted towards values somewhat lower than $\hat{n}_\varphi$ of equation (2). We also recall that the launched $k_\parallel$ spectrum is further broadened by the fact that each poloidal Fourier component has a different $k_\parallel$: in all our simulations the poloidal wavenumber spanned the range $-31 \leq m_\vartheta \leq +31$. Both these effects are automatically taken into account in TORIC .

Finally, increasing the number of straps N_g improves the directionality of the antenna and increases the driven current; this number, however, is limited by the available access and by the fact that to control the phases each strap must be individually fed. Figure 4 shows how the total driven current varies with $(w+d)$ and N_g for DEMO1 (left) and DEMO2 (right). In figure 4 the trend of $I_{\rm RF}$ with $(w+d)$ can be understood by taking into account also the dependence on $(w+d)$ of the fraction of RF power absorbed by electrons shown in figure 5. At 70 MHz in DEMO1 and 55 MHz in DEMO2 the electrons suffer competition only by tritium with its first IC harmonic on the high field side. Electron absorption is larger in DEMO2 because of the higher T_e. For both DEMO1 and DEMO2 the results obtained with flat profiles show on average higher total driven current: this is again due to the higher T_e necessary to balance the density flattening in order to keep constant the plasma pressure and the pedestal. To explore

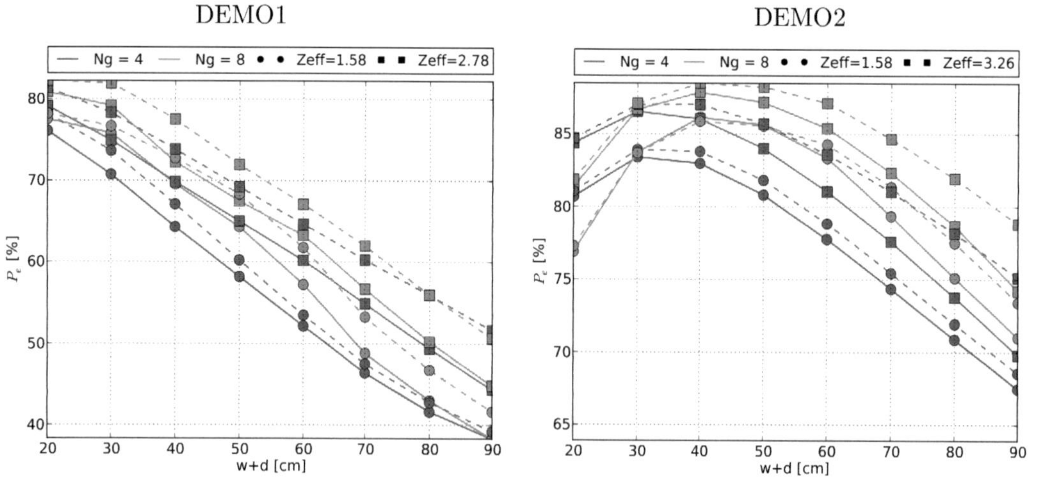

Figure 5. Power absorbed by electrons for the cases considered in figure 4.

the sensitivity of CD efficiency on $Z_{\rm eff}$, we have considered $Z_{\rm eff} = 1.58$, which is lower than the values of table 1, and closer to the values used in [5, 6, 7, 4, 15]. For this purpose, we have reduced the iron concentration, which entails also an increase of deuterium-tritium fuel and thus of α particles. As expected, the total driven current with the lower value of $Z_{\rm eff}$ is systematically higher than $I_{\rm RF}$ with higher $Z_{\rm eff}$.

For completeness, figure 6 shows the driven currents $I_{\rm RF}$ for a low frequency scenario ($f \lesssim 20$ MHz). These values are comparable with those of figure 4, although at these low frequencies the RF power is mainly absorbed by electrons. However, if the antenna parameters are kept unchanged, the CD efficiency decreases with frequency, since x_e decreases. Finally, figure 7 shows how $I_{\rm RF}$ varies with the antenna phasing $\Delta\phi$. In DEMO2 the highest predicted $I_{\rm RF}$ is around $\Delta\phi \approx 60^o$ (figure 7(right)), lower than 90° where one expects the highest antenna directionality, as it is for DEMO1 (figure 7(left)). Also in this case, the sensitivity of the amount of RF power absorbed by electrons on n_φ plays a major role on $I_{\rm RF}$. To allow a

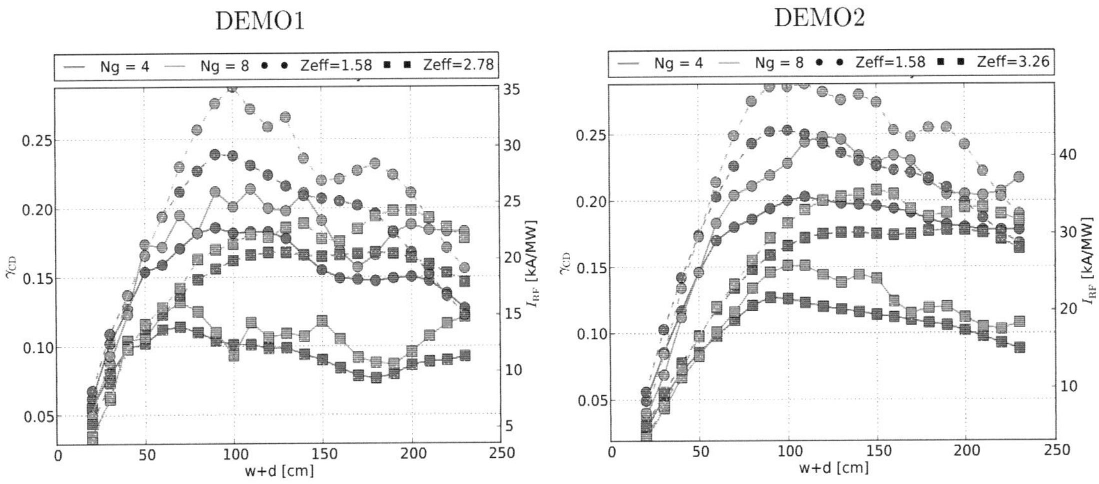

Figure 6. Same of figure 4 with 20 MHz for both DEMO1 and DEMO2.

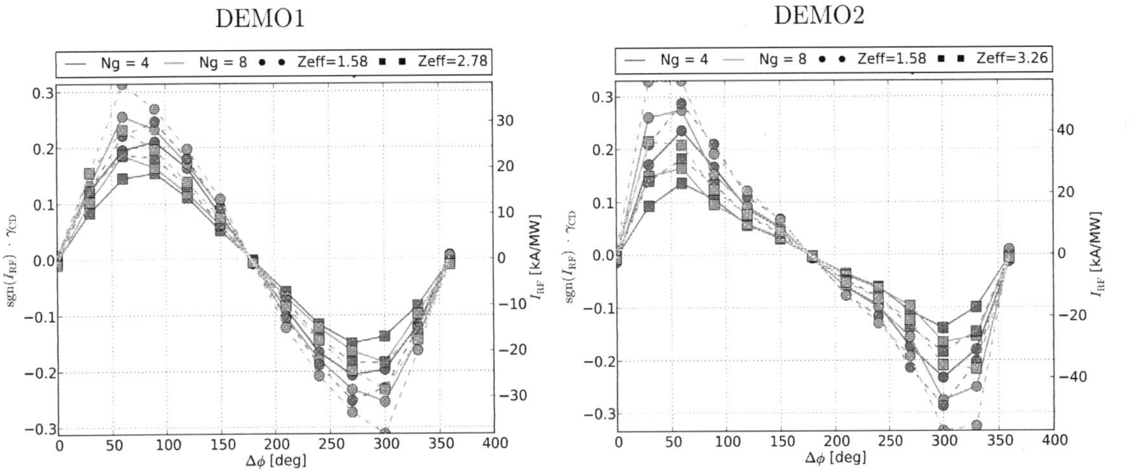

Figure 7. Same of figure 4 when the phasing $\Delta\phi$ is varied. In all these runs it holds $(w + d) = 30$ cm.

straightforward comparison with other CD systems, in figure 4,6, and 7 the left ordinate axis reports the corresponding values of the global figure of merit γ_{CD} defined as

$$\gamma_{\mathrm{CD}} = \langle n_e \rangle_{[10^{20}\mathrm{m}^{-3}]} \, R_{0[\mathrm{m}]} \, \frac{I_{\mathrm{RF[A]}}}{P_{\mathrm{RF[W]}}} \tag{3}$$

with $\langle n_e \rangle$ the volume-average density. The γ_{CD} values are in the same range as those predicted for EC [15] and for NBI CD at low Z_{eff} [4].

5. Discussion

Our simulations for the two presently proposed DEMOs show that the performances of a FWCD system will substantially depend not only on plasma parameters but also on the antenna geometry. With an antenna located at mid plane on the low-field side detrimental absorption by fusion-born α particles can be avoided at very low frequencies ($f \lesssim 20$ MHz) and in a window at intermediate frequencies, 70 MHz for DEMO1 and 55 MHz for DEMO2. For low values of Z_{eff} and for reasonable antenna parameters, the values of γ_{CD} are comparable with those of other current drive systems [4], provided the antenna spectrum is opportunely tailored. Since absorption by electrons is strongest in the plasma core where the electron temperature is higher, and reduction of the efficiency by toroidal trapping of resonant electrons is also weakest near the magnetic axis, FWCD is most efficient for on-axis CD [16]. Thus, FWCD seems to be insufficient to replace the ohmic contribution to off-axis current profile, and it must be complemented with an other CD system [4]. Undesirable synergies must also be taken into account: for example, in the presence of simultaneous NBI a reduction of FWCD performances might occur as a non negligible fraction of the RF power can be efficiently absorbed by the fast ions born from NBI neutrals [17].

FWCD in the range of frequencies considered in this study is presently the only CD method for which reliable, robust and economically appealing sources already exist. The critical element of an ICRF system will be the antenna. Nevertheless, many of the important issues in antenna design, including optimization of coupling and electron absorption, avoidance of parasitic absorption in the scrape-off plasma and sputtering in passive elements and the wall, can already be further investigated in present devices, in which the frequencies used are in the same range as those proposed fore FWCD in DEMO [5, 6, 7, 16].

Acknowledgment

We are grateful to C. Angioni, Y. Kazakov, E. Poli, G. Tardini, and H. Zohm for useful discussions.

[1] ZOHM, H., ANGIONI, C., FABLE, E., et al., Nuclear Fusion **53** (2013) 073019.
[2] FEDERICI, G., GIRUZZI, G., LOWRY, C., et al., in *2013 IEEE 25th Symposium on*, pp. 1–8, 2013.
[3] FRANKE, T., BARBATO, E., CARDINALI, A., et al., AIP Conference Proceedings **1580** (2014) 207.
[4] ZOHM, H., BARBATO, E., JENKINS, I., et al., in *40th EPS conference*, p. O3.108, 2013.
[5] HANNAN, A., HELLSTEN, T., and JOHNSON, T., Nuclear Fusion **53** (2013) 043005.
[6] LERCHE, E., VAN EESTERA, D., MESSIAEN, A., et al., AIP Conference Proceedings **1580** (2014) 338.
[7] KAZAKOV, Y., EESTER, D. V., LERCHE, E., et al., in *40th EPS conf.*, p. 2.031, 2014.
[8] BRAMBILLA, M., Plasma Physics and Controlled Fusion **41** (1999) 1.
[9] POST, D., MIKKELSEN, D., HULSE, R., et al., Journal of Fusion Energy **1** (1981) 129.
[10] EHST, D. and KARNEY, C., Nuclear Fusion **31** (1991) 1933.
[11] KNIGHT, P., A user's guide to process systems code, UKEA Fusion, 1996.
[12] A.A.IVANOV, R.R.KHAYRUTDINOV, S.YU.MEDVEDEV, et al., in *32th EPS conf.*, pp. P–5.063, 2005.
[13] FABLE, E., ANGIONI, C., IVANOV, A. A., et al., Plasma Physics and Controlled Fusion **55** (2013) 074007.
[14] BILATO, R., BRAMBILLA, M., HARTMANN, D., and PARISOT, A., Nuclear Fusion **45** (2005) L5.
[15] POLI, E., TARDINI, G., ZOHM, H., et al., Nuclear Fusion **53** (2013) 013011.
[16] BRAMBILLA, M, BILATO, R., submitted to Nuclear Fusion.
[17] BILATO, R., BRAMBILLA, M., MAJ, O., et al., Nuclear Fusion **51** (2011) 103034.

Journal of Physics: Conference Series **561** (2014) 012002 doi:10.1088/1742-6596/561/1/012002

Self-consistent quasi-linear modelling of Lower Hybrid Current Drive in ITER and DEMO

A. Cardinali, R. Cesario, F. Santini, L. Amicucci, C. Castaldo, S. Ceccuzzi, L. Panaccione, F. Mirizzi, A.A. Tuccillo

ENEA, Unità Tecnica Fusione, Via E Fermi 45 Rome, Italy

E-mail: alessandro.cardinali@enea.it

Abstract. First pass absorption of the Lower Hybrid waves in thermonuclear devices like ITER and DEMO is modelled by coupling the ray tracing equations for the wave phase and amplitude with the quasi-linear evolution of the electron distribution function. A system of coupled ordinary differential equations for each Fourier component of the spectrum radiated by the LH antenna is derived and solved when considering both 1D/2D Fokker-Planck model for the electron distribution function. This allows to reconstruct and to evolve the quasi-linear diffusion coefficient consistently with the wave propagation, to calculate the power deposition profile and the amount of current driven by the wave. As usually assumed, the Lower Hybrid Current Drive is not effective in a plasma of a tokamak fusion reactor like ITER or DEMO, because the high electron temperature would enhance the wave absorption and then limit the RF power deposition to the very periphery of the plasma column (near the separatrix). In this work by extensively using this self-consistent modelling for the propagation and absorption of the LH wave, a parametric study on the wave spectrum (and consequently on the antenna design) as spectrum width, peak value, secondary lobes etc. has been performed very accurately. Such a careful investigation aims at controlling the power deposition layer possibly in the external half radius of the plasma, thus providing a valuable aid to the solution of how to control the plasma current profile in a toroidal magnetic configuration. This analysis is useful not only for exploring the possibility of profile control of a pulsed operation reactor, but also in order to reconsider the feasibility of steady state regime.

1. Introduction

The reactor–oriented research, now in progress, is based on Deuterium-Tritium plasmas in a Tokamak configuration, and is aimed at demonstrating the scientific, technological and economic feasibility of fusion power. Although a reactor should in principle operate in steady-state, a more conservative pulsed option is now receiving major consideration, because tools for active and fine control of the radial current density profile are unavailable. This control of the current density, necessary for preventing the q-profile to evolve towards conditions deleterious for stability [1], is mandatory for steady state operation, as well as for pulsed operations that require profile control by non-inductive current to be added to the self-generated bootstrap current. The lower hybrid current drive (LHCD) [2,3] is attractive for a reactor, owing to the non-inductive current produced with high efficiency by microwave power at a few GHz coupled to plasma by waveguide antennas. However, two problems have persisted for decades which prevented the LHCD tool to be considered of practical use for a reactor: *i*) the difficult extrapolation of the LHCD effect at reactor graded high densities, due to parasitic effect of plasma edge [4] and, *ii*) the strong wave damping which would make the LH power deposition in hot plasmas too peripheral [5]. The first problem (i) was solved by a method assessed on FTU [4], and based on previous theoretical predictions that high plasma edge temperature in reducing parasitic effects as Parametric Instabilities (PI) at the edge, should allow the penetration of the LH power to the plasma core [6,7]. This favourable condition should be naturally met in reactor plasma, as shown later on. Concerning the second problem (ii), the strong dependence of the LH wave damping on the electron temperature is considered problematic for DEMO reactor, as the power should be deposited not deeper than the pedestal radial layer, where $T_e \approx 8$ keV [5]. In this paper, we propose a self-consistent modelling based on the simultaneous solution of the ray tracing equation, which involve all the components of the power spectrum, and coupled with the evolution in the velocity space of the electron distribution function (Quasi-Linear Fokker-Planck Equation). The ray tracing system that includes the evolution along the trajectory of the power associated to the single ray at the antenna, enables the reconstruction of the quasi-linear diffusion coefficient and consequently the determination of the electron distribution function due to LH. This method is particularly suitable for the first pass absorption where apart from a very short transient time τ_{ql}, where the incoming wave is linearly absorbed, at the steady state the quasi-linear effects are prevailing and determine the establishment of an equilibrium distribution function (tail) which is responsible of the subsequent damping of the wave. In reactor plasma, full LH power deposition takes place in less than half radial pass and application of this self-consistent method appears well justified. This feature of the modelling clarifies better the crucial role of power spectrum (the nominal one delivered by the antenna) and that of the broadening (possibly produced by PI at the plasma edge) in determining the radial LH deposition profile [6,7,8,9]. Moreover, by controlling the shape of the spectrum, its width, the n|| peak and the electric field intensity (surface power density at the antenna), it is possible to control the deposition layer and to recover the analytical formula obtained by F Santini in Ref. [10]. This approach has been indeed routinely used in the "LHstar" code, which after calculating the effects of spectral broadening produced at the edge by PI, integrate the ray tracing equations simultaneously with the Quasi-Linear Fokker-Planck evaluation of the electron distribution function in 2D velocity space. This code was successfully employed earlier for producing LH deposition profiles in agreement with available data of important experiments of JET [6], and for assessing on FTU the method for producing LHCD effects at reactor relevant high plasma density [4]. In this work in order to elucidate the calculation mechanism we will use the quasi-analytical solution for the 1D QL Fokker-Planck equation, by considering the quasi-linear diffusion coefficient which stems out from ray-tracing and power damping equations and then we will give an example by calculating the LHCD for DEMO reactor plasma parameter as resulting from Ref. [11]. In this context it should be also useful to give some evaluation of LH damping on the non-negligible fraction of α-particles present in DEMO, which results of about <5% of the total absorbed power.

2. Pertinent equations and modelling

The electron distribution function satisfies the quasi-linear Fokker-Planck equation and in 1D velocity space we can write

$$\frac{\partial \hat{f}_e(\tau,u)}{\partial \tau} = \frac{\partial}{\partial u}\left(D(u)\frac{\partial \hat{f}_e(\tau,u)}{\partial u}\right) + \alpha \frac{\partial}{\partial u}\left(\frac{1}{u^3}\frac{\partial \hat{f}_e(\tau,u)}{\partial u} + \frac{\hat{f}_e(\tau,u)}{u^3}\right) \tag{1}$$

where $\hat{f}_e(u)$, is the normalized to $n_0 u_{the}^3$ electron distribution function, $\tau = \nu_{e,e} t$, is the normalized time (to the electron-electron collision frequency $\nu_{e,e} = 2.91\times 10^{-6} \ln \Lambda n_0 T_{eV}^{-3/2}$), $u = \frac{v_\parallel}{u_{the}}$ is the

Journal of Physics: Conference Series **561** (2014) 012002 doi:10.1088/1742-6596/561/1/012002

normalized parallel velocity (to the thermal velocity), $\alpha = \left(\frac{2+Z_i}{2}\right)$, $u_{the} = 4.2 \times 10^7 \sqrt{T_e}$, Z_i is the ion charge and

$$D(u,x) = \frac{D_{RF}(u,x)}{u_{the}^2 \nu_{e,e}} = \left(\frac{1}{u_{the}^2 \nu_{e,e}}\right) \frac{e^2}{m_e^2} \frac{\pi}{\omega} \left| E_{\parallel}(x,k_{\parallel}) \right|^2 \Big|_{k_{\parallel} = \frac{\omega}{u_{the} u}} \tag{2}$$

is the normalized quasi-linear diffusion coefficient. Note that the dependence of the quasi-linear diffusion coefficient from the space coordinate is only a local dependence where the effects of the 2D wave propagation are accounted. An estimate of the quasi-linear diffusion time in which the fast electron tail will be restored in the layer where the LH is absorbed can be analytically evaluated. The following expression is obtained by considering the Fokker-Planck equation Eq. (1)

$$\tau_{ql} \approx \tau_{e,e} \frac{u_1^3 \Delta u}{D(u)} \tag{3}$$

where Δu is the resonance width of the coupled LH spectrum in the velocity space normalized to the electron thermal speed, u_1 is the lower bound of the spectrum, i.e. $u_1 \sim 3.5 u_{the}$ to satisfy the electron Landau absorption criterion, $n_{\parallel} \sqrt{T_{keV}} \sim 5-7$ [12]. D is the normalized quasi-linear diffusion coefficient Eq. (2) that is correlated with the absorbed power. In a typical DEMO LH scenario, we have that the absorbed lower hybrid power is 80MW for the case under consideration; thus, D is about 10^3 and the launched $n_{\parallel peak} \approx 1.8$. Consequently, by inserting these values in Eq. (3) the quasi-linear time is $\tau_{ql} \approx 2.8ms$. This time is very short with respect to the RF pulse life-time, this means that a steady state in the quasi-linear LH absorption is suddenly formed just after the switch on of the RF generator. At the steady state the solution of Eq. (1) is

$$\hat{f}_e(u) = \frac{1}{(2\pi)^{3/2}} \exp\left\{ -\int_u \frac{\alpha u}{\left(\alpha + u^3 D(u)\right)} du \right\} \tag{4}$$

where C is a constant which can be determined by the integral of the distribution function over the velocity space. The equation for the spatial damping rate can be written as [13]

$$\frac{dP(x)}{dx} = -2 \frac{\Gamma_{QL}^{es}\left(\omega, k_{\perp}, k_{\parallel}, x, P(x)\right)}{\partial H\left(\omega, k_{\perp}, k_{\parallel}, x\right) / \partial k_{\perp}} P(x) \tag{5}$$

where

$$\Gamma_{QL}^{es} = \sqrt{\frac{\pi}{2}} \frac{\omega_{pe}^2}{\left(k_{\perp}^2 + k_{\parallel}^2\right) u_{the}^2} \left(\frac{\omega}{k_{\parallel} u_{the}}\right) \left(\frac{\alpha}{\alpha + \left(\omega / (k_{\parallel} u_{the})\right)^3 D\left(\omega / (k_{\parallel} u_{the})\right)}\right) \exp\left\{ -\int_u \frac{\alpha u}{\left(\alpha + u^3 D(u)\right)} du \Big|_{u = \frac{\omega}{k_{\parallel} u_{the}}} \right\} \tag{6}$$

which in the case of Maxwellian plasma reduces to the well-known expression

$$\Gamma^{es}_{Lin} = \sqrt{\frac{\pi}{2}} \frac{\omega^2_{pe}}{\left(k^2_\perp + k^2_\parallel\right) u^2_{the}} \frac{\omega}{k_\parallel u_{the}} \exp\left\{-\frac{1}{2}\left(\frac{\omega}{k_\parallel u_{the}}\right)^2\right\} \tag{7}$$

By using the power instead of the electric field we can write the quasi-linear diffusion coefficient in Eq. (6) as

$$D(x,u) = \frac{4\pi^2}{\omega c} \frac{e^2}{m^2_e} \frac{1}{u^2_{the}(x) \nu^{ee}_{coll}(x)} \left(\frac{\omega}{c k_\perp}\right) \frac{P(x,u)}{\left|\Sigma_{antenna}\right|_x} = D_{coeff}(x,u) P(x,u) \tag{8}$$

Upon using Eq. (8) in (5), and giving the power spectrum at the edge $P(x=1,k_\parallel) = P_0 \exp\left(\frac{k_\parallel - k_{\parallel c}}{\Delta k_\parallel}\right)$, Eq. (5) transforms in a non-linear system of coupled differential equations, after changing the integral in a summation over all the "N" components in which we have decomposed the wave spectrum. Defining the integrand in Eq. (6) as

$$G(P(x,u),u,x)\Big|_{u=\frac{\omega}{k_\parallel u_{the}}} = \frac{\alpha u}{\left(\alpha + u^3 D_{coeff}(x,u) P(x,u)\right)}\Bigg|_{u=\frac{\omega}{k_\parallel u_{the}}} \tag{9}$$

we have

$$\int_u G(P(u,x),u,x)\,du\Bigg|_{u=\frac{\omega}{k_\parallel u_{the}}} = \sum_{n=1,N} G(P_n(u_n,x),u_n,x)\Bigg|_{u_n=\frac{\omega}{k_{\parallel n} u_{the}}} = \sum_{n=1,N} G\left(\left(\frac{\omega}{k_{\parallel n}(x) u_{the}(x)}\right), P_n(k_{\parallel n}(x))\right) \tag{10}$$

and

$$\frac{dP_n(x)}{dx} = -2F(\omega,k_\perp,k_{\parallel n},x)\exp\left\{-\sum_{n=1,N} G\left(\left(\frac{\omega}{k_{\parallel n} u_{the}}\right), P_n(k_{\parallel n})\right)\right\} P_n(x) \tag{11}$$

where

$$F(\omega,k_\perp,k_{\parallel n},x) = \frac{1}{2}\sqrt{\frac{\pi}{2}} \frac{\omega^2_{pe}}{k_\perp u^2_{the}} \left(\frac{\omega}{k_\parallel u_{the}}\right)\left(1 - \frac{\omega^2_{pe}}{\omega^2 - \Omega^2_{ce}} - \frac{\omega^2_{pi}}{\omega^2 - \Omega^2_{ci}}\right)\left(\frac{\alpha}{\alpha + \left(\omega/\left(k_\parallel u_{the}\right)\right)^3 D\left(\omega/\left(k_\parallel u_{the}\right)\right)}\right) \tag{12}$$

Eqs. (11) is the relevant equation system to be solved in order to have the quasi-linear self-consistent power deposition profile, and current drive evaluation. We recall that the 2D effects in the wave propagation (mainly affecting the evolution of the parallel wavenumber, as well as the shape of the antenna image inside the plasma) are kept into account by analytical integration of the propagation as given in Ref. [14]. For LH application to tokamak devices of thermonuclear interest like ITER and DEMO where the LH is characterized by first pass absorption, this method is particularly effective because the wave spectrum behaves coerently during the propagation and the rays characterizing the antenna both in the shape and spectrum do not scatter too much. In the multipass regime (not interesting for reactor application) this type of approach does not work correctly at least for more than two passes, and other approach based on statistical ray-tracing must be used [15].

Journal of Physics: Conference Series **561** (2014) 012002 doi:10.1088/1742-6596/561/1/012002

3. Numerical results for DEMO

DEMO can operate in pulsed and Steady State regime both characterized by flat or peaked kinetic profiles. We have applied the above theory to calculate the QL power deposition profile for DEMO in flat and peaked density profiles with the two kinds of electron temperature profiles, as established in Ref. [16]. The LH power deposition profiles has been accounted by studying the effect of the wave spectrum on the deposition layer, in particular the width of the spectrum (considering an analytical Gaussian shape) and the strength of the admissible electric field surface density expressed in kW/cm^2 (for the LH <10kW/cm^2). The results proposed here refers to the 1D QL theory as shown in the previous section, but it has been confirmed also by a more complete (full numerical approach) which uses the ray-tracing equations in a realistic DEMO flux-magnetic configuration and a much more sophisticated 2D full numerical integration for the relativistic Fokker-Planck equation, showing the correctness of our semi-analytical 1D QL self-consistent approach. The plasma data we have used in our simulation for DEMO flat scenario are the following: plasma radius a=2.25m, major radius R_0=9m, magnetic field on axis B_0=6.8T, f_{LH}=5GHz, n_{e0}=8.8X10^19m^-3, n_{eS}=5.9X10^19m^-3 with a very rapid decrease (e-folding 22cm), we considered also a model of the scrape off plasma (from the separatrix at the wall of about 20cm) in this layer the density exponentially drops of two order of magnitude. Concerning the electron temperature -which can indeed plays a role in the LH absorption-we have that at the plasma centre reaches the value of 25keV down to 2.4keV at the separatrix, being of about 9keV at the pedestal. From the separatrix to the wall in the scrape-off layer the temperature also drops of a couple of order of magnitudes reaching the value of 25eV at the antenna mouth.

In Fig. (1), a plot which shows the power deposition profile in W/m^3 for two wave spectra (Gaussian shape) characterized by a different width $\Delta n_{\|} = \frac{c}{\omega}\Delta k_{\|}$, is depicted vs x (the normalized radius).

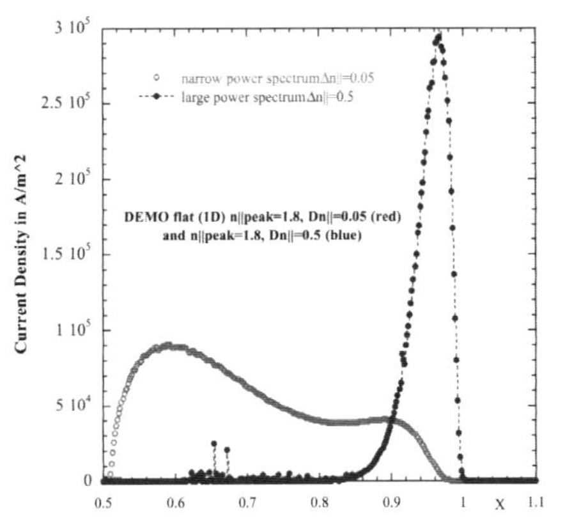

Figure 1. Current Driven Profile in A/m^2 for two wave spectra: (red line) Dn||=0.05, (blue line) Δn||=0.5.

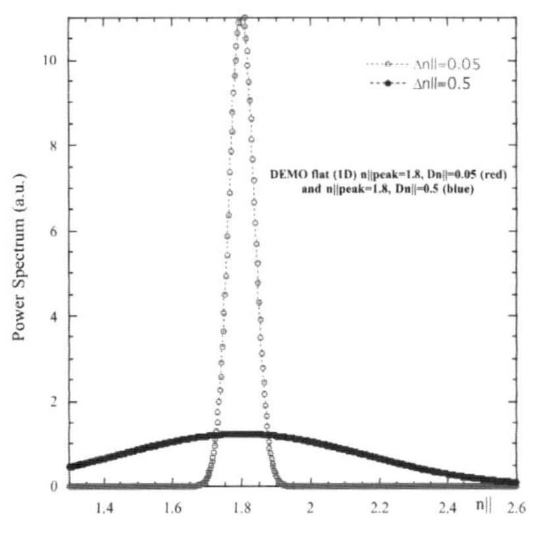

Figure 2. Power spectra in a.u. for the cases considered in Fig. 1.

The red curve with a sufficiently on-axis deposition is related to the thin spectrum $\Delta n_{\|} = 0.05$, with a peak value of n||=1.8; the blue curve with deposition very peripheral, is related to the same spectrum but much larger $\Delta n_{\|} = 0.5$ than the previous. In Fig. 2 the power spectra for both cases are compared. In Fig. 3 the power damping rate $\frac{1}{P}\frac{dP}{d\hat{x}}$ is reported vs x (normalized radius) for both spectra shown in

Fig. 2 (red (thin) and blue (large)), and also when quasi-linear effects are neglected (Maxwellian distribution function (green line)). In the case of linear damping (Maxwellian plasma) it is worth to note that the wave absorption is always peripheral independently of the spectrum shape. This leads to argue that the quasi-linear effects accounted by a formation of the tail in the distribution function as can be seen in Fig. (4) where the Maxwellian f_e is compared to the tailored-LH distribution function, are as important as the spectrum is narrower.

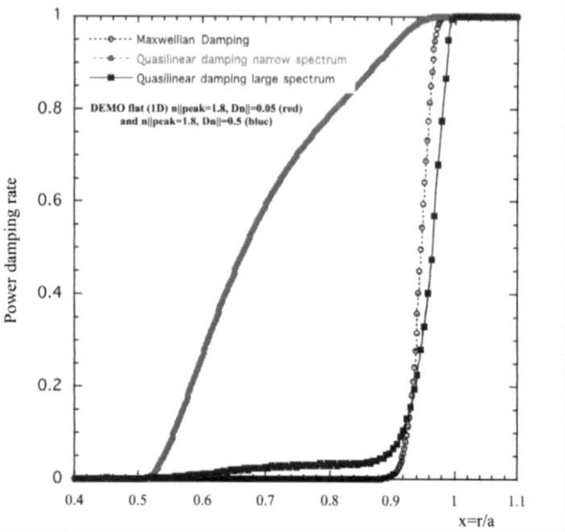

Figure 3. Power Damping Rate for the cases considered in Fig.1 and the Maxwellian damping (green line).

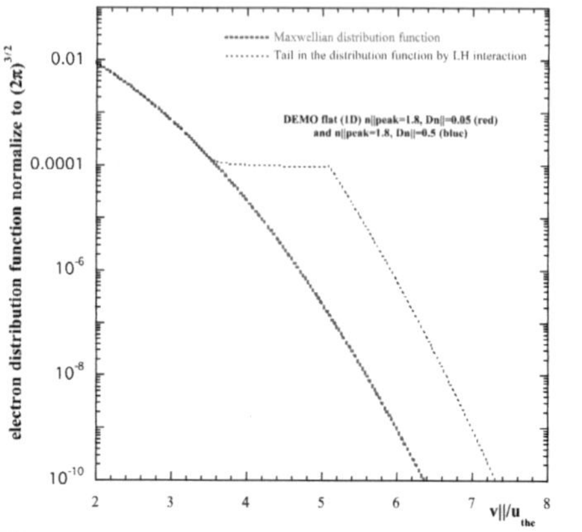

Figure 4. Electron distribution function Maxwellian compared to the LH quasi-linear tailored distribution function.

The quasi-linear effect, in fact, by smoothing the distribution function, reduces the damping factor $\left(\Gamma \sim \frac{\partial f_e}{\partial u} \to 0\right)$, at least for lower resonant velocities, allowing the wave to penetrate deeper. Using the linear Maxwellian damping the absorption is fulfilled when $T_{keV} = 7.7keV$ (linear) to $T_{keV} = 15keV$ (quasi-linear) [17], which is coherent with the numerical results which shows that for $n_{\parallel peak} = 1.8$ and narrow spectrum the linear damping is located around x=0.95 while the quasi-linear around x=0.6. This fact can be quantitatively understood by analytical inspection of the Eq. (6), and (8); looking at Eq. (8), in fact, we can assume that $D(u)$ behaves like $D(u) \approx D_0 \exp\left\{-\left[(u-u_c)/\Delta u\right]^2\right\}$: the Gaussian form of the wave spectrum in the velocity space. When the quasi-linear diffusion coefficient is very strong $D_0 >> 1$, and the spectrum is very narrow $(\Delta u/u_c << 1)$, by expanding the exponential function in Taylor series we have that the integral in Eq. (6) reduces to

$$\int_u \frac{\alpha u}{\left(\alpha + u^3 D(u)\right)} du \Bigg|_{u=\frac{\omega}{k_\parallel u_{the}}} \approx \frac{\alpha}{D_0} \frac{\Delta u}{u^2}\Bigg|_{u=\frac{\omega}{k_\parallel u_{the}}} + \ldots\ldots\ldots\ldots \quad (13)$$

and consequently approximating the exponential function in Eq. (6) we have

$$\exp\left\{-\int_u \frac{\alpha u}{\left(\alpha+u^3 D(u)\right)} du \Big|_{u=\frac{\omega}{k_\| u_{the}}}\right\} \approx \exp\left\{\frac{\alpha}{D_0}\left(\frac{u_{the}k_{\|peak}}{\omega}\right)^2 \Delta k_\| + ...\right\} \approx 1+\frac{\alpha}{D_0}\left(\frac{u_{the}k_{\|peak}}{\omega}\right)^2 \Delta k_\| + ... \quad (14)$$

This means that the quasi-linear damping Eq. (6) at the lowest order has the following form

$$\Gamma_{QL}^{es} = 2\sqrt{\pi}\frac{\omega_{pe}^2}{\left(k_\perp^2+k_\|^2\right)v_{the}^2}\left(\frac{\omega}{k_\| v_{the}}\right)\left(\frac{\alpha}{\alpha+\left(\omega/\left(k_\| u_{the}\right)\right)^3 D\left(\omega/\left(k_\| u_{the}\right)\right)}\right)\left[1+\frac{\alpha}{D_0}\left(\frac{u_{the}k_{\|peak}}{\omega}\right)^2 \Delta k_\|\right], \quad (15)$$

It is obvious that Γ_{QL}^{es} appears much weaker with respect to the linear damping Γ_{Lin}^{es} (given in Eq. (7)) in which the exponential reduction dominates:

$$\frac{\Gamma_{QL}^{es}}{\Gamma_{Lin}^{es}} = \left(\frac{\alpha}{\alpha+\left(\omega/\left(k_\| u_{the}\right)\right)^3 D\left(\omega/\left(k_\| u_{the}\right)\right)}\right)\left[1+\frac{\alpha}{D_0}\left(\frac{u_{the}k_{\|peak}}{\omega}\right)^2 \Delta k_\|\right]. \quad (16)$$

when $\Delta k_\|$ increases this condition is no more fulfilled and Γ_{QL}^{es} does not differ so much from the linear Γ_{Lin}^{es}. In Fig. (5) we have evaluated and plotted, in equal condition of narrow spectrum $\left(\Delta k_\| = 0.05\right)$, and absorbed power (50MW), the effects due to the reduction of D and increase of $\Delta k_\|$. As is possible to notice on the plot, where the peak of the power density profile x_p is plotted (bottom axis and red bullets) vs the surface power density (at the antenna in kW/cm^2), by reducing P_d at the antenna (or equivalently D), the peak of the deposition tends to move toward the plasma edge. On the same plot (top axis and blue diamonds) x_p is plotted vs $\Delta k_\|$; by increasing $\Delta k_\|$ the peak of the power deposition moves towards the plasma edge. This behaviour is perfectly accounted by Eq. (15), which shows that the damping rate Γ_{QL}^{es} increases by decreasing D (or equivalently the surface power density at the antenna P_d), and by increase $\Delta k_\|$. It is worth to note that this analytical calculation explains the formulae derived by F Santini in Ref. [10]. In Fig. 5 the red bullets and the blue diamonds are the result of the numerical code while the black shaded line is the interpolation line which in the first case scales like $\left(x_p \sim C/P_d^{0.1}\right)$ (where C≈0.5), while in the second case like a straight line as prescribed by Eq. (15). Finally in Fig. 6 the dynamical evolution of the power spectrum from the antenna to the absorption layer is depicted. It is possible to note, in addition to the wave absorption which reduces the size of the spectrum, also the propagation effects that lead to the increase of the parallel wavenumber and consequently the peak of the spectrum moves from 1.8 at the antenna to almost n||=2 near the radius of total wave absorption.

3. Conclusions

In conclusion, an analytical inspection of the quasi-linear damping for the LH wave in conjunction with a self-consistent approach in the integration of the ray-trajectories, in particular the power damping equation system, has shown that, by properly considering QL effect of a strong LH power coupled to high temperature plasma, the lower hybrid current drive should be indeed used for active profile control in a reactor like DEMO and/or ITER by overcoming the difficulties connected to a peripheral absorption of the wave. The active control is operated by the antenna design, involving a suitable power spectrum and wavenumber control. The semi-analytical considerations have also been supported by full 2D numerical approach (not reported here but in Ref. [8]). The above results could be tested in an experiment using a small $\Delta k_{||}$ ($\approx$0.05) but relatively large $n_{||0}$ ($\approx$2.8) to compensate the lower electron temperatures of the today's tokamaks, e.g., EAST [18], ranging at around 5 keV in the core. Furthermore, a small value of $\Delta k_{||}$ should be achieved in reactor devices since the large toroidal dimensions allows to a have a grill with a large number of elementary waveguides.

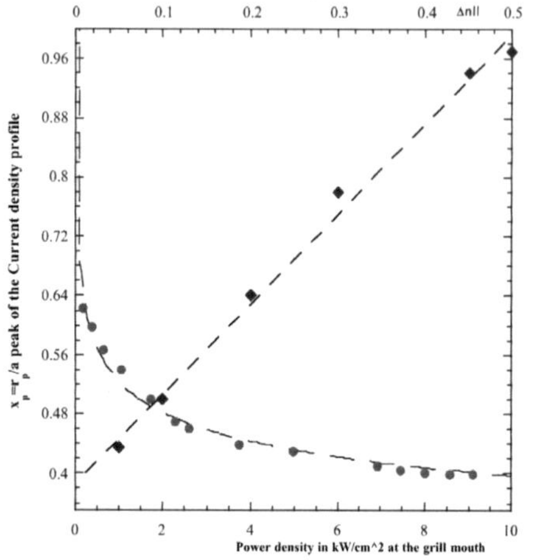

Figure 5. Location of the peak of the deposition profile vs the antenna-surface power density (bottom), and the width of power spectrum (top). The dashed lines represent the interpolation of the numerical data (bullets).

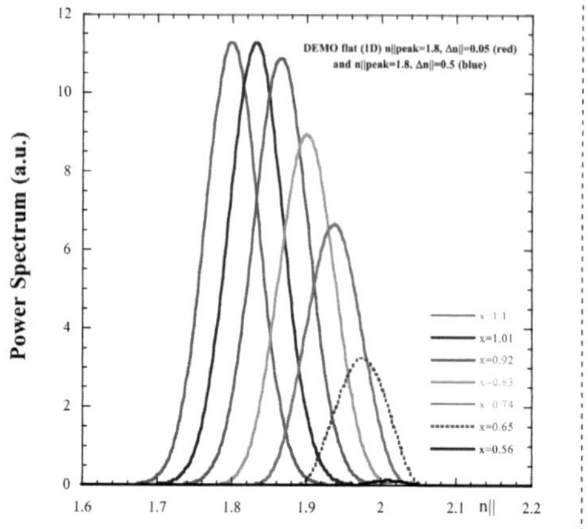

Figure 6. Power spectrum (a.u.) vs n|| in dynamical evolution inside the plasma. The red curve corresponds to x=1.1 (antenna location). The absorption in the internal layers is related to the reduction of Gaussian size.

References

[1] Giruzzi G et al., Final Report on EFDA Task "WP13-SYS01-B: Scenario modelling", EFDA_D_2LJPY6 v1.0 (Dec. 2013)
[2] Fisch N 1987 *Rev. Mod. Physics* **59** 175
[3] Bernabei S et al. 1982 *Phys. Rev. Letters* **49** 1255
[4] Cesario R et al. 2010 *Nature Communications* **1** 55
[5] Garcia J et al. 2008 *Nucl. Fusion* **48** 075007
[6] Cesario R et al. 2004 *Phys. Rev. Letters* **92** 175002
[7] Cesario R et al. 2006 *Nucl. Fusion* **46** 462
[8] Cardinali A et al. 2014 Submitted to *Nature Communications*
[9] Cesario R et al. 2014 *Nucl. Fusion* **54** 043002
[10] Santini F 1991 *Comments Plasma Phys. Controlled Fusion* **14** 199
[11] Tardini G EFDA 3PT H & CD Final Meeting, November 28th 2013
[12] Brambilla M 1978 *Nucl. Fusion* **18** 4
[13]M. Brambilla, A. Cardinali 1982 *Plasma Phys. Control. Fusion* **24** 1187
[14] Cardinali A and Romanelli F 1986 *Phys. of Fluids* **29** 810
[15] Peysson Y, Decker J and Morini L 2012 *Plasma Phys. Control. Fusion* **54** 045003
[16] Ceccuzzi S et al. 2013 Fusion Science and Technology **64** 748
[17] Brambilla M 1999 *Kinetic Theory of Plasma Waves*, Oxford Press
[18] Ding BJ et al. 2013 *Nucl. Fusion* **53** 113027

Staircase temperature profiles and plasma transport self-organisation in a minimum kinetic model of turbulence based on the trapped ion mode instability

T. Cartier-Michaud, P. Ghendrih, Y. Sarazin, G. Dif-Pradalier, T. Drouot*, D. Estève, X. Garbet, V. Grandgirard, G. Latu, C. Norscini, C. Passeron

CEA/DSM/IRFM, Cadarache, 13108 Saint-Paul-Lez-Durance, France
* IJL, Nancy, France

E-mail: `t.cartiermichaud@gmail.com`

Abstract. A minimum model of plasma turbulence in a kinetic framework is presented. It is based on trapped ion turbulence, gyro and bounce averaged, and implemented in the versatile and efficient code TERESA. Zonal flow - streamer interplay are readily shown to be key players that govern the confinement properties of the model. The parameter space of the model is explored with brute force numerics. A generic result is either a streamer dominated pattern with large transport, or a staircase temperature profile with very marked corrugations and quenched transport. A case with off-axis heating is found to exhibit quasiperiodic relaxation events relevant to investigate dynamical turbulence self-organisation.

1. Introduction

Although the issue of confinement performance of burning plasmas could be considered as solved, at least empirically, the JET results with the ITER like wall have eroded the performance capability of this device [1]. The possibility of stepping up the fusion gain on JET with respect to the 1997 DT campaign [2] is clearly questioned by the latter results. Theoretical investigation of confinement performance then appears to be required to help in clarifying the situation and drawing appropriate research lines to master plasma performance with this new wall environment. Of particular importance is the issue of transport barriers, their onset, their width and their confinement properties [3]. In major gyrokinetic simulations such transport barriers have been triggered by external plasma polarisation [4, 5]. However, self-generated barriers have only been observed with micro-barriers leading to corrugations in the temperature gradient [6]. Such a regulation of turbulence can lead to key regimes for more favourable confinement performance with either short lived distributed corrugations that govern an enhanced confinement time, or the seed to the transition towards stronger barriers, and consequently improved confinement regimes. In this paper we analyse the onset and evolution of such corrugations with a reduced kinetic model of turbulence.

State of the art simulations show complex confinement properties combining all scales of the system and in particular on going self-organisation and interplay between different turbulence scales [7, 8, 9]. However, such models are very demanding to run and analyse. We have thus developed a simple kinetic code, the TERESA code [10], modelling trapped ion turbulence [11, 12, 13, 14, 15, 16, 17]. This model is the minimum kinetic model of plasma turbulence. It is derived in the framework of Hamiltonian dynamics and action-angle variables [17], and benefits from a versatile and efficient parallelisation implementation in the code TERESA [10].

First simulations have led to staircase temperature profiles and thus strong corrugations of the temperature gradient. This signature of self-organisation of plasma transport is reminiscent of that obtained with the GYSELA $5D$ kinetic code [6] and the TOKAM2D fluid code [18]. TERESA simulations thus offer the possibility to address the key elements of turbulence self-organisation whilst retaining the efficiency of toy models like TOKAM2D [19]. Furthermore, it addresses kinetic features that are considered mandatory [20] but still require appropriate investigation to assess their actual effect. The paper is built as follows. The model and numerical settings are presented in Section 2. In the case where the plasma is driven out of equilibrium by a contact with two thermal baths, the long-time behaviour is observed to settle in either a streamer or in a zonal flow dominated regime, Section 3. In Section 4, a more systematic exploration of the space of control parameters shows that this feature is rather stiff. With off-axis heating, therefore a flux-driven case, quasi-periodic relaxation of the transport barriers have been observed, hence with time dependent self-organisation properties. Finally, Section 5 closes the paper.

2. A reduced model for trapped ion dynamics

Compared to the gyrokinetic case that is based on averaging out the fast gyrophase [21], the bounce average reduction consists of averaging out another fast phase related to the bounce motion. This model is thus restricted to trapped ion turbulence. The kinetic model is then reduced from the original $6D$ phase space $(r, \theta, \varphi, v_r, v_\theta, v_\varphi)$ in toroidal geometry, with given magnetic equilibrium, to $4D$ $(\alpha, \psi, \kappa, E)$ [11, 12, 13, 14, 15, 16], with two invariants, $\kappa \in [0, 1[$ the trapping parameter, and the normalised particle energy $E \in [0, +\infty[$ [17]. Given the safety factor q, the variable $\alpha = \varphi - q\theta$ labels a trapped orbit and is comparable to the toroidal angle φ at fixed poloidal angle θ. It is the conjugate coordinate [17] of the normalised poloidal magnetic flux $\psi \in [0, 1]$ that also labels the radial position of a trapped orbit. Note that ψ is normalised by the radial extent of the box, according to the convention used in the experiments with $\psi = 0$ on the magnetic axis and $\psi = 1$ at the outer boundary. Similarly, α is defined according to the φ variation $\alpha \in [0, 2\pi]$ when the reference poloidal angle is set at $\theta = 0$ (midplane low field side)

With such a reduction, the new particles are not local but extended in space, the extent being the Larmor radius in the α-direction and the so-called banana width in the ψ-direction [17, 22]. Regarding the trapping parameter, $\kappa \to 0$ for highly trapped particles and $\kappa \to 1$ for barely trapped particles. It is interesting to note that averaging out the last angle α would then yield a fully integrable system, hence without transport. A large part of the interest of the model is therefore that it is a minimum model for turbulent transport with a kinetic description.

The two fast angle averaging procedures yield a low pass filter in time, by construction, and in space via Finite Size Effects due to the spatial extent of the averaged trajectories. The gyrophase average introduces the Larmor radius, here ρ_0, as characteristic scale for the FSE in both the radial and poloidal directions, while the bounce average introduces the banana width δ_0 for the FSE in the radial direction. Removing the two fast angle dependences, restricts the physics of the model to low frequency turbulence comparable to the drift frequency of the trapped orbit, $\Omega_d E$. This frequency depends on both invariants, E as made explicit here, and κ. For the sake of simplicity the latter can be ignored, which is the case for deeply trapped particles corresponding to harmonic oscillations, so that the model is effectively reduced to $3D$.

The characteristic frequency is then $\Omega_d/\Omega_0 \propto q_0\rho_*^2/A_0$ where $\rho_* = \rho_0/r_0$, r_0 being a reference minor radius, $q_0 = q(r_0)$ the reference value of the safety factor, and A_0 the reference aspect ratio $A_0 = r_0/R_0$ and Ω_0 the reference ion cyclotron frequency.

Given the description in terms of action-angle variables, the Vlasov equation is readily recovered given the Hamiltonian $\mathcal{H}$ describing the dynamics of the trapped orbits [17], namely:

$$\mathcal{H} = E(1 + \Omega_d\psi) + \Phi \tag{1}$$

Here Φ is the fast angle averaged of the actual electric potential ϕ. It comprises therefore a gyroaverage and a bounce average, hence $\Phi = \mathcal{J}\phi$, where $\mathcal{J}$ is the fast angle averaging operator. For numerical efficiency, the FSE governed by this averaging process is taken into account by a Pade approximation, hence:

$$\mathcal{F} = \mathcal{J}F = \left(1 - \frac{E\delta_0^2}{4}\frac{\partial^2}{\partial\psi\partial\psi}\right)^{-1}\left(1 - \frac{E\rho_0^2}{4}\frac{\partial^2}{\partial\alpha\partial\alpha}\right)^{-1} F \tag{2}$$

The code then evolves the trapped orbit distribution function $\mathcal{F}(\alpha, \psi, \kappa, E)$ given the Vlasov equation with possible volumetric source term $\mathcal{S}$, and diffusion transport $\mathcal{D}$ in buffer regions, Fig.(1):

$$\frac{\partial}{\partial t}\mathcal{F}\Big(\alpha, \psi, \kappa, E\Big) - \Big[\Phi, \mathcal{F}\Big(\alpha, \psi, \kappa, E\Big)\Big] + E\,\Omega_d\frac{\partial}{\partial\alpha}\mathcal{F}\Big(\alpha, \psi, \kappa, E\Big) = \mathcal{D}(\mathcal{F}) + \mathcal{S}(\mathcal{F}) \tag{3}$$

In this expression the Poisson bracket is defined by $[\Phi, \mathcal{F}] = \partial_\psi\Phi\partial_\alpha\mathcal{F} - \partial_\alpha\Phi\partial_\psi\mathcal{F}$, which also readily takes the form of the divergence of a flux, $[\Phi, \mathcal{F}] = \partial_\alpha\Big(\partial_\psi(\Phi)\,\mathcal{F}\Big) + \partial_\psi\Big(-\partial_\alpha(\Phi)\,\mathcal{F}\Big)$. The initial condition is defined as an α-dependent perturbation to an initial equilibrium distribution function:

$$\mathcal{F}(t = 0) = \mathcal{F}_{eq}(\psi, \kappa, E)(1 + \varepsilon\,\zeta(\alpha, \psi)) \tag{4}$$

where $\mathcal{F}_{eq}(\psi, \kappa, E)$ is an equilibrium solution at $\Phi = 0$, since it does not depend on α and where ε is a small parameter that sets the magnitude of the initial perturbation $\zeta(\alpha, \psi)$, a random function, of the equilibrium solution. Unless specified, $\mathcal{F}_{eq}$ is a non-shifted Maxwellian with density and temperature depending on ψ in such a way that this steady state is unstable. In particular, the temperature ranges from T_1 at $\psi = 1$, $\mathcal{F}_{eq}(\psi = 1, \kappa, E) = \mathcal{F}_1$, to T_0 at $\psi = 0$, $\mathcal{F}_{eq}(\psi = 0, \kappa, E) = \mathcal{F}_0$, with constant temperature e-folding length L_T. Conversely the density profile is usually flat. In the $3D$ version of TERESA, no dependence on κ is taken into account. The perturbation yields a modification of the density of trapped orbits.

$$\delta_n(\alpha, \psi) = \varepsilon\int_0^{1^-} J_\kappa d\kappa \int_0^{+\infty} J_E dE\;\; \mathcal{F}_{eq}(\psi, \kappa, E)\zeta(\alpha, \psi) \tag{5}$$

where J_κ and J_E are the Jacobians associated to the change of coordinates.

Boundary conditions of eq.(3) are Dirichlet on the cold side $\mathcal{F}(\psi = 1) = \mathcal{F}_1$ and Dirichlet on the hot side $\mathcal{F}(\psi = 0) = \mathcal{F}_0$ for the Thermal Bath (TB) driven case or Neumann conditions $\partial_\psi\mathcal{F}(\psi = 0) = 0$ for the Flux Driven (FD) case [23]. The diffusion operator $\mathcal{D}(\mathcal{F}) = \partial_\psi(D(\psi)\partial_\psi\mathcal{F})$ with $D(\psi)$ different from zero only close to ψ boundaries, Fig.(1), is a term which locally changes the nature of transport, converting turbulent convection into diffusive transport. In these buffer regions the Trapped Ion Modes are stable (large damping) so that a constant electric potential can be imposed at the boundaries. Furthermore, these buffer regions ensure a thermal contact with the thermal bath conditions with a heat flux:

$$Q_D = -D(\psi)\partial_\psi \int_0^{1^-} J_\kappa d\kappa \int_0^{+\infty} J_E dE \mathcal{F} \tag{6}$$

that depends on the radial gradient of the distribution function. Conversely, in the FD case this flux is vanishing at $\psi = 0$ due to the symmetry boundary condition.

In the Thermal Bath driven case, the control parameter is the temperature difference ΔT between the two boundaries. The heat flux is not constrained and characterises the confinement regime. In the Flux Driven case, the mean heat flux is imposed and the temperature profile then characterises the confinement regime. The volumetric source term $\mathcal{S}(\mathcal{F}) = S_0 S_\psi(\psi) L_1(E) e^{-E}$ is a heating operator that transfers sub-thermal particles into supra-thermal particles. Using the first Laguerre polynomial $L_1(E)$ in E, energy is injected without particles [22, 23]. At each time step, the magnitude of the source S_0 is monitored to avoid generating negative values of the distribution function. This corresponds to some form of saturation of the heating efficiency. Two ψ-profiles $S_\psi(\psi)$ of the source term have been used in FD simulations, Fig.(1): central heating with a gaussian shape centred on $\psi = 0$ in the core buffer region, and off-axis heating with a gaussian shape shifted radially out of the core buffer region. The initial condition of the distribution function is an α-independent function (hence an equilibrium function) with given exponential temperature profile, therefore constant gradient length L_T, perturbed by a random function $\zeta(\alpha, \psi)$ of amplitude ε.

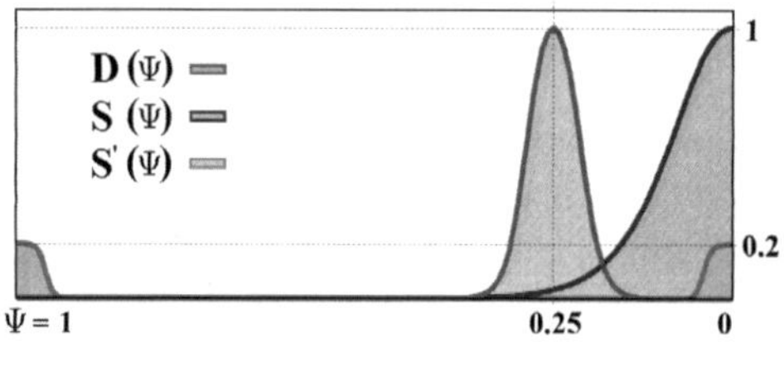

Figure 1. Profiles of the diffusion operator (red), core heating within the core buffer region (blue) and off-axis heating outside the buffer region (green).

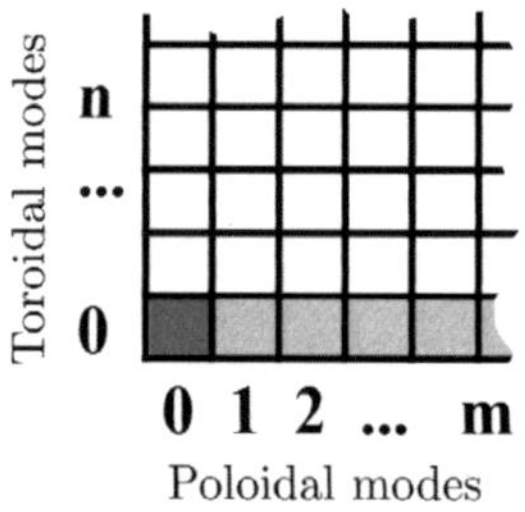

Figure 2. Cancelled modes of the adiabatic response: $5D$ with gyro-averaging only the $(n = 0, m = 0)$ mode is removed (red point) and $4D$ with gyro and bounce averages where one cannot separate the poloidal modes at given toroidal mode, red plus orange points for $n = 0$.

The Maxwell-Gauss equation, in the quasineutral asymptotic limit, is used to compute the electric potential. The finite size effects yield the non-isotropic operator Δ_{FSE} [17].

$$C_e(\phi - \lambda \langle \phi \rangle_\alpha) - C_i \Delta_{FSE}\phi = \frac{1}{\sqrt{\pi} n_{eq}} \int_0^{1^-} J_\kappa d\kappa \int_0^{+\infty} J_E dE \; \mathcal{J}\mathcal{F} - 1 \tag{7}$$

$$\Delta_{FSE} = \rho_0^2 \frac{\partial^2}{\partial \alpha^2} + \delta_0^2 \frac{\partial^2}{\partial \psi^2} \tag{8}$$

C_e and C_i are constant depending on T_i and T_e the ion and electron temperatures as well as on the fraction of trapped particles. The standard adiabatic electron response takes the form

$\delta n_e/n_0 = e(\phi - \langle\phi\rangle_{\theta,\varphi})/T_e$ where the average is a magnetic flux surface average, hence on both the poloidal angle θ and the toroidal angle φ. This enforces that $\langle\delta n_e\rangle_{\theta,\varphi} = 0$. In Fourier space, it corresponds to $\hat{n}_{m,n} = 0$ for $(m,n) = (0,0)$, Fig.(2). Given the electric potential, the only average available after the reduction is $\langle\phi\rangle_\alpha$. For the sake of simplicity, let us assume that the α-average corresponds to the φ-average. The electron response in eq.(7) is then such that $\langle\delta n_e\rangle_\varphi \approx C_e(1-\lambda)\langle\delta\phi\rangle_\varphi$. Hence for $\lambda = 1$, $\langle\delta n_e\rangle_\varphi = 0$. Such a constraint is much stronger that the actual one for adiabatic electrons. In Fourier space it corresponds to $\hat{n}_{m,n} = 0$ for $n = 0$, $\forall m$, Fig.(2). Conversely, the other limit $\lambda = 0$ removes altogether the cancellation of the electron density fluctuation averaged on a magnetic surface. Since the model does not allow one to properly compute the adiabatic electron response, we introduce λ as a free parameter, $0 \leq \lambda \leq 1$, and assume the form chosen in eq.(7). Boundary conditions for the quasineutrality equation are set with $\Phi(\psi = 0) = 0$ which cancels the radial component of the electric drift velocity, $\partial_\alpha\Phi(\psi = 1) = 0$ and $\partial_\alpha\Phi(\psi = 1) = 0$.

The numerical implementation of TERESA $4D$ is similar to that of GYSELA [24]. It is a semi-Lagrangian code using spectral method on α [10]. Several convergence studies have been performed to ensure momentum and energy conservation. The parallelisation scheme is based on MPI and OpenMP with high efficiency from 10 to $10k$-cores. A recent upgrade of TERESA $4D$ allows one to perform simulations using several ion species and electrons [25].

With respect to previously published papers on Trapped Ion turbulence in the kinetic collisionless regime, the present work is based on several novel aspects. The code has been completely rewritten to meet the high standard required by massively parallel computing, as indicated above. As in [22, 23] the system is flux driven and thus departs significantly from the initial work [17]. However, a major difference has been introduced. Indeed, the volumetric source term implemented in the code is an operator taking subthermal particles and changing them into suprathermal particles at constant particle number. The possibility of maintaining a constant heating power then relies on the transport properties of the system that must replenish the subthermal particle population in the heating volume via transport mechanisms. When the latter are too weak, the heating operator generates negative values of the distribution function. This issue was identified in [22, 23] and taken into account by assuming that the code only addressed part of the trapped population, as would a δf code. In fact, this working assumption is not appropriate. We have therefore introduced a saturation mechanism in the heating efficiency as the number of available subthermal particles tends to zero. This prevents negative values of the distribution function at the cost of a time dependent heating power. This more complex but correct approach to the heating mechanism is used to address the critical issue of self-generated transport barriers [6].

Regarding the physics at hand, it is interesting to note that this model is in fact a kinetic model of the Rayleigh-Bénard instability [26] in the 2D approach. Indeed the equations of the the latter, in a fluid approach, combine a vorticity equation for the stream function and a heat transport equation. The coupling between the two equations is governed by the fluid convection of heat on the one hand and on the g-term, a buoyancy force due to the fluid expansion with temperature on the other hand. At lowest order this force is proportional to gravity and the temperature variation. The Trapped Ion turbulence model can be changed by computing the evolution equation of the integral of the distribution function $\mathcal{F}$ using the Vlasov equation (3). Then using the quasineutrality condition, eq.(7), one obtains a vorticity equation for the electric potential. In this equation the drift term proportional to $\Omega_d E$ plays the same role as the g-term in the fluid vorticity equation. One can then draw an equivalence between the stream function and the electric potential. Finally, the Vlasov equation is similar to the heat transport equation of neutral fluids, the distribution function replacing the temperature variation. In the latter, the differences stem from the molecular collision closure of the fluid equation that yields the heat diffusion term. This term competes with heat convection and leads to a threshold for the Rayleigh-Bénard instability. Regarding the vorticity equation, one finds another difference which corresponds to the electron response to the electric potential fluctuations. This term introduces a volumetric loss term. When this term is not null it will tend to stabilise the larger modes and modify the inverse cascade properties.

3. Thermal Bath regimes: Streamers versus Zonal Flows

Zonal Flows [27] and Streamers [28] are two different kinds of structures present in the electric potential that govern large scale transport. Zonal Flows are aligned in the α-direction and thus convect in the direction perpendicularly to the radial one. They do not yield radial transport but generate shear [27]. Conversely streamers are aligned in the ψ direction that directly connect the source to the sink regions of the plasma [28]. The control parameter λ [22] governs the transition from zonal flow dominated transport, $\lambda = 1$, to streamer dominated transport, $\lambda = 0$, see Fig.(3).

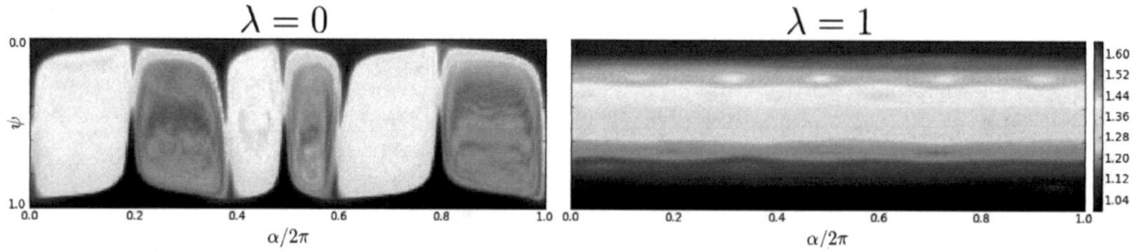

Figure 3. Large scale patterns of the temperature for different values of λ. Left hand side: streamer regime with $\lambda = 0$. Right hand side zonal flow regime with $\lambda = 1$.

Let us monitor the state of the plasma by the α-averaged heat flux $< Q_D >_\alpha$, eq.(6), the zonal flow amplitude measured by $ZF = \int_0^1 |\partial_\psi < \Phi >_\alpha| d\psi$ and the zonal flow shear magnitude ZF' defined by , $ZF' = \int_0^1 |\partial_\psi^2 < \Phi >_\alpha| d\psi$. One then finds that the simulation evolution in the TB case is organised in 5 characteristic steps, Fig.(4):

- a) Linear phase. Streamer structures develop according to the most unstable mode k_α. The heat flux increases exponentially.
- b) Nonlinear saturation phase. The heat flux reaches its maximum value just before the development of zonal flows, increase of ZF.
- c) Relaxation phase. Fast dynamics, apparently chaotic, take place with strong zonal flow and streamer interplay. This step is the most demanding from a numerical point of view because of the fast dynamics and sharp gradients in both (α, ψ) directions.
- d) Bifurcation phase. As the fast dynamics slow down, the system evolves either towards a zonal flow or a streamer dominated case. In some simulations, a slowly decaying streamer regime governs a long transient before switching finally towards a zonal flow dominated regime.
- e) final streamer dominated phase or e') final zonal flow dominated phase. In the latter, the heat flux is quenched while in the former state the heat flux is strong. In both cases the variation of $< Q_D >_\alpha$, ZF and ZF' are small compared to the prior phases. For the streamer dominated state some reorganisation in the number of vortices can take place.

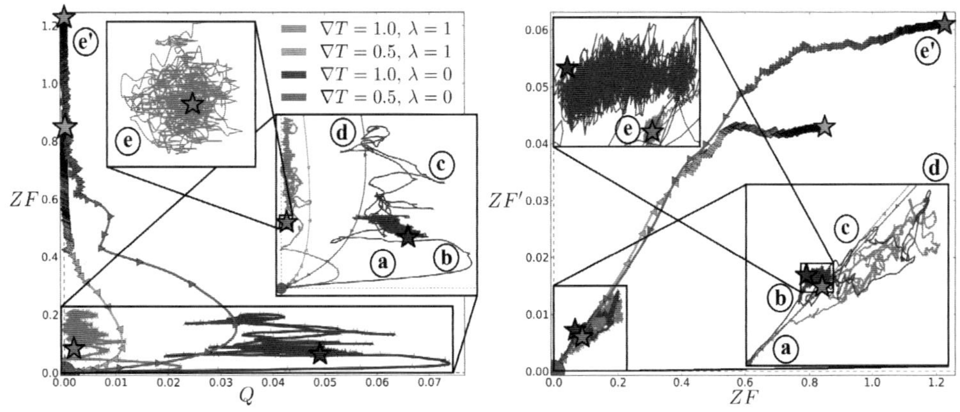

Figure 4. Phase portraits in the planes heat flux, Zonal Flow ($< Q_D >_\alpha, ZF$) and Zonal Flow, Zonal Flow shear (ZF, ZF') of the 5 chronological steps of typical simulations. Blue and red: $\lambda = 0$, green and cyan: $\lambda = 1$. Blue and green: large temperature difference ΔT, red and cyan: small temperature difference ΔT.

While the transition from the linear state a) to the nonlinear quasi-steady state e) is chaotic, it is possible that some memory effects take place. This aspect is related to the streamer pattern and the k_α selection which stems both from the initial noise, which defines the spectrum of injected k_α-modes, and the k_α-dependent growth rate during the linear phase. After that phase, the energy cost to reorganise the k_α pattern can be too large to take place, so that different final states can be reached for otherwise comparable initial conditions.

4. Phase portrait of the TERESA simulation domain

4.1. The numerical investigation

The equations, eqs.(3, 7), initial condition, eq.(5) and boundary conditions determine the parameter space associated to the TERESA toy system. This must be completed by the finite size effects that appear in eq.(8) and the averaging operators $\mathcal{J}$, including the relation $\Phi = \mathcal{J}\phi$. The reduction process that is followed to obtain the TERESA system has led us to introduce several arbitrary control parameters such as λ, that can play a role in the behaviour of the system and lead to differences with respect to the reference case discussed in Section 3.

We aim at describing the properties of the non-linear state, in particular possible interplay between zonal flow pattern and streamer pattern [29], hence between state e) and state e'). The exploration of the impact of 7 control parameters is based on a brute force analysis with some 650 simulations using 85000 CPU.h. For each selected control parameter, between 3 and 10 values distributed in geometric progressions have been tested over specified ranges, see table 1 and table 2. Scans in $1D$ and $2D$ have been performed in order to recover a $7D$ reconstruction of the pattern behaviour. Such an exploration is now possible given TERESA numerical and computational optimisation and the increasing performance of supercomputers.

From a numerical point of view, a wide range of values, table 3, have been used to ensure that numerical error on the energy conservation of the distribution function would not exceed 1%. The number of points in the (α, ψ) space, is N_α and N_ψ respectively. Several runs with different resolution have been performed to check the numerical convergence. A similar check has been performed for the time step Δt and regarding the energy E, depending on the maximum value E_{max} and the number of energy values N_E, which determine the precision of the computation

Parameters	λ	ρ_0	δ_0	Ω_d
Ref case	0.5	0.06	0.03	1
Interval	[0 : 1]	[0.0003 : 0.12]	[0.003 : 0.12]	[0.1 : 1]

Table 1. Values used for the scan of the "internal" control parameters

Parameters	ΔT	S_0	ϵ
Ref case	0.5	TB	10^{-4}
Interval	[0.125 : 2]	TB/FD [0 : 2]	$[10^{-6} : 10^{-2}]$

Table 2. Values used for the scan of the "external" control parameters

of the trapped orbit density in eq.(7). In the present analysis the trapping parameter is kept constant.

Parameters	N_α	N_ψ	N_κ	N_E	E_{max}	Δt
Ref case	257	129	1	48	25	$[10^{-5} : 10^{-2}]$
Interval	[129 : 4097]	[129 : 257]	[1]	[48 : 144]	[25 : 50]	$[10^{-8} : 10^{-2}]$

Table 3. Range of values used in the numerical tests, and retained values of the parameters that control the numerical precision of TERESEA.

4.2. Changing the response of the adiabatic electrons

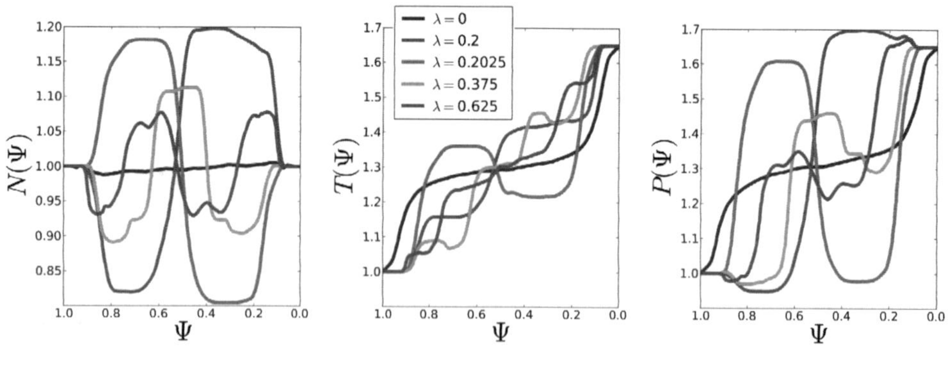

Figure 5. Density, temperature and pressure of trapped orbits for different values of λ.

In Section 3, the two extreme values of λ are used, namely 0 and 1. Extending this analysis to intermediate values allows one to modify the step like pattern of the profiles. The trapped orbit density and pressure are computed as the zero and first energy moment of the distribution function $\mathcal{F}$, the ratio of these moments defining the temperature, see Fig.(5). First, one can observe non-monotonic profiles with temperature inversion which underlines the very efficient insulation property of the zonal flow shear layers, red curve for $\lambda = 0.2025$. A small variation of λ governs a strong change of the profile, see $\lambda = 0.2$, which exhibits a monotonic temperature profile, Fig.(5). Furthermore, one also finds a variation in the number of steps, essentially leading to a doubling of the number steps as λ increases. Odd numbers of steps have not been observed: this is likely governed by symmetry and momentum conservation. The control parameter C_i and C_e eq.(7) can play a similar role as the parameter λ or modify the finite size effects that are discussed in the following section. However, their impact is not addressed in the present study.

4.3. Finite Size Effects

The number of degrees of freedom is determined by ρ_0 and δ_0. The first one determines the number of structures in the α direction while the second constrains the ψ direction. Comparing TERESA simulations to more standard gyrokinetic simulations one can state that δ_0 is equivalent to the standard ρ_* parameter while δ_0/ρ_0 is a measure of shaping properties, in particular the elongation [30]. Small values of ρ_0 and δ_0 yield a large number of structures within the simulation box. Note that the radial size of the simulation region is used to normalise ψ so that: $L_\alpha \times L_\psi = [0 : 2\pi] \times [0 : 1]$.

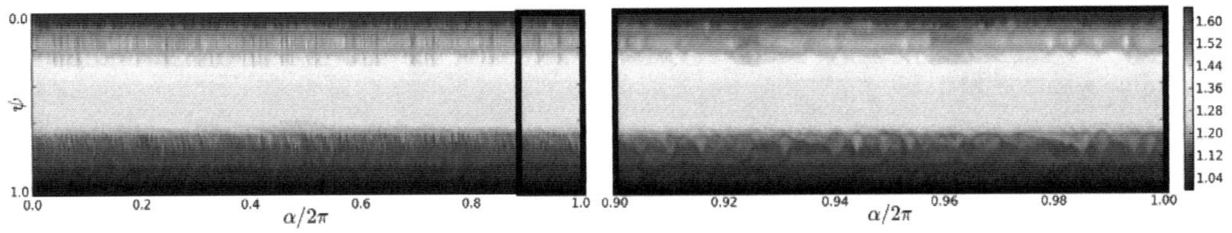

Figure 6. Temperature of trapped orbits for $2\pi/\rho_0 = 20500$ during phase e'), factor 10 zoom on the right hand side.

Since ρ_0 controls the number of modes in the α-direction, it determines the number of streamers Fig.(6) and consequently the heat flux from one thermal bath to the other. However the threshold value for switching from streamers to zonal flows does not appear to depend on ρ_0, Fig.(7 (top row)). The parameter δ_0 modifies the zonal flow pattern. It yields a comparable effect as the parameter λ on the transition between streamer and zonal flow dominated states, Fig.(7(bottom row)). This result is in line with the fact that both parameters play similar role regarding the zonal flow, mode 0, in the quasi-neutrality equation.

4.4. Heat source effect on transport

As readily expected for the present Rayleigh-Bénard setting, the temperature difference ΔT governs the instability threshold. Otherwise it appears to essentially change the steps height, and consequently the gradients with standard consequences on the numerical robustness of the simulation. Similarly, Ω_d is a key parameter in determining the critical temperature difference $\Delta T_c \propto \Omega_d/(1 + \mathcal{O}(\delta_0^2) + \mathcal{O}(\rho_0^2))$. This is readily expected since, the energy normalisation, and consequently the temperature, are equivalent control parameters in the Vlasov equation.

The last part of this study is based on two series of simulations exploring the FD approach for on axis and off-axis heating, see Fig.(1). For on-axis heating, the diffusion buffer region ensures the contact between the source and the turbulent plasma and preserves the symmetry in the α-direction. This heating leads to the zonal flow dominated regime. The transport is quenched and no steady state regime can be achieved. Similar observations have been made in ELM-free H-modes [31, 32, 33]. In the streamer dominated regime, with off-axis heating, standard large scale convection can be achieved. However, as in the ELM-free experiments, stable conditions

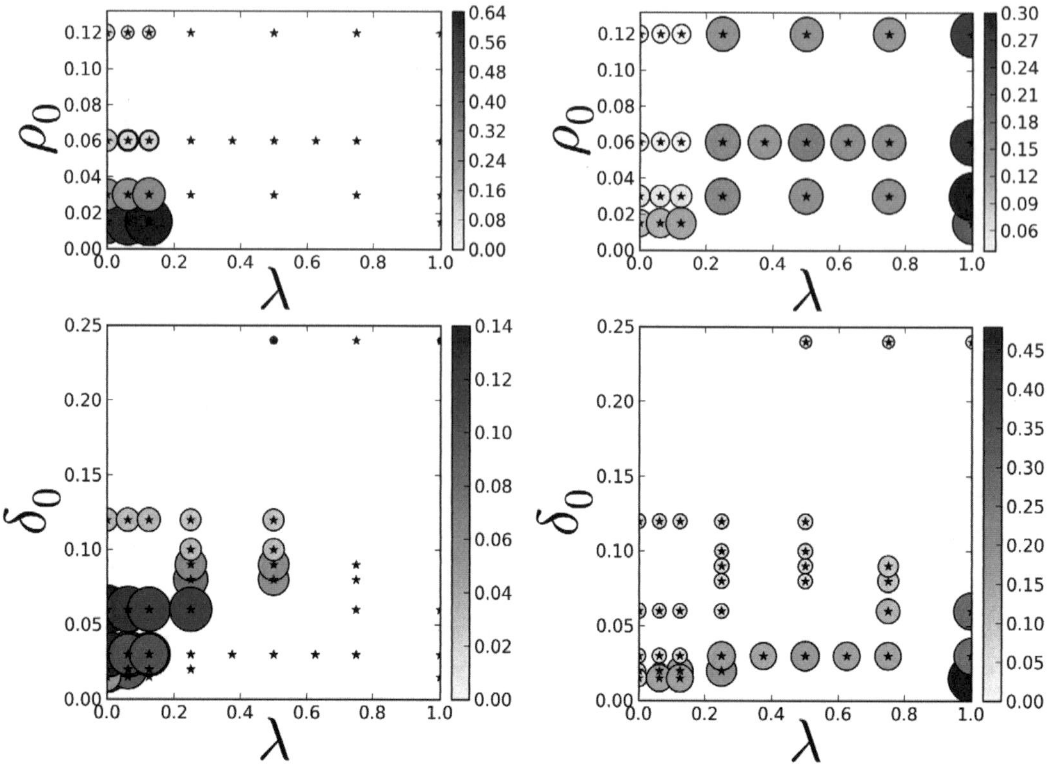

Figure 7. Heat flux $< Q_D >_\alpha$ (left column) and ZF (right column) depending on λ and ρ_0 (top row) or δ_0 (bottom row). Colour and radius of circles are proportional to the magnitude of $< Q_D >_\alpha$ or ZF respectively. Black stars locate simulations that have been analysed.

are only achieved with a power step-down procedure [33].

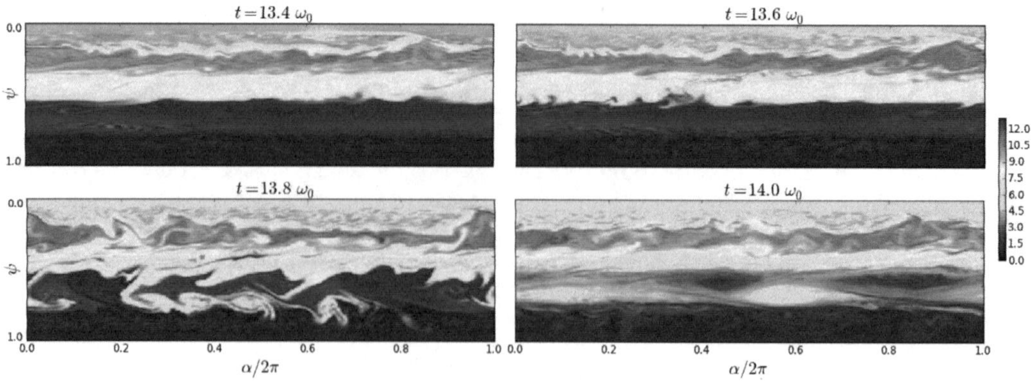

Figure 8. Pressure of trapped particle orbits during a barrier relaxation: before (top left), relaxation onset (top right), relaxation (bottom left) and barrier restoration (bottom right).

Off-axis heating is observed to have less efficient coupling to the plasma. Indeed, on-axis

heating is combined to the strong diffusion in the core buffer region that replenishes efficiently the distribution function, hence preventing negative values. In the off-axis case, turbulent transport is too weak to prevent the self regulation of the magnitude of the heating source to come into play. The coupling of the heating source is then switched-off to avoid generating negative values in the distribution function.
The interesting point with these TERESA simulations is to understand why the corrugation results are particularly stiff, generally preventing steady state regimes in Flux Driven simulations. Indeed, the corrugation dynamics observed in GYSELA [9] do not lead to such strong barriers [3]. Furthermore, regarding simulations with barriers, secondary instabilities occur leading to Barrier Relaxation Modes, as in GYSELA kinetic simulations [5, 4] and in edge fluid simulations [34, 35]. The case with off-axis heating is still under investigation to understand what process unlocks the transport barrier, leading to quasi-periodic relaxation events, Fig.(8). Although the overall heat transport remains small compared to the streamer regime, these relaxation events contribute to sufficient energy transfer to reach steady state regimes. As observed in the fluid simulations, a Dupree like mechanism, with competition between the growth of instabilities and large damping processes [35] can delay the relaxation mechanism so that a weak energy source is mandatory to allow the perturbation to grow and trigger the relaxation before too large gradients lock the barrier. This mechanism would then be reminiscent of the power step-down experiments [33]. However, further analysis is required to identify the actual mechanism at hand.

5. Discussion and Conclusion

The turbulence governed by trapped ion instabilities provides a minimum kinetic model when averaging out two fast angles, that of the gyro-motion and that of the bounce-motion. In the limit of deeply trapped ions, when the trapping parameter is nearly constant, one can further reduce the model to 3D: 2D space conjugate variables and the energy –setting the trapping parameter to a single value. This interchange instability model is akin to a kinetic Rayleigh Bénard model. It provides a toy model, reduced to $3D$, which is implemented in the TERESA code.

In the reduction procedure, several free parameters are introduced in the model leading to a large parameter space where turbulence can be investigated. The zonal flow streamer interaction is addressed by using the control parameter that determines the response of the adiabatic electrons. One then finds that the self-organisation of the system leads either to streamer dominated solutions, with large heat flux, and thus low confinement, or to zonal flow dominated solutions, where the radial transport is locked. In the latter limit, staircase temperature profiles develop with marked corrugations. Tuning the control parameter allows one to change the number of steps in the temperature profiles and even to achieve temperature reversal.

A brute force exploration of the parameter space of the model has allowed one to draw a first series of phase portraits of the system. On the overall, the system is observed to be rather stiff, switching between streamer and zonal flow states, sometimes after very long transients, but then remaining in steady state. Quasiperiodic relaxation cycles between these two states, as characterised by predator-prey models, is most often not observed. It appears that off-axis heating seems to allow the onset of a secondary instability and subsequent relaxation events. Further analysis of the model is under-way to complete the description of the parameter space. This reduced model, with versatile and efficient numerics, is a powerful tool to investigate the physics at play in more relevant and complete models of plasma turbulence. Furthermore, it provides a means to investigate in great detail the kinetic features, an aspect that is most often overlooked.

Acknowledgments

The authors acknowledge the fruitful and lively discussions held during the Festival de Théorie in Aix-en-Provence. They are most indebted to Pat Diamond for his strong support and very active scientific participation to this event. This work is carried out in the framework of the ANR-GYPSI and ANR-ESPOIR projects (ANR-10-BLAN-941 and ANR-09-BLAN-0035-01). The computation has been performed at the Mésocentre d'Aix-Marseille Université.

References

[1] M. Beurskens, L. Frassinetti, C. Challis, C. Giroud, S. Saarelma, B. Alper, C. Angioni, P. Bilkova, C. Bourdelle, S. Brezinsek, P. Buratti, G. Calabro, T. Eich, J. Flanagan, E. Giovannozzi, M. Groth, J. Hobirk, E. Joffrin, M. Leyland, P. Lomas, E. de la Luna, M. Kempenaars, G. Maddison, C. Maggi, P. Mantica, M. Maslov, G. Matthews, M.-L. Mayoral, R. Neu, I. Nunes, T. Osborne, F. Rimini, R. Scannell, E. Solano, P. Snyder, I. Voitsekhovitch, P. de Vries, and J.-E. Contributors, "Global and pedestal confinement in jet with a be/w metallic wall," *Nuclear Fusion*, vol. 54, no. 4, p. 043001, 2014.

[2] J. Jacquinot and the JET team, "Deuterium-tritium operation in magnetic confinement experiments: results and underlying physics," *Plasma Physics and Controlled Fusion*, vol. 41, no. 3A, p. A13, 1999.

[3] E. Floriani, G. Ciraolo, P. Ghendrih, R. Lima, and Y. Sarazin, "Self-regulation of turbulence bursts and transport barriers," *Plasma Physics and Controlled Fusion*, vol. 55, no. 9, p. 095012, 2013.

[4] A. Strugarek, Y. Sarazin, D. Zarzoso, J. Abiteboul, A. S. Brun, T. Cartier-Michaud, G. Dif-Pradalier, X. Garbet, P. Ghendrih, V. Grandgirard, G. Latu, C. Passeron, and O. Thomine, "Unraveling quasiperiodic relaxations of transport barriers with gyrokinetic simulations of tokamak plasmas," *Phys. Rev. Lett.*, vol. 111, p. 145001, Oct 2013.

[5] A. Strugarek, Y. Sarazin, D. Zarzoso, J. Abiteboul, A. S. Brun, T. Cartier-Michaud, G. Dif-Pradalier, X. Garbet, P. Ghendrih, V. Grandgirard, G. Latu, C. Passeron, and O. Thomine, "Ion transport barriers triggered by plasma polarization in gyrokinetic simulations," *Plasma Physics and Controlled Fusion*, vol. 55, no. 7, p. 074013, 2013.

[6] G. Dif-Pradalier, P. H. Diamond, V. Grandgirard, Y. Sarazin, J. Abiteboul, X. Garbet, P. Ghendrih, A. Strugarek, S. Ku, and C. S. Chang, "On the validity of the local diffusive paradigm in turbulent plasma transport," *Phys. Rev. E*, vol. 82, p. 025401, Aug 2010.

[7] X. Garbet, Y. Idomura, L. Villard, and T. Watanabe, "Gyrokinetic simulations of turbulent transport," *Nuclear Fusion*, vol. 50, no. 4, p. 043002, 2010.

[8] M. Nakata and Y. Idomura, "Plasma size and collisionality scaling of ion-temperature-gradient-driven turbulence," *Nuclear Fusion*, vol. 53, no. 11, p. 113039, 2013.

[9] P. Ghendrih, "Phase space structures in gyrokinetic simulations of fusion plasma turbulence." to be published in *Eur. Phys. J. D.*

[10] Cartier-Michaud, Thomas, Ghendrih, Philippe, Grandgirard, Virginie, and Latu, Guillaume, "Optimizing the parallel scheme of the poisson solver for the reduced kinetic code teresa*," *ESAIM: Proc.*, vol. 43, pp. 274–294, 2013.

[11] B. B. Kadomtsev and O. P. Porgutse, "Turbulence in toroidal systems," *Reviews of Plasma Physics*, vol. 5, p. 249, 1970.

[12] M. Tagger, G. Laval, and R. Pellat, "Trapped ion mode driven by ion magnetic drift resonance in a fat torus," *Nuclear Fusion*, vol. 17, no. 1, p. 109, 1977.

[13] W. M. Tang, J. C. Adam, and D. W. Ross, "Residual trapped-ion instabilities in tokamaks," *Physics of Fluids (1958-1988)*, vol. 20, no. 3, pp. 430–435, 1977.

[14] H. Biglari, P. H. Diamond, and M. N. Rosenbluth, "Toroidal ion-pressure-gradient-driven drift instabilities and transport revisited," *Physics of Fluids B: Plasma Physics (1989-1993)*, vol. 1, no. 1, pp. 109–118, 1989.

[15] X. Garbet, L. Laurent, F. Mourgues, J. P. Roubin, A. Samain, X. L. Zou, and J. Chinardet, "Ionic instability thresholds in tokamaks," *Physics of Fluids B: Plasma Physics (1989-1993)*, vol. 4, no. 1, pp. 136–144, 1992.

[16] M. Artun, W. M. Tang, and G. Rewoldt, "Trapped ion mode in toroidally rotating plasmas," *Physics of Plasmas (1994-present)*, vol. 2, no. 9, pp. 3384–3400, 1995.

[17] G. Depret, X. Garbet, P. Bertrand, and A. Ghizzo, "Trapped-ion driven turbulence in tokamak plasmas," *Plasma Physics and Controlled Fusion*, vol. 42, no. 9, p. 949, 2000.

[18] Norscini, "Turbulent transport close to marginal instability: role of the source driving the system out of equilibrium." 2014.

[19] Y. Sarazin and P. Ghendrih, "Intermittent particle transport in two-dimensional edge turbulence," *Physics of Plasmas (1994-present)*, vol. 5, no. 12, pp. 4214–4228, 1998.

[20] A. M. Dimits, G. Bateman, M. A. Beer, B. I. Cohen, W. Dorland, G. W. Hammett, C. Kim, J. E. Kinsey,

M. Kotschenreuther, A. H. Kritz, L. L. Lao, J. Mandrekas, W. M. Nevins, S. E. Parker, A. J. Redd, D. E. Shumaker, R. Sydora, and J. Weiland, "Comparisons and physics basis of tokamak transport models and turbulence simulations," *Physics of Plasmas (1994-present)*, vol. 7, no. 3, pp. 969–983, 2000.

[21] A. J. Brizard and T. S. Hahm, "Foundations of nonlinear gyrokinetic theory," *Rev. Mod. Phys.*, vol. 79, pp. 421–468, Apr 2007.

[22] Y. Sarazin, V. Grandgirard, E. Fleurence, X. Garbet, P. Ghendrih, P. Bertrand, and G. Depret, "Kinetic features of interchange turbulence," *Plasma Physics and Controlled Fusion*, vol. 47, no. 10, p. 1817, 2005.

[23] G. Darmet, P. Ghendrih, Y. Sarazin, X. Garbet, and V. Grandgirard, "Intermittency in flux driven kinetic simulations of trapped ion turbulence," *Communications in Nonlinear Science and Numerical Simulation*, vol. 13, no. 1, pp. 53 – 58, 2008. Vlasovia 2006: The Second International Workshop on the Theory and Applications of the Vlasov Equation.

[24] V. Grandgirard, Y. Sarazin, P. Angelino, A. Bottino, N. Crouseilles, G. Darmet, G. Dif-Pradalier, X. Garbet, P. Ghendrih, S. Jolliet, G. Latu, E. Sonnendrcker, and L. Villard, "Global full- f gyrokinetic simulations of plasma turbulence," *Plasma Physics and Controlled Fusion*, vol. 49, no. 12B, p. B173, 2007.

[25] D. et al., "Competition between trapped ion and trapped electron instabilities," in *41st EPS Conference on Plasma Physics, P2.039*, 2014.

[26] M. C. Cross and P. C. Hohenberg, "Pattern formation outside of equilibrium," *Rev. Mod. Phys.*, vol. 65, pp. 851–1112, Jul 1993.

[27] P. H. Diamond, S.-I. Itoh, K. Itoh, and T. S. Hahm, "Zonal flows in plasma: a review," *Plasma Physics and Controlled Fusion*, vol. 47, no. 5, p. R35, 2005.

[28] A. M. Dimits, J. F. Drake, A. B. Hassam, and B. Meerson, "Formation of streamers in plasma with an ion temperature gradient," *Physics of Fluids B: Plasma Physics (1989-1993)*, vol. 2, no. 11, pp. 2591–2599, 1990.

[29] S. Champeaux and P. Diamond, "Streamer and zonal flow generation from envelope modulations in drift wave turbulence," *Physics Letters A*, vol. 288, no. 34, pp. 214 – 219, 2001.

[30] P. Angelino, X. Garbet, L. Villard, A. Bottino, S. Jolliet, P. Ghendrih, V. Grandgirard, B. F. McMillan, Y. Sarazin, G. Dif-Pradalier, and T. M. Tran, "Role of plasma elongation on turbulent transport in magnetically confined plasmas," *Phys. Rev. Lett.*, vol. 102, p. 195002, May 2009.

[31] T. Shoji, H. Tamai, Y. Miura, M. Mori, H. Ogawa, A. Leonard, T. Jensen, A. Hyatt, A. Howald, G. Fuchs, N. Ohyabu, N. Asakura, T. Fujita, M. Shimada, S. Tsuji, H. Maeda, H. Aikawa, K. Hoshino, S. Kasai, T. Kawakami, H. Kawashima, M. Maeno, T. Matsuda, T. Ogawa, T. Seike, N. Suzuki, K. Uehara, T. Yamamoto, T. Yamauchi, T. Hamano, K. Hasegawa, A. Honda, M. Kazawa, Y. Kashiwa, K. Kikuchi, H. Okano, E. Sato, N. Seki, T. Shibata, T. Shiina, S. Suzuki, T. Tani, T. Tokutake, Y. Tomiyama, and T. Yamasato, "Effects of ergodization on plasma confinement in jft-2m," *Journal of Nuclear Materials*, vol. 196198, no. 0, pp. 296 – 300, 1992. Plasma-Surface Interactions in Controlled Fusion Devices Proceedings of the Tenth International Conference on Plasma-Surface Interactions in Controlled Fusion Devices.

[32] H. Tamai, T. Shoji, K. Nagashima, Y. Miura, T. Yamauchi, H. Ogawa, H. Kawashima, T. Matsuda, M. Mori, K. Ida, and S. Ohdachi, "Improvement of the density limit with an external helical field on jft-2m tokamak," *Journal of Nuclear Materials*, vol. 220222, no. 0, pp. 365 – 369, 1995. Plasma-Surface Interactions in Controlled Fusion Devices.

[33] F. Marcus, B. Balet, N. Deliyanakis, R. Giannella, C. Gowers, O. Jarvis, T. Jones, R. Koenig, P. Lomas, F. Rimini, P. Smeulders, and P. Thomas, "A power step-down approach to extended high performance of elm-free h modes in jet," *Nuclear Fusion*, vol. 37, no. 8, p. 1067, 1997.

[34] P. Beyer, S. Benkadda, G. Fuhr-Chaudier, X. Garbet, P. Ghendrih, and Y. Sarazin, "Turbulence simulations of transport barrier relaxations in tokamak edge plasmas," *Plasma Physics and Controlled Fusion*, vol. 49, no. 4, p. 507, 2007.

[35] P. Beyer, *Transport Barrier Relaxations in Tokamak Edge Plasmas*, vol. Relaxation Dynamics in Laboratory and Astrophysical Plasmas of *Reviews of the Theory of Magnetized Plasmas*, ch. 8, pp. 255–268. world scientific ed., 2009.

Formation of plasmoid chains in fusion relevant plasmas

L Comisso[1], D Grasso[1] and F L Waelbroeck[2]

[1] Dipartimento Energia, Politecnico di Torino, Corso Duca degli Abruzzi 24, 10129, Torino, and Istituto dei Sistemi Complessi - CNR, Via dei Taurini 19, 00185, Roma, Italy
[2] Institute for Fusion Studies, The University of Texas at Austin, Austin, TX 78712-1060, USA

E-mail: `luca.comisso@polito.it`

Abstract. The formation of plasmoid chains is explored for the first time within the context of the Taylor problem, in which magnetic reconnection is driven by a small amplitude boundary perturbation in a tearing-stable slab plasma equilibrium. Numerical simulations of a magnetohydrodynamical model of the plasma show that for very small plasma resistivity and viscosity, the linear inertial phase is followed by a nonlinear Sweet-Parker evolution, which gives way to a faster reconnection regime characterized by a chain of plasmoids instead of a slower Rutherford phase.

1. Introduction

Magnetic reconnection is a fundamental plasma process that changes the topology of the magnetic field lines and has profound consequences in a wide variety of different phenomena in laboratory, space and astrophysical plasmas [1]. It can be spontaneous and/or forced. In the latter case, one of the most important paradigms is the so-called Taylor problem, which consists in the study of the response of a tearing-stable slab plasma equilibrium to a small amplitude boundary perturbation that drives magnetic reconnection. This problem was investigated for the first time in a seminal work by Hahm and Kulsrud [2]. Adopting a resistive magnetohydrodynamic (MHD) model of the plasma, they showed that the reconnection process evolves through five dynamical phases, the first four dominated by linear layer physics, and the fifth one characterized by a nonlinear Rutherford phase [3].

The analysis by Hahm and Kulsrud was reconsidered some years later by Wang and Bhattacharjee, who showed that under certain conditions the reconnection process passes into the nonlinear regime according to a Sweet-Parker evolution which gives way, on the long time scale of resistive diffusion, to the Rutherford regime [4]. Subsequently, an expression for the threshold perturbation amplitude required to trigger the Sweet-Parker phase in a visco-resistive plasma was derived in a careful work by Fitzpatrick [5].

The Taylor problem is of great interest within the fusion community, mainly because it represents a convenient model to describe magnetic reconnection processes due to resonant magnetic perturbations. However, despite the numerous works related to this problem (see, e.g., [6-14]), in the framework of resistive and visco-resistive MHD the nonlinear possible scenarios of the Taylor paradigm have not changed from those outlined in Refs. [2, 4, 5].

In this paper we show a different possible scenario, in which the Sweet-Parker phase identified by Wang and Bhattacharjee does not lead to a slower Rutherford evolution, but to a faster regime characterized by the formation of plasmoid chains. Actually, it is known from quite a long time that elongated current sheets can become unstable to the formation of plasmoids (secondary islands) [15-18]. Nevertheless, the eventuality that the current sheet may be unstable has never been taken into account whithin the context of the Taylor problem.

In the last years there has been a growing interest to investigate the plasmoid instability of thin current sheets due to the fact that the plasmoid formation facilitates faster reconnection [19-30]. Here we show that the development of plasmoids is possible also whithin the Taylor paradigm and that strongly affects the reconnection rate. Interestingly, a recent work by Dewar and coworkers [31] have found the existence of MHD equilibria with plasmoids as solutions of the Taylor geometry. However, this work was concerned only with finding static equilibrium solutions.

2. Basic equations and geometry of the viscous Taylor problem

We consider a visco-resistive MHD description of the plasma. Within this description the equation of motion of the plasma takes the form

$$\rho\left(\partial_t + \mathbf{v}\cdot\nabla\right)\mathbf{v} = -\nabla p + \frac{1}{c}\mathbf{j}\times\mathbf{B} + \rho\nu\nabla^2\mathbf{v}\,, \tag{1}$$

whereas the induction equation reads

$$\partial_t\mathbf{B} = \nabla\times(\mathbf{v}\times\mathbf{B}) + D_\eta\nabla^2\mathbf{B}\,, \tag{2}$$

where $\mathbf{v}$ and p are the velocity and pressure of the plasma, respectively, $\mathbf{B}$ is the magnetic field, $\mathbf{j} = (c/4\pi)\nabla\times\mathbf{B}$ denotes the electric current density, and ρ stands for the mass density. The kinematic viscosity can be recognized as ν, while $D_\eta = \eta c^2/4\pi$ is the magnetic diffusivity. If we further assume that the plasma is incompressible and homogeneous, then the mass density is constant in space and time, and the conservation of mass $\partial_t\rho + \nabla\cdot(\rho\mathbf{v}) = 0$ reduces to the condition $\nabla\cdot\mathbf{v} = 0$.

We also consider a two-dimensional dynamics with $\partial_z = 0$ for all the fields. In the presence of a strong and nearly constant guide field B_z, and if $\beta = p/\left(B_z^2/8\pi\right) \ll 1$, the components v_z and B_z decouple from $\mathbf{v}_\perp$ and $\mathbf{B}_\perp$. In this case, after the normalization

$$\left(L\nabla,\,\frac{t}{\tau_A},\,\frac{\mathbf{B}}{B_0}\right) \to (\nabla,\,t,\,\mathbf{B})\,, \tag{3}$$

where $\tau_A = L/(B_0/\sqrt{4\pi\rho})$, and after introducing the stream and flux functions $\phi(x,y,t)$ and $\psi(x,y,t)$ such that the normalized in-plane velocity can be written as $\mathbf{v}_\perp = \mathbf{e}_z\times\nabla\phi$ and the normalized in-plane magnetic field as $\mathbf{B}_\perp = \nabla\psi\times\mathbf{e}_z$, the equations (1) and (2) may be reduced to [32]

$$\partial_t\omega_z + [\phi,\omega_z] = [j_z,\psi] + \nu\nabla^2\omega_z\,, \tag{4}$$

$$\partial_t\psi + [\phi,\psi] = -\eta j_z + E_0\,. \tag{5}$$

The canonical Poisson brackets are given as usual by $[f,g] = \partial_x f\partial_y g - \partial_y f\partial_x g$. Furthermore, $\omega_z = \nabla_\perp^2\phi$, $j_z = -\nabla_\perp^2\psi$, and $\nabla_\perp = \mathbf{e}_x\partial_x + \mathbf{e}_y\partial_y$. Notice also that in Eq. (5) we have added a term representing an externally applied electric field $E_0 = \eta j_z^{(0)}$, which is required to maintain the equilibrium magnetic field in the presence of a small but finite electrical resistivity η.

In accordance with the Taylor problem, we assume a slab plasma which is bounded by perfectly conducting walls at $x = \pm L_x/2$ and is periodic in the y direction with period L_y. Furthermore, we consider a tearing-stable equilibrium given by

$$\psi^{(0)}(x) = -x^2/2\,, \qquad \phi^{(0)}(x) = 0\,. \tag{6}$$

Therefore, ${B_y}^{(0)}(x) = x$, ${j_z}^{(0)}(x) = 1$, and ${B_x}^{(0)}(x) = {v_x}^{(0)}(x) = {v_y}^{(0)}(x) = {\omega_z}^{(0)}(x) = 0$. For this set-up we have (in unnormalized units) $L = L_x$ and $B_0 = \Delta B_y^{(0)} = {B_y}^{(0)}(L_x/2) - {B_y}^{(0)}(-L_x/2)$.

Then, we suppose that the conducting walls are subject to a sudden displacement

$$x_w = \pm 1/2 \mp \Xi(t)\cos(ky)\,, \tag{7}$$

where $k = 2\pi/L_y$ and $\Xi(t) = \Xi_0\left(1 - e^{-t/\tau} - (t/\tau)e^{-t/\tau}\right)$ for $t \geq 0$ [8]. Since the displacement is small, i.e. $\Xi_0 \ll 1/2$, we can adopt the following boundary conditions at the flux conserving walls

$$\psi(\pm 1/2, y, t) = -1/8 - \Psi(t)\cos(ky)\,, \qquad j_z(\pm 1/2, y, t) = 1\,, \tag{8}$$

$$\phi(\pm 1/2, y, t) = \pm\frac{d\Xi(t)}{dt}\frac{\sin(ky)}{k}\,, \qquad \omega_z(\pm 1/2, y, t) = 0\,, \tag{9}$$

where $\Psi(t) = \left.\dfrac{d\psi^{(0)}}{dx}\right|_{x=-1/2} \Xi(t)$.

3. Reconnection rate and plasmoid formation

We solve Eqs. (4) and (5) numerically by splitting all the fields in the time independent equilibrium and an evolving perturbation, which is advanced in time according to a third order Adams-Bashforth algorithm. We employ a compact finite difference algorithm on a non-equispaced grid to compute the spatial operations in the x direction, while a pseudospectral method is used for the periodic direction. We adopt a space discretization of 512×8192 grid points, so that the minimum step size in the x direction is $dx_{\min} = 8.04 \times 10^{-4}$, whereas in the y direction we have $dy = 3.07 \times 10^{-3}$ ($L_y = 8\pi$). Convergence studies have been conducted to ensure that the results do not change when increasing the resolution.

In the following, we consider the results of two simulations with the same boundary perturbation $\Psi_0 = d\psi^{(0)}/dx|_{x=-1/2}\Xi_0$, but different values of the plasma resistivity and viscosity. These values correspond to $\eta = 10^{-5}$, $\nu = 10^{-4}$, in one case, and $\eta = 10^{-8}$, $\nu = 10^{-7}$, in the other case. Therefore, while the Lundquist number $S = 1/\eta$ is different, the Prandtl number $P_m = \nu/\eta$ does not change.

Fig. 1 shows clearly that the time evolution of the reconnection rate, defined as

$$R(t) = \frac{d}{dt}\Big(\max\left(\psi(0, y, t)\right) - \min\left(\psi(0, y, t)\right)\Big)\,, \tag{10}$$

is very different in the two simulations. As we shall see later, this is because the forced reconnection process evolves into a regime characterized by the formation of plasmoid chains when the plasma resistivity and viscosity are very small.

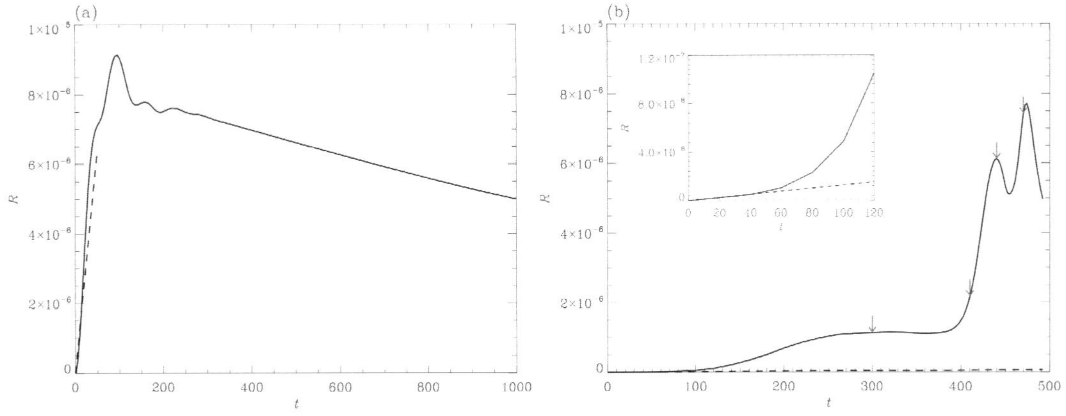

Figure 1. Reconnection rate $R(t)$ for (a) $\eta = 10^{-5}$, $\nu = 10^{-4}$, and (b) $\eta = 10^{-8}$, $\nu = 10^{-7}$. Furthermore, $\Psi_0 = 1 \times 10^{-2}$, $\tau = 1$, and $L_y = 8\pi$. Therefore, the usual tearing stability parameter for the equilibrium (6) and wave number $k = 2\pi/L_y$ is $\Delta' = \ 2k/\tanh{(k\,L_x/2)} \approx -4.02$. The theoretical reconnection rates in the inertial regime are represented by dashed lines. The inset shows a zoom at early times. Red arrows correspond to the time when the current density contours are shown in Fig. 2.

In both cases there is an initial inertial phase for $t \ll \nu^{-1/3}k^{-2/3}$. Neglecting the very small initial transient, the perturbed flux function at the X-point is [2]

$$\delta\psi_{\rm i}(t) = \frac{\eta}{\pi}\frac{2k^2\Psi_0}{\sinh{(k\,L_x/2)}}\,t^2\,. \tag{11}$$

Therefore, since in this phase the X and O points of the driven magnetic island are located in correspondence of the maximum and minimum of the perturbed flux function, the reconnection rate defined by Eq. (10) can be evaluated as

$$R_{\rm i}(t) = \frac{\eta}{\pi}\frac{8k^2\Psi_0}{\sinh{(k\,L_x/2)}}\,t\,. \tag{12}$$

From Fig. 1 we can see that in both simulations the initial phase evolves in accordance with the above formula represented in the plots by dashed lines. During this inertial phase, which is characterized by a nonconstant-ψ behaviour, a current layer builds up along the resonant surface $x = 0$.

After the inertial phase, the evolution of the system can be very different depending on the magnitude of the boundary perturbation with respect to that of the relevant plasma parameters. Fig. 1(a) shows the reconnection evolving as predicted by Hahm and Kulsrud [2]. In this case, the inertial phase is followed by a linear constant-ψ phase. For $P_m = \nu/\eta \gg 1$ this phase is known as the visco-resistive regime, which occurs for $t \gg \nu^{1/3}\eta^{-2/3}k^{-2/3}$ [5]. In our simulation the system remains in the visco-resistive regime until full reconnection is achieved. However, if the amplitude of the boundary perturbation is sufficiently large to allow the system to enter into the nonlinear regime, the constant-ψ phase is followed by a Rutherford evolution [3, 2, 5].

On the contrary, if the driven magnetic island is comparable or exceeds the linear layer width while it is still in the nonconstant-ψ phase, the system passes into the nonlinear regime (at $t \approx 200$ for the simulation shown in Fig. 1(b)) according to a Sweet-Parker evolution instead of a Rutherford one [4]. In this case the magnetic island is characterized by a helicity-conserving dynamics that produces an elongated current sheet [33] and leads to a reconnection rate $R \propto \eta^{3/4}\nu^{-1/4}$ for $P_m \gg 1$ [34]. The threshold perturbation amplitude required to trigger the Sweet-Parker phase is [5]

$$\Psi_0 \gtrsim \frac{2k}{\sinh(k\,L_x/2)}\frac{(\eta\nu)^{1/6}}{k^{1/3}}, \tag{13}$$

which tells us that in high-performance tokamaks the Sweet-Parker phase is facilitated because of the smallness of the plasma resistivity and viscosity.

Wang and Bhattacharjee concluded that the Sweet-Parker phase gives way to a slower Rutherford regime [4]. However, this is only one possible scenario. In fact, it is important to point out that in their analysis it was implicitly assumed a stable Sweet-Parker current sheet. This is a strong assumption, which may be violated for small values of plasma resistivity and viscosity. In this case, as shown in Fig. 2, the thin current sheet of width $\delta_g \propto \eta^{1/2}(\nu/\eta)^{1/4}$ [34] becomes unstable to a chain of plamoids, leading to a strong increase of the reconnection rate.

Fig. 2(a) shows the Sweet-Parker current sheet before the onset of the instability, which occurs at $t \approx 350$. The plasmoids closer to the center of the current sheet grow at a faster rate than those more distant from it; therefore, the four central plasmoids pass into the nonlinear phase at $t \approx 410$ while the other plasmoids are still in the linear growth phase. As shown in Fig. 1(b), approximately at this time there is a sudden increase of the reconnection rate. After $t \approx 440$, when many plasmoids are well into the nonlinear phase, the reconnection rate seems to fluctuate around a constant mean value, as expected on the basis of previous numerical simulations of nonlinear reconnection with plasmoids (see, e.g., [20, 23, 26]). The growth of the plasmoids is so fast that they start to merge before being ejected in the downstream region, and the simulation is stopped at $t = 490$, when our truncated Fourier expansion can no longer resolve the large gradients in the y direction that are due to the coalescence of the plasmoids in this stage of the reconnection process.

Fig. 3(a) shows a zoom around the central plasmoids of j_z at $t = 440$. The initial Sweet-Parker current sheet is broken into smaller sheets separated by plasmoids. The sheets closer to the center are characterized by higher current density than those more distant from it. Furthermore, Fig. 3(b) shows that the field v_x is characterized by a multipolar structure close to $x = 0$, indicating that the presence of plasmoids leads to a temporary deflection of the plasma from the outflow channel. This yields a non-monotonic velocity v_y at $x = 0$, as shown in Fig. 3(c). It is also worth noting that, apart for the central microscopic current sheet, the other sheets are characterized by an asymmetric outflow that may lead to a kind of single wedge shape [35]. Finally, Fig. 3(d) shows that the plasmoids do not develop net vorticity, differently from what happens when the reconnecting magnetic fields are not symmetric [29].

It is possible to obtain a rough estimation of the reconnection rate in the highly nonlinear regime by assuming a chain of plasmoids connected by various marginally stable current sheets of length L_c and width δ_c. Following Huang and Bhattacharjee [26], L_c and δ_c may be deduced by supposing a statistical steady state in which the interplasmoid current layers follow the same Sweet-Parker scaling of the primary (global) current sheet, but with reduced length. Therefore, for $P_m \gg 1$ one can obtain

$$L_c \sim \eta\frac{S_c}{v_{A,up}} \sim L_g\frac{S_c}{S_g}, \tag{14}$$

and

$$\delta_c \sim \frac{L_c}{S_c^{1/2}}P_m^{1/4} \sim L_g\frac{S_c^{1/2}}{S_g}P_m^{1/4}, \tag{15}$$

where $S_g = v_{A,up}L_g/\eta$ is the Lundquist number based on the global current sheet lenght L_g, $v_{A,up}$ is the Alfvén speed based on the reconnecting component of the magnetic field $B_{y,up}$, and

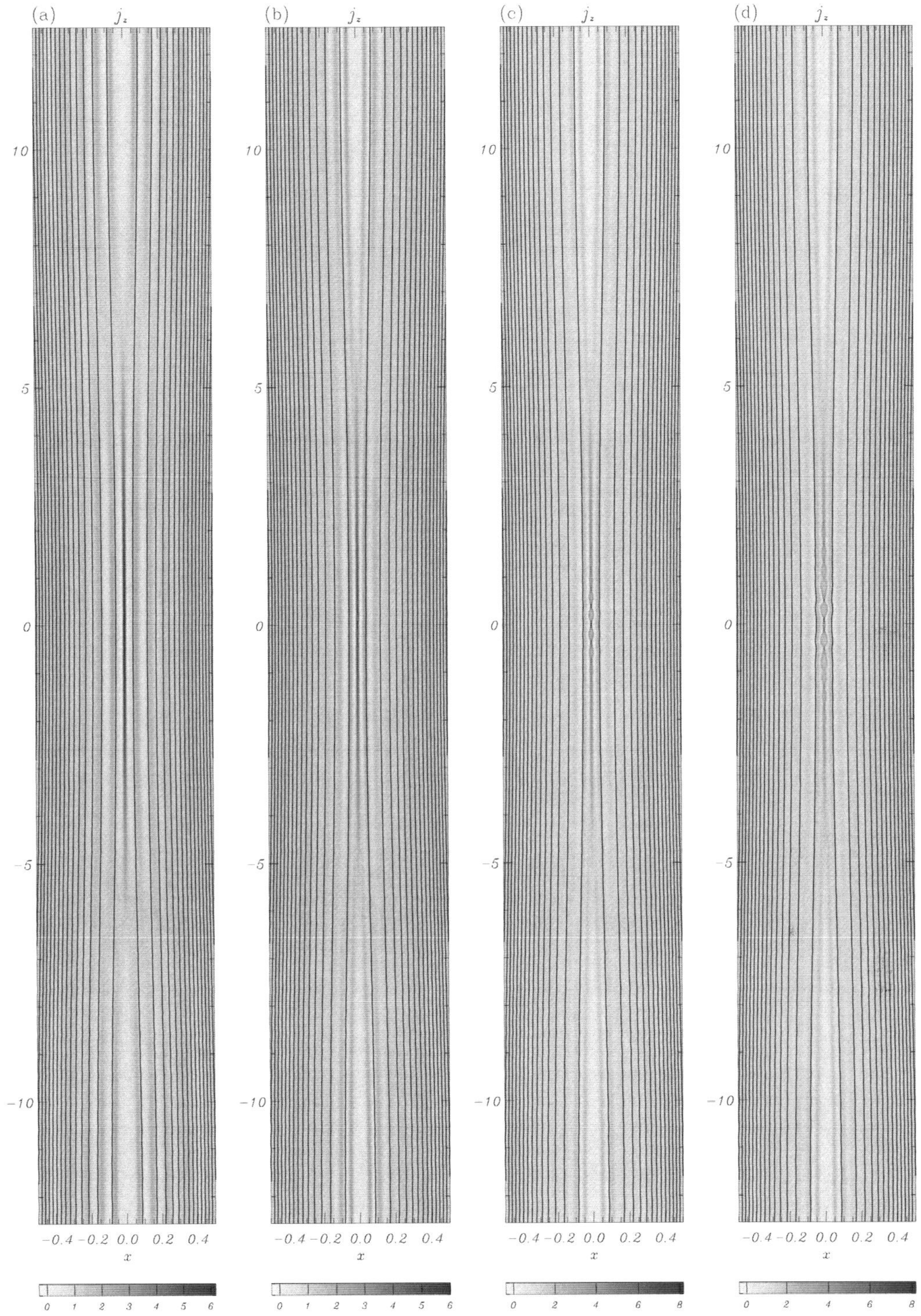

Figure 2. From the numerical simulation shown in Fig. 1(b), contour plots of the out-of-plane current density j_z with the in-plane component of some magnetic field lines (black lines) superimposed at (a) $t = 300$, (b) $t = 410$, (c) $t = 440$ and (d) $t = 470$.

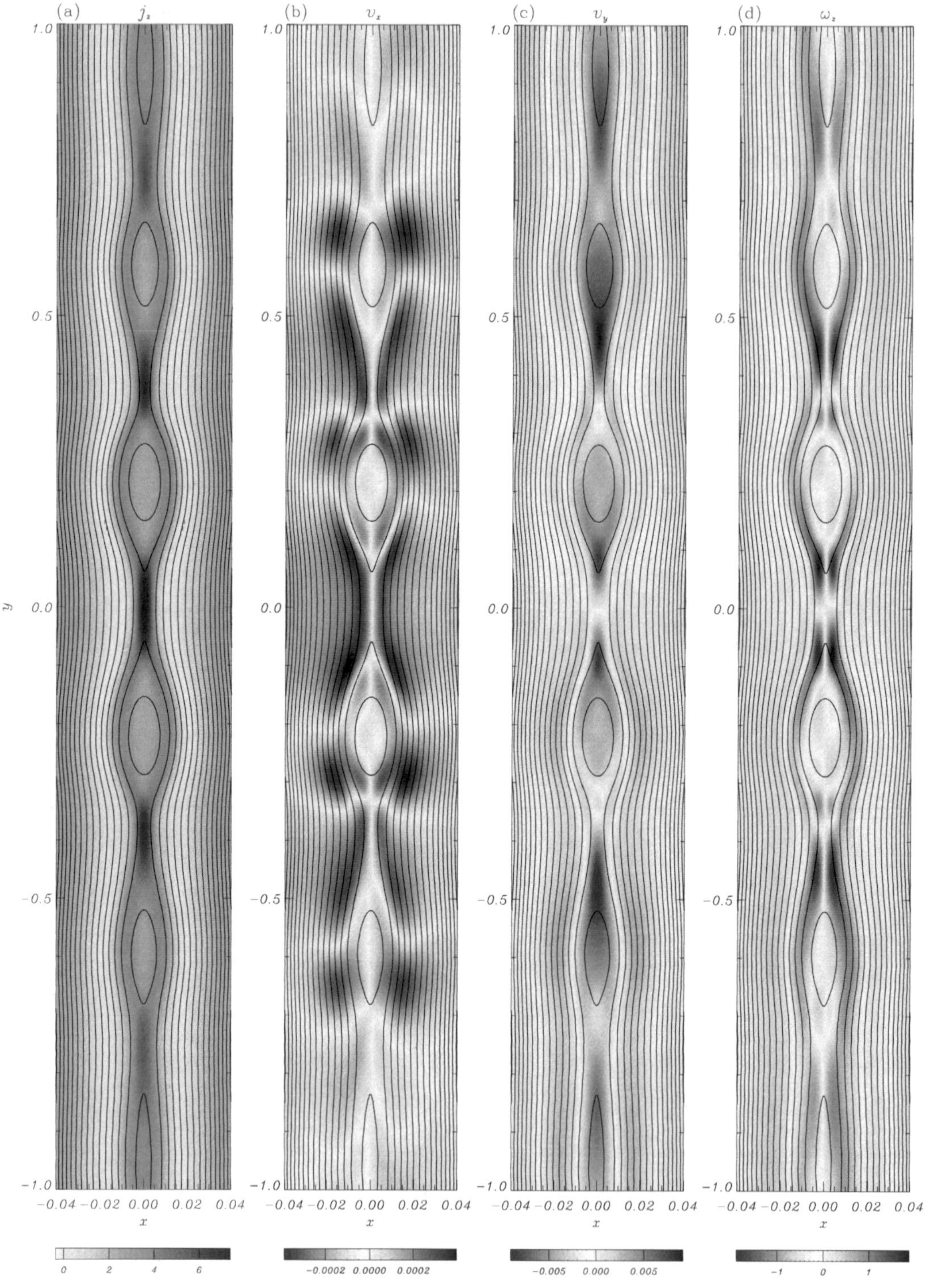

Figure 3. From the numerical simulation shown in Fig. 1(b), blowup around the central plasmoids of the (a) out-of-plane current density j_z, (b) velocity v_x, (c) velocity v_y and (d) vorticity ω_z at $t = 440$. The in-plane component of some magnetic field lines have been superimposed (black lines).

S_c is the critical Lundquist number above which the current sheet is unstable to the plasmoid instability. The reconnection rate may be estimated simply as the rate of change of the flux reconnected via the most central layer [25]. Hence it follows that

$$R \sim \eta \frac{B_{y,up}}{\delta_c} \sim \frac{B_{y,up}\, v_{A,up}}{S_c^{1/2} P_m^{1/4}}, \tag{16}$$

showing that R is intimately linked to the aspect ratio of the elementary current sheet of the plasmoid chain. When the plasma viscosity is negligibe, i.e. $P_m \ll 1$, numerical simulations indicate that $S_c \sim 10^4$ [16]. Thus, since in the absence of viscosity Eq. (16) has to be replaced by [26] $R \sim S_c^{-1/2} B_{y,up}\, v_{A,up}$, in this regime the statistical steady state reconnection rate is independent of the microscopic plasma parameters. For $P_m \gg 1$ there are not yet numerical indications for S_c. However, heuristic arguments by Loureiro and coworkers suggest $S_c \sim 10^4 {P_m}^{1/2}$ [28]. Therefore, the reconnection rate in the visco-resistive regime can be estimated as

$$R \sim 10^{-2} {P_m}^{-1/2} B_{y,up}\, v_{A,up}\,. \tag{17}$$

A numerical check of the above formula would require to perform several numerical simulations at different magnetic Prandtl numbers. We leave such study for a future work. However, it is worth to compare the prediction of this formula with the reconnection rate of the highly nonlinear plasmoid regime shown in Fig. 1(b). In this stage of the reconnection process the reconnecting component of the magnetic field just upstream of the diffusion region is $B_{y,up} \approx 0.032$. Then, the relation (17) gives $R \sim 3.24 \times 10^{-6}$, which is reasonably close to the numerical reconnection rate. According to the scalings of Park *et al.* [34], in the absence of plasmoids the value of the reconnection rate for this small-resistivity small-viscosity case would be much lower than the one observed in Fig. 1(a); in contrast, the plasmoids increase the reconnection rate so strongly that the maximum reconnection rate is essentially the same despite the large difference in the values of plasma resistivity and viscosity.

4. Conclusions

In this paper we have found a new nonlinear scenario of the Taylor problem which complements those of Hahm and Kulsrud [2] and Wang and Bhattacharjee [4]. For very small boundary perturbations Hahm and Kulsrud [2] demonstrated that the driven magnetic island evolves into the nonlinear phase through a slow Rutherford regime, whereas Wang and Bhattacharjee [4] showed that larger boundary perturbations lead to a nonlinear evolution characterized by a Sweet-Parker phase that subsequently gives way to the Rutherford regime. Here instead we have shown that larger boundary perturbations may give rise to a Sweet-Parker phase with a current sheet that is unstable to the plasmoid instability [19] if very small values of plasma resistivity and viscosity are considered. As a consequence, the advanced nonlinear phase is characterized by the development of a plasmoid chain that is responsible for a substantial speed up of the reconnection process. When the plasmoid chain is fully developed, the reconnection rate in the large magnetic Prandtl number regime is estimated to be $R \sim 10^{-2} {P_m}^{-1/2} B_{y,up}\, v_{A,up}$. A detailed scaling study of the reconnection rate in the visco-resistive regime is the subject of ongoing work, as well as the determination of an expression for the threshold perturbation required to trigger the plasmoid phase.

Acknowledgments

The authors would like to acknowledge fruitful conversations with Dario Borgogno, Richard Fitzpatrick, Enzo Lazzaro and Fulvio Militello. One of us (L.C.) is grateful for the hospitality of the Institute for Fusion Studies at the University of Texas at Austin, where part of this work was done. This work was carried out under the Contract of Association Euratom-ENEA and was also supported by the U.S. Department of Energy under Contract No. DE-FG02-04ER-54742.

References

[1] Yamada M, Kulsrud R and Ji H 2010 *Rev. Mod. Phys.* **82**, 603
[2] Hahm T S and Kulsrud R M 1985 *Phys. Fluids* **28**, 2412
[3] Rutherford P H 1973 *Phys. Fluids* **16**, 1903
[4] Wang X and Bhattacharjee A 1992 *Phys. Fluids B* **4**, 1795
[5] Fitzpatrick R 2003 *Phys. Plasmas* **10**, 2304
[6] Rem J and Schep T J 1998 *Plasma Phys. Controlled Fusion* **40**, 139
[7] Valori G, Grasso D and de Blank H J 2000 *Phys. Plasmas* **7**, 178
[8] Fitzpatrick R 2004 *Phys. Plasmas* **11**, 937
[9] Cole A and Fitzpatrick R 2004 *Phys. Plasmas* **11**, 3525
[10] Bian N and Vekstein G 2005 *Phys. Plasmas* **12**, 072902
[11] Birn J, Galsgaard K, Hesse M, Hoshino M, Huba J, Lapenta G, Pritchett P L, Schindler K, Yin L, Büchner J, Neukirch T and Priest E R 2005 *J. Geophys. Res.* **106**, 3715
[12] Vekstein G and Bian N 2006 *Phys. Plasmas* **13**, 122105
[13] Lazzaro E and Comisso L 2011 *Plasma Phys. Control. Fusion* **53**, 054012
[14] Hosseinpour M 2013 *J. Plasma Physics* **79**, 519
[15] Matthaeus W H and Lamkin S L 1985 *Phys. Fluids* **28**, 303
[16] Biskamp D 1986 *Phys. Fluids* **29**, 1520
[17] Lee L C and Fu Z F 1986 *J. Geophys. Res.* **91**, 6807
[18] Fu Z F and Lee L C 1986 *J. Geophys. Res.* **91**, 13373
[19] Loureiro N F, Schekochihin A A and Cowley S C 2007 *Phys. Plasmas* **14**, 100703
[20] Bhattacharjee A, Huang Y-M, Yang H and Rogers B 2009 *Phys. Plasmas* **16**, 112102
[21] Samtaney R, Loureiro N F, Uzdensky D A, Schekochihin A A and Cowley S C 2009 *Phys. Rev. Lett.* **103**, 105004
[22] Daughton W, Roytershteyn V, Albright B J, Karimabadi H, Yin L and Bowers K J 2009 *Phys. Rev. Lett.* **103**, 065004
[23] Cassak P A, Shay M A and Drake J F 2009 *Phys. Plasmas* **16**, 120702
[24] Shepherd L S and Cassak P A 2010 *Phys. Rev. Lett.* **105**, 015004
[25] Uzdensky D A, Loureiro N F and Schekochihin A A 2010 *Phys. Rev. Lett.* **105**, 235002
[26] Huang Y-M and Bhattacharjee A 2010 *Phys. Plasmas* **17**, 062104
[27] Ni L, Ziegler U, Huang Y-M, Lin J and Mei Z 2012 *Phys. Plasmas* **19**, 072902
[28] Loureiro N F, Schekochihin A A and Uzdensky D A 2013 *Phys. Rev. E* **87**, 013102
[29] Murphy N A, Young A K, Shen C, Lin J and Ni L 2013 *Phys. Plasmas* **20**, 061211
[30] Yu Q, Günter S and Lackner K 2104 *Nucl. Fusion* **54**, 072005
[31] Dewar R L, Bhattacharjee A, Kulsrud R M and Wright A M 2103 *Phys. Plasmas* **20**, 082103
[32] Strauss H R 1976 *Phys. Fluids* **19**, 134
[33] Waelbroeck F L 1989 *Phys. Fluids B* **1**, 2372
[34] Park W, Monticello D A and White R B 1984 *Phys. Fluids* **27**, 137
[35] Murphy N A 2010 *Phys. Plasmas* **17**, 112310

Journal of Physics: Conference Series **561** (2014) 012005 doi:10.1088/1742-6596/561/1/012005

An Eulerian Vlasov code for plasma-wall interactions

David Coulette and Giovanni Manfredi

Institut de Physique et Chimie des Matériaux de Strasbourg, CNRS and Université de Strasbourg, 23 rue du Loess, BP 43, F-67034 Strasbourg, France

E-mail: david.coulette@ipcms.unistra.fr, giovanni.manfredi@ipcms.unistra.fr

Abstract. In this work we present a new Eulerian kinetic 1D3V code devoted to the study of plasma-wall interactions and sheath structure. The code solves the Vlasov-Poisson system for two or more kinetic species on a one-dimensional spatial grid between two limiting plates, in the presence of a uniform magnetic field tilted with respect to the normal to the plates. A source-sink term in the Vlasov equation leads to a stationary solution comprising a magnetic and a collisional presheath, as well as the usual Debye sheath in front of the wall. Thanks to a nonuniform spatial grid the sheath structure can be resolved accurately. Several advection schemes are implemented and the code is parallelized. Here, the code is used to illustrate the problems of a stationary sheath-presheath structure in front of a negatively biased wall, with and without a tilted magnetic field.

1. Introduction

Despite the constant improvement of power and accessibility of computing resources, the accurate numerical simulation of plasma-wall interactions remains a challenging task. The intermixing of various time and space scales, dynamical regimes (from collisional to ballistic) and mutually-interacting multiple species, along with their nontrivial interactions with materials, lead to highly complex and costly modeling. As of today, the most comprehensive simulations are massively parallel PIC/Monte Carlo codes [1]. Such simulations have the advantage that atomic processes can be modeled in a straightforward way, but require heavy computational resources, since large numbers of macro-particles must be used to obtain reasonable accuracy. Accuracy is particularly hard to maintain in low-density regions such as the Debye sheaths. In contrast, Eulerian (Vlasov, Fokker-Planck) codes have the potential to fare better in those situations, as the distribution function is sampled with the same accuracy in all phase-space regions. However, they are computationally more demanding, and their treatment of collisions is more costly and less intuitive compared to PIC codes.

In this paper, we report on the latest advancements in the development of an Eulerian code for the study of the plasma-wall transition. We adopt a simplified geometry restricted to one dimension in space, with a uniform and constant magnetic field, while retaining full 3D velocity space description. The current implementation is a parallelized extension and generalization of the codes used in [2, 3, 4] to arbitrary number of species. As an illustration, we will show a few numerical results on the stationary sheath-presheath structure in front of a negatively biased wall, with and without a tilted magnetic field.

2. Model and code structure

2.1. Physical model

We consider a plasma bounded by two material plates (Fig. 1). The system is assumed to be invariant along the two coordinates parallel to the plates. The x direction, normal to the plates, may either represent the distance along the principal axis of a small linear machine, or the curvilinear coordinate along an open magnetic field line between divertor plates in a tokamak. The system is magnetized by a constant and uniform magnetic field $\mathbf{B}_0 = B_0(\sin\alpha\mathbf{e}_x + \cos\alpha\mathbf{e}_y)$. For each species, the dynamics is governed by the collisional Vlasov equation

$$\partial_t f_s + v_x \partial_x f_s + \frac{q_s}{m_s}(\mathbf{E} + \mathbf{v} \times \mathbf{B}_0) \cdot \boldsymbol{\nabla}_v f = \mathcal{C}_s + \mathcal{S}_s, \tag{1}$$

where $f_s(x, \mathbf{v})$ is the sth species distribution function, $\mathcal{C}_s$ stands for the sum of all contributions from collisional processes, $\mathcal{S}_s$ is the sum of all independent sources, and $\mathbf{E} = -\partial_x \phi$ is the electrostatic field obtained self-consistently from the Poisson equation

$$-\partial_{xx}\phi = \sum_s \frac{q_s}{\epsilon_0} \int f_s d^3v. \tag{2}$$

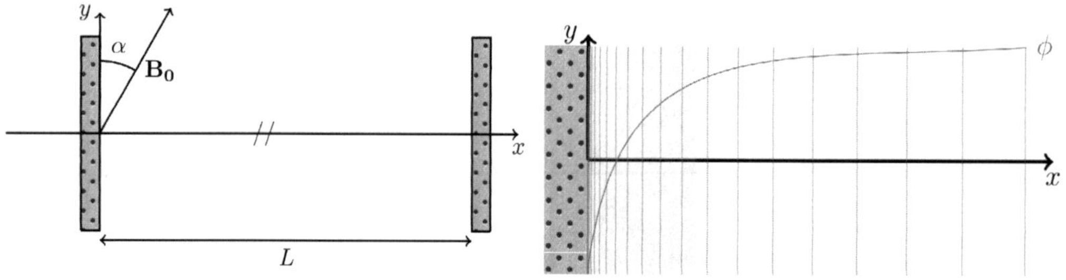

Figure 1. Model geometry (a) and non-uniform grid near the wall (b)

The distribution functions solution of Eq. (1) are assumed to be zero at all boundaries in velocity space ($\pm v_{\max}$). In the x direction, it is assumed that the incoming flux at the plates is zero (perfectly absorbing plate), but more complex boundary conditions can also be implemented to simulate particle-wall interactions such as physical sputtering or secondary electron emission. In some cases, only a half-infinite plasma will be considered, between one plate and a bulk position where the plasma is supposed to be at equilibrium.

Boundary conditions for the electrostatic potential at the plates may either be mixed Neumann/Dirichlet or Dirichlet/Dirichlet, which allows for the treatment of both floating and externally biased walls. When a full kinetic description is not necessary, the dynamics of some species (most often electrons) can follow a simplified model such as a Boltzmann law $n_s(x) = n_{s0} \exp[-(q_s(\phi - \phi_{ref})/k_B T_s]$, in which cases the Poisson equation becomes nonlinear and is solved iteratively.

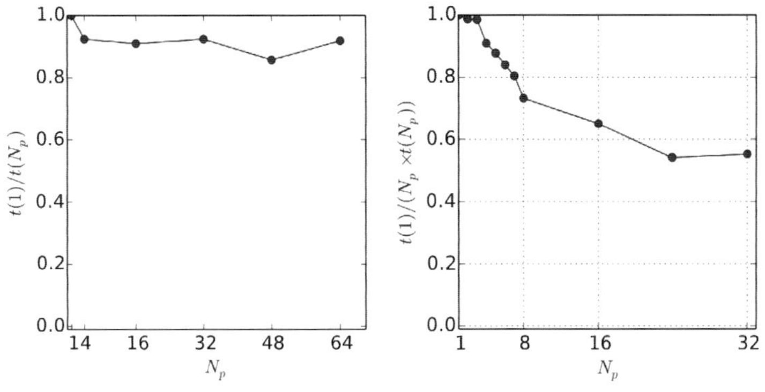

(a) (b)

Figure 2. Parallel efficiency of x domain decomposition (MPI): N_p is the number of MPI processes, $t(N_p)$ is the run duration. Efficiency $t(1)/t(N_p)$ scaling with growing global domain size $n_x = 64 \times N_p$ and $(n_{vx}, n_{vy}, n_{vz}) = (64, 64, 64)$ (a) and efficiency $t(1)/[N_p \times t(N_p)]$ scaling with a constant-size global domain $n_x = 1024$, $(n_{vx}, n_{vy}, n_{vz}) = (96, 96, 96)$ (b).

2.2. Numerical methods and implementation

Equation (1) is solved using a the following time-splitting scheme:

$$
\begin{array}{lr||l}
1: & \partial_t f_s + v_x \partial_x f_s = 0 & x \text{ advec. } (\Delta t) \\
2: & -\Delta\phi - \frac{1}{\epsilon_0}\rho = 0 & \text{Poisson solver} \\
3: & \partial_t f_s + \frac{q_s}{m_s}(E_x - B_y v_z)\partial_{v_x} f_s = 0 & v_x \text{ advec}(\Delta t) \\
4: & \partial_t f_s + \frac{q_s}{m_s}(B_x v_z)\partial_{v_y} f_s = 0 & v_y \text{ advec}(\Delta t) \\
5: & \partial_t f_s + \frac{q_s}{m_s}(B_x v_y - B_y v_x)\partial_{v_z} f_s = 0 & v_z \text{ advec.}(\Delta t) \\
6: & \partial_t f_s + \nu(f_s - F_0) = 0 & \text{BGK relaxation } (\Delta t)
\end{array}
\quad (3)
$$

where the advections in all four phase-space directions (steps 1, 3–5) and the source and collision terms (step 6) are treated sequentially.

Collisional processes are modeled using a single BGK relaxation operator acting as a restoring force that drives the distribution function towards a fixed Maxwellian $F_0 = n_{0s}\exp(-m_s v^2/2T_{0s})$. Thus $\mathcal{C}_s = -\nu(f_s - F_0)$, where ν is the relaxation rate.

All of the conservative advection equations take the simple linear form $\partial_t f + u_\xi \partial_\xi f = 0$, where $\xi \in \{x, v_x, v_y, v_z\}$ and u_ξ does not depend on ξ. They are integrated in time using a finite volume scheme [5] with reconstruction by primitive. Earlier versions of the code [2, 3, 4] used the flux-balanced scheme [6]. Several higher-order schemes have now been implemented and tested: (i) a fixed stencil polynomial interpolation with a choice of classical limiter functions, (ii) an ENO [7] scheme of arbitrary order, (iii) and a version of the PFC3 scheme [8] whose limiters have been modified to be used on a nonuniform grid. The latter is the default scheme in all results presented hereafter. It was selected as a good compromise between accuracy and computing effort.

The parallelization strategy relies on domain decomposition in the x dimension, implemented using MPI, and parallelization over species on each spatial sub-domain using OpenMP. This design does not fully exploit the parallelism that could be extracted by domain-decomposing the velocity dimensions. The rationale for this choice is to allow for the possibility of implementing more realistic collision operators. In that context, the full non-local coupling between velocity spaces of interacting species could lead to detrimental communication costs. Parallelization over velocity dimensions, either by domain decomposition or fine-grained OpenMP parallelism, will thus be considered only once the collision operators framework is implemented.

Scalability tests were performed to evaluate the MPI parallel performance of the current design. The first (Fig. 2a) measures the so-called "weak" scaling: starting from a given spatial sub-domain size $n_x = 64$ with one MPI process $N_p = 1$, global domain size is constrained to grow linearly with the number of processes ($n_x = 64N_p$). The efficiency $e = t(1)/t(N_p)$ is measured against the ideal case, for which the runtime would be constant. The results show an acceptable scaling: the communication overhead is nearly constant in the tested N_p range. From a user point of view, it means that global domain size can be extended with no penalty.

The second test measures the "strong" scaling of efficiency. The global problem size is kept constant, while the number of sub-domains (and MPI processes) grows. The efficiency $e = t(1)/[N_p \times t(N_p)]$ is measured against the ideal case where runtime would scale as $1/N_p$. The results (Fig. 2b) show that the efficiency degrades and saturates with growing number of processes. The cost the Poisson solver [step 2 in Eq. (3)], which is parallelized using a block-decomposition scheme [9], stays negligible. Scalability is thus mostly limited by the x advection (step 1), for which the overhead brought forth by buffer exchange between sub-domains, redundant computation on overlapping regions, and all-to-all communications required by the PFC3 scheme become non-negligible with respect to the ideally parallel steps 3–6. From a user's perspective, and for "small" spatial domain sizes (typically a few hundred points), one can expect a maximal speed-up of about 10 to 20 without wasting computing resources.

3. Numerical results

We tested our code on two well-documented steady-state problems: the plasma-wall transition with and without a tilted magnetic field. In both cases, we considered a half-infinite plasma between an absorbing plate (located at $x = 0$) and a bulk plasma at equilibrium ($x = L$). The domain length is chosen to encompass several mean-free paths relatively to the BGK relaxation operator, so as to provide a smooth transition in the pre-sheath. At $t = 0$, the ion distribution is set to a spatially homogeneous Maxwellian with bulk parameters. During the evolution, governed by the Vlasov-Poisson equation, the ions are lost to the wall and the sheath forms rapidly. The system then evolves slowly until the balance between particle transport and the BGK relaxation is established in the presheath. The results shown in the forthcoming sections are all taken in this asymptotic steady-state regime.

3.1. Static sheath in a linear machine

First, we consider the steady state of a small linear machine filled with a non-magnetized argon plasma, leading to a simple 1D1V model. An analytical kinetic treatment of similar systems in a collisionless regime can be found in Refs. [11, 12]. The bulk plasma parameters ($n_e = n_i = 5.5 \times 10^{15}\mathrm{m}^{-3}$, $k_B T_i = 0.027\mathrm{eV}$, $k_B T_e = 2.5\mathrm{eV} \approx 100\mathrm{k_B T_i}$) were chosen to match those of Ref. [10]. In the actual machine, the wall potential is not driven by the bulk electrons, but by the highly energetic primary electrons emitted by the heated filament and then accelerated by a fixed discharge potential (-50V to -100V). The wall bias then reaches values of the order of the discharge potential. As the density of primary electrons is three orders of magnitude lower than that of bulk electrons, their direct contribution to the space charge density is neglected. In order to mimic their impact on wall bias, we force the wall potential to values of the order of the discharge potential. This assumption makes it possible to reproduce the ion velocity profile observed in the experiment [10].

The bulk electrons are treated using a Boltzmann model. Additional simulations with kinetic electrons showed good agreement with the results obtained using the Boltzmann law. All collisional processes affecting the ions are modeled using a single BGK operator that forces

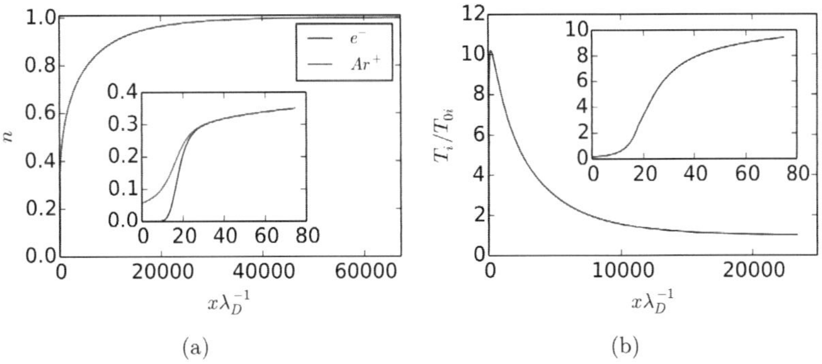

Figure 3. Argon discharge steady-state with 50V wall bias and collision rate $\nu = 10^{-4}\omega_{pi}$: particles densities (a); ion temperature profile (b). The insets show zooms of the respective quantities.

the ion velocity distribution function (IVDF) to relax towards a Maxwellian with bulk plasma parameters, with a relaxation rate ν ranging from $\nu = 10^{-4}\omega_{pi}$ to $\nu = 10^{-2}\omega_{pi}$.

Although the model is simple, the actual parameters impose quite stringent requirements on the phase space resolution if one wants to describe accurately the evolution of the ion distribution function up to the wall. The required velocity range is governed by the wall potential, which is rather large compared to the ion bulk temperature. In the sheath, the ion distribution becomes very narrow, which requires even higher resolution, as both cell averaging and numerical diffusion tend to smear out the distribution profile. For this set of test cases, a $(n_x, n_{vx}) = (1024, 1024)$ grid has been used. The cell width in x ranges from $\Delta x = 0.01\lambda_D$ near the wall at $x = 0$ to $\Delta x = 100\lambda_D$ at the right boundary ($x \approx 2 \times 10^5\ \lambda_D$) where the plasma is at equilibrium

The density and temperature profiles are shown in Fig. 3. In the quasi-neutral pre-sheath, the ion behavior is dominated by the balance between plasma flow towards the wall and relaxation driven by the BGK term. The electric field tends to accelerate the ions towards the wall, whereas the BGK term tries to drag the ion distribution towards the equilibrium Maxwellian. The net result is that the ion velocity distribution widens in the presheath, leading to an increased kinetic temperature (Fig. 3b), defined as the velocity spread of the distribution function. The widening of the ion distribution in the presheath is clearly visible in Fig. 4a at position $x = 385\lambda_D$.

In the Debye sheath, characterized by strong charge separation (Fig. 3a) and strong electric fields (Fig. 4b), the effect of the BGK term is negligible, so that the ion trajectories are essentially ballistic. The ion distribution reduces to a cold beam that is strongly accelerated towards the wall, as can be seen in Fig 4a between $x = 18\lambda_D$ and $x = 0$. The thinning of the ion distribution is reflected in a drop of the ion kinetic temperature (Fig. 3b, inset).

In order to check that the chosen grid and interpolation scheme do not introduce any artificial widening of the distribution function, either by cell averaging or numerical diffusion, the simulation results are compared against an analytical ballistic model in the sheath (Fig. 5). Starting from a reference distribution function at the entrance of the sheath ($x = 34.3\lambda_D$), the ion distribution functions closer to the wall are obtained by conservation of the distribution along the characteristics. Although a slight erosion of the IVDF maximum can be seen (mostly due to the action of the limiters), the IVDF geometry, and most importantly its asymmetry, are well preserved by the PFC3 scheme, and spurious oscillations are controlled.

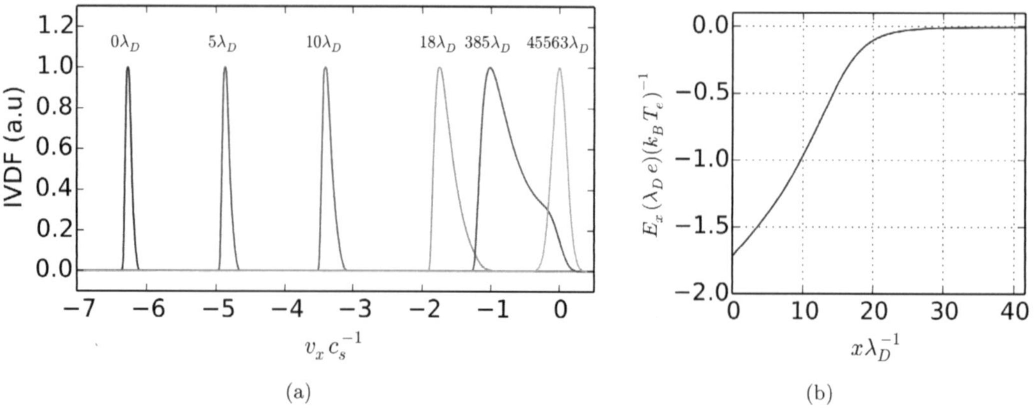

Figure 4. Argon discharge with -50V wall bias and collision rate $\nu = 10^{-4}\omega_{pi}$. Evolution of the IVDF (normalized to its peak value) from the bulk plasma to the wall (a); distances from the wall are indicated above each peak. Electric field profile near the wall (b).

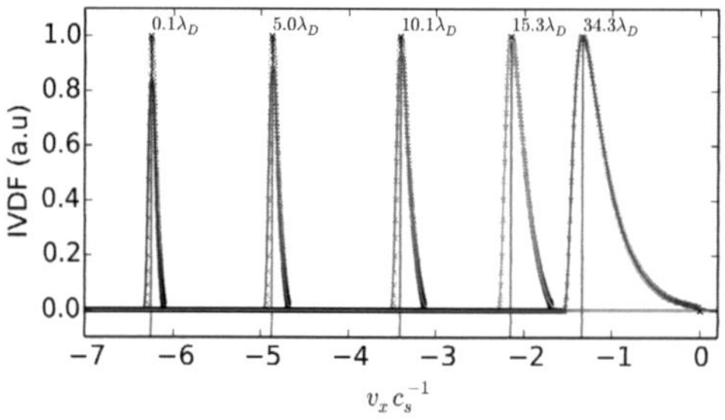

Figure 5. Argon discharge with a -50V wall bias and $\nu = 10^{-4}\omega_{pi}$. Evolution of the IVDF in the sheath: numerical results (continuous lines); IVDF computed from the reference profile at $x = 34.3\lambda_D$ using a ballistic model (crosses). The vertical lines mark the ballistic model peak velocity. All distributions are normalized to the peak value of the reference distribution at $x = 34.3\lambda_D$.

3.2. Magnetized deuterium plasma with tilted magnetic field

In order to test the operation of the code with a full 3D description in velocity space, we consider a deuterium plasma with kinetic ions and Boltzmann electrons. The setup is similar to the cases studied in Refs. [3, 4, 13, 14], with the geometry depicted in Figure 1. The discharge parameters are $T_e = T_i = 5$ eV, $\omega_{Ci} = 0.05\ \omega_{pi}$, $\alpha = 15°$, $\nu = 0.002\ \omega_{pi}$. The phase space is discretized on a $(256, 96, 96, 96)$ grid. The cell width ranges from $\Delta x = 0.5\lambda_D$ to $\Delta x = 500\lambda_D$, and the ion velocity range is $[-6v_{thi}; 6v_{thi}]$ in all directions. Albeit rather coarse, particularly in the sheath, this resolution is sufficient to obtain the main features of the collisional pre-sheath/magnetic pre-sheath/sheath transitions described in Ref. [13].

In a non-magnetized (or with normal incidence, $\alpha = 90°$) and weakly collisional plasma, the transition from the equilibrium bulk plasma to the quasi-ballistic regime in the sheath occurs in the collisional pre-sheath, where collisional/ionization processes compete with the ion flow towards the wall. In the magnetized plasma considered here, the magnetic field prevents the ions to flow directly towards the wall, particularly for grazing incidence.

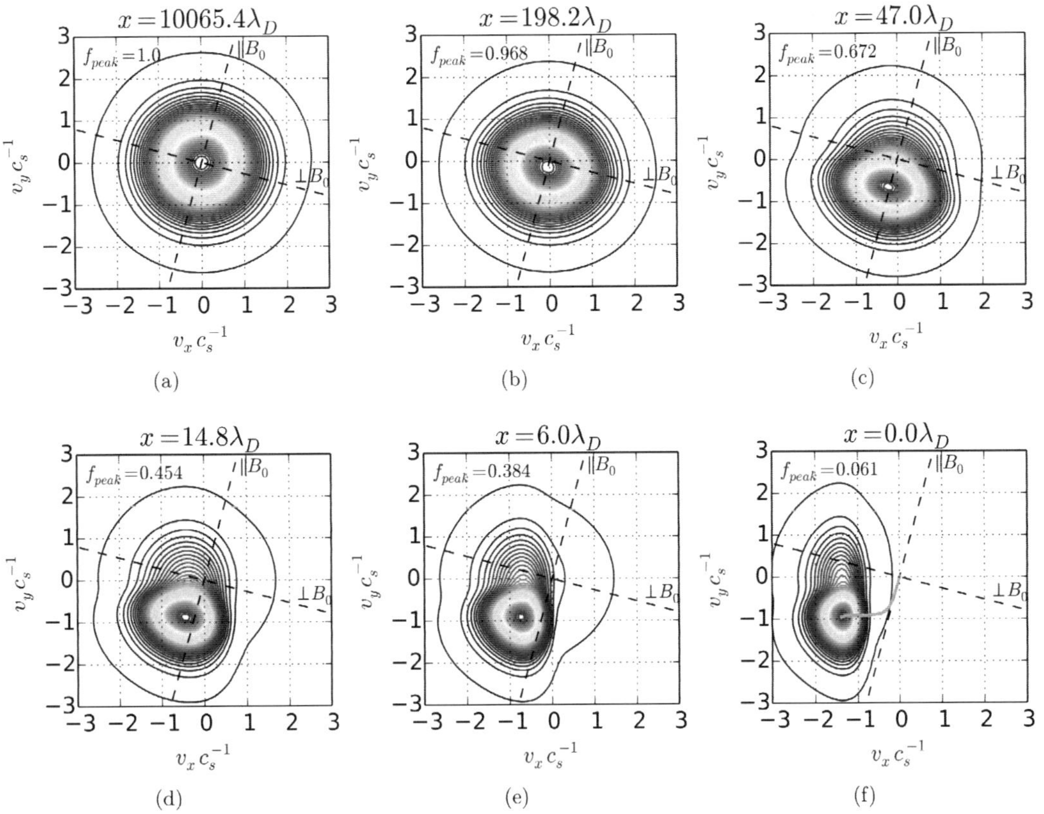

Figure 6. Contour plots of the v_z-averaged ion distributions normalized to their peak value (indicated in the upper left corner); dashed lines indicate the $\mathbf{B}_0$ and $\mathbf{e}_z \times \mathbf{B}_0$ directions. Bulk plasma (a); collisional pre-sheath (b-c); magnetic pre-sheath (d); Debye sheath (e); wall (f). The orange curve in (f) retraces the evolution of the peak of the distribution from the bulk plasma to the wall.

The overall transition from bulk to wall can be seen by looking at the evolution of the steady-state ion distribution function in the (v_x, v_y) plane (Fig. 6). In the collisional pre-sheath, as the ion mean velocity is slowly accelerated up to the sound speed, the magnetic field maintains the flow aligned with its own direction (Fig. 6a–c). In the magnetic pre-sheath, while the plasma is still quasi-neutral (Fig. 7b), the magnitude of the electric field is sufficient to break the alignment of the flow with the magnetic field (Fig. 6d). Eventually, the electric field dominates in the Debye sheath (Fig. 6e–f), where the ions are strongly accelerated in the direction normal to the wall.

The transition from the **B**-dominated pre-sheath to the **E**-dominated sheath is clearly visible in Fig. 7a, where we compare the magnitude of the various forces acting in the x direction. The Debye sheath entrance corresponds not only to the beginning of the nonneutral region (Fig. 7b), but also to the point where the electric force starts dominating over the Lorentz force $\mathbf{v} \times \mathbf{B}$.

4. Conclusion and perspectives

We presented the first developments of an Eulerian 1D3V code devoted to the study of plasma-wall interactions. Building up from earlier versions [3, 4], significant modifications were made

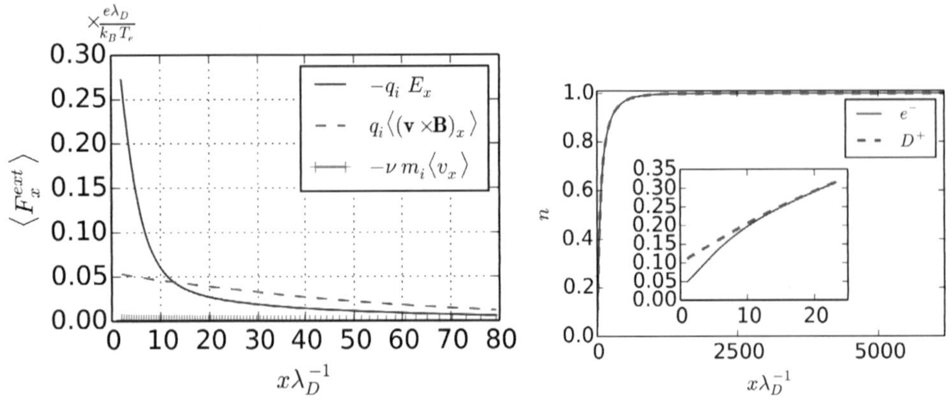

(a) (b)

Figure 7. Deuterium plasma with tilted magnetic field, $\alpha = 15°$. Profiles of the various forces in the momentum balance equation (a). Note that the friction force $-\nu m_i \langle v_x \rangle$ caused by the BGK term is negligible over the whole domain. Density profiles for electrons and ions (b).

in order to adapt the code to its planned applications. Particle fluxes reaching the wall are the source of complex dynamics, including non-trivial feedback mechanisms, so that an accurate description of the distribution function must be preserved up to the wall. This necessity triggered the switch from the flux balance method to the PFC3 advection scheme. In addition, the present version can also address dynamical transient, and not only steady-state solutions, as was the case before [3, 4]. Finally, the parallelization over spatial sub-domains and different species opens the way to more demanding applications with acceptable run times.

Some standard cases, both in 1D1V and 1D3V, were used to check the validity and parallel performance of the code. The next steps will concentrate on the fully kinetic regime, in particular the comparison of Boltzmann and kinetic electron models for near-grazing incidence ($\alpha = 1-5°$) of the magnetic field. It is also planned to incorporate more accurate collision models than the simple BGK relaxation used so far. This is a major step, as it will introduce nontrivia integral terms in velocity space. Finally, more general plasma-wall effects should also be modeled, such as secondary-electron emission and sputtering from the wall [4].

Acknowledgements

The authors acknowledge the support of the French Agence Nationale de la Recherche (ANR) under reference ANR-12-BS09-028-01.

References

[1] Tskhakaya D and Kuhn S 2004 *Contrib. Plasma Phys.* **44** 564.
[2] Manfredi G and Valsaque F 2004 *Comp. Phys. Comm.* **164** 262.
[3] Devaux S and Manfredi G *Phys. Plasmas* **13** 083504.
[4] Devaux S and Manfredi G *Plasma Phys. Control. Fus.* **50** 025009.
[5] Leveque R.J 2004 *Finite Volume Methods for Hyperbolic Problems* ed. Cambridge University Press 387-95.
[6] Fijalkow E 1999 *Comp. Phys. Comm.* **116** 319.
[7] Abgrall R 1994 *J. Comp. Phys.* **114** 45.
[8] Filbet F, Sonnendrücker E and Bertrand P 2001 *J. Comp. Phys.* **172** 166.
[9] Polizzi E, Sameh A H 2006 *Parallel Computing* **32** 177.
[10] Claire N 2006 *Phys. Plasmas* **13** 062103.
[11] Emmert G A, Wieland R M Mense A T and Davidson J N 1980 *Phys. Fluids* **23** 803.

[12] Bissel R C, Johnson P C Stangeby PC 1989 *Phys. Fluids B* **1** 1133.
[13] Chodura R 1982 *Phys. Fluids* **25** 1628.
[14] Riemann K U 1994 *Phys. Plasmas* **1** 552.

Joint Varenna-Lausanne International Workshop 2014 IOP Publishing
Journal of Physics: Conference Series **561** (2014) 012006 doi:10.1088/1742-6596/561/1/012006

Modeling of ion-cyclotron resonant heating in Wendelstein 7-X equilibrium

J.M. Faustin, W.A. Cooper, J.P. Graves, D. Pfefferlé
Ecole Polytechnique Fédérale de Lausanne (EPFL), Centre de Recherches en Physique des Plasmas (CRPP), CH-1015 Lausanne, Switzerland

Abstract. W7X stellarator 3D equilibrium has been computed with the equilibrium code ANIMEC (Anisotropic Neumann Inverse Moments Equilibrium Code). This equilibrium was used to model ICRH minority heating in ^{4}He(H) plasma with the 3D full-wave code LEMan (Low frequency ElectroMagnetic wave propagation). The coupled power spatial distribution is shown for different resonance positions within the range of frequencies foreseen for the ICRH antenna. It is found that for the high mirror equilibrium examined, the antenna frequency can be chosen to optimise the power deposition in the plasma core while limiting the absorption at the edge.

1. Introduction

Wendelstein 7-X (W7X) is a large superconducting quasi-isodynamic stellarator under construction in Greifswald, Germany. This fully optimised device aims at demonstrating the feasibility of operating steady state plasmas relevant for a fusion powerplant while assuring the confinement of fast particles. Several sources of heating will be mixed in W7X in order to reach fusion plasma relevant temperatures. An assessment of the viability of ion-cyclotron resonance heating (ICRH) applied to W7X constitutes an important step to determine whether this approach is feasible. The design of antennas for this method of heating relies heavily on ICRH simulations in quasi-isodynamic stellarator systems.

Numerical simulations using three-dimensional codes were performed to compute a W7X relevant equilibrium and wave propagation in the ICRH range of frequency. A high-mirror equilibrium was produced using the equilibrium code ANIMEC (Anisotropic Neumann Inverse Method Equilibrium Code) [1]. The full-wave code LEMan (Low frequency ElectroMagnetic wave propagation) [2] uses this equilibrium to solve the wave equation to 0th order in Larmor radius including warm contributions to the dielectric tensor. LEMan is therefore limited to the calculation of fundamental minority heating scenarios. According to Ref. [3] in particular one of the antenna designs foreseen for W7X cover a range of frequencies valid for fundamental minority heating of H in D or ^{4}He, which are scenarios LEMan can compute. However the neutrons generated by D-D fusion reactions may activate the machine which renders deuterium a less desirable majority species for ICRH scenarios for machine maintenance purposes. Consequently we will focus here on ^{4}He(H) plasmas.

This paper is organised as follow. Section 2 describes the main features of the equilibrium case on which the wave propagation is based. In section 3 we discuss the plasma parameters and the scan perfomed in the wave calculation. In section 4 we present the power deposition on the fast ion fraction of the minority species.

2. 3D Equilibrium

The ANIMEC code is used to compute an anisotropic W7X fixed boundary 3D geometry plasma equilibrium. Many toroidal mirror configurations are accessible with a proper use of the modular and planar coils. In this study, we relied on the so-called high mirror configuration which offers the potential for α-particle confinement in a reactor-scaled device. This configuration is of particular interest insofar as ICRH will be used in W7X not only to investigate plasma heating but also to generate a fast particle population to emulate the behaviour of α-particles in a reactor.

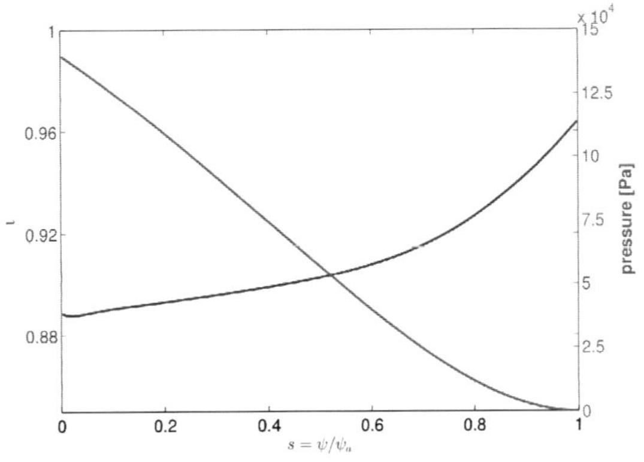

Figure 1. Rotational transform $\iota(s)$ and pressure $p(s)$ profiles used for the equilibrium calculation.

The rotational transform, and the plasma pressure used in input is shown in Figure 1, where s refers to the radial variable. Note that s is proportional to the toroidal magnetic flux ψ and proportional to the enclosed plasma volume. The equilibrium magnetic field amplitude for different toroidal angles, displayed in Figure 2, is obtained and gives a toroidal mirror ratio defined as [4] :

$$\frac{B_{\varphi=0} - B_{\varphi=\pi/5}}{B_{\varphi=0} + B_{\varphi=\pi/5}} = 8.5\% \quad (1)$$

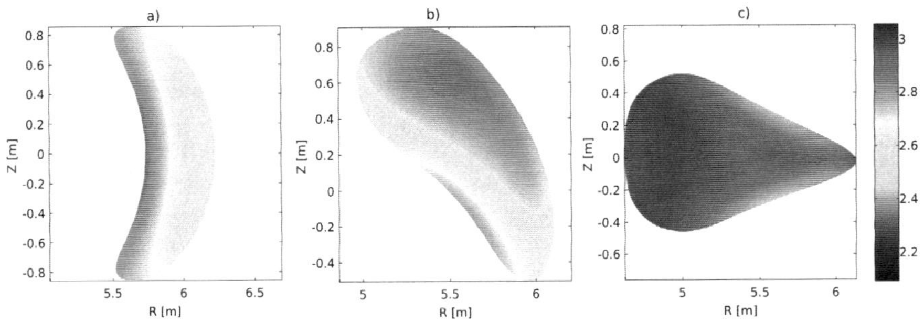

Figure 2. Magnetic field amplitude [T] for three cross section between the two symmetry planes : a) $\varphi = 0$, b) $\varphi = \pi/10$, c) $\varphi = \pi/5$.

An important feature of this high mirror equilibrium for the wave propagation is that the minimum magnetic field in the bean-shaped cross section is higher that the maximum magnetic field in the triangular cross section (respectively a) and c) in Figure 2), i.e. there is no available wave frequency which corresponds to a resonant magnetic field at every toroidal angle. This will be clearly illustrated in the next section. Finally, The ANIMEC code uses a bi-Maxwellian model [1] to incorporate the contribution of hot particles to the equilibrium. The anisotropy was chosen to be constant and set to $T_{\perp}/T_{\|} = 3$, thus envisaging strong perpendicular ICRH heating.

Now that the plasma equilibrium has been presented we will focus on the wave field calculation.

3. 3D Wave field

On the basis of the ANIMEC equilibrium converted to Boozer coordinates, the 3D full-wave code LEMan computes the wave field. LEMan solves the Maxwell's equations in terms of scalar and vector potentials ϕ and $\mathbf{A}$ respectively, which take the following form in fourier space :

$$\nabla^2\mathbf{A} + \left(\tfrac{\omega}{c}\right)^2 \hat{\epsilon}\cdot\mathbf{A} + i\tfrac{\omega}{c}\hat{\epsilon}\cdot\phi = -\tfrac{4\pi}{c}\mathbf{j}_{ant} \quad (2)$$

$$\nabla\cdot(\hat{\epsilon}\cdot\nabla\phi) - i\tfrac{\omega}{c}\nabla\cdot(\hat{\epsilon}\cdot\mathbf{A}) = -4\pi\rho_{ant} \quad (3)$$

Here c stands for the velocity of light and $\hat{\epsilon}$ is the warm dielectric tensor. The elements of this tensor are derived from a Maxwellian distribution function for the electrons and the background species and the same bi-Maxwellian model that in the ANIMEC code for the fast fraction of the minority species [6]. Note that this set of equations has been obtained by introducing the Coulomb Gauge ($\nabla\cdot\mathbf{A} = 0$). The problem is then discretised using finite elements in the radial direction and Fourier harmonics in the poloidal and toroidal directions [2]. The subscripts *ant* refers to the source contributions, namely the charge density and the current density flowing in the antenna. The geometrical description of the antenna is a crucial aspect of the wave calculation. In LEMan, the current density takes a divergence-free form :

$$\mathbf{j}_{ant} = \nabla s \times \nabla\sigma(s,\theta,\varphi) \quad (4)$$

The function $\sigma(s,\theta,\varphi) = \sigma_s(s)\sigma_\theta(\theta)\sigma_\varphi(\varphi)$ contains the geometrical aspect of the antenna and gives a degree of freedom concerning its localisation and extension around the torus. A realistic antenna would be localised between two poloidal and toroidal angles and would even introduce phase shifts between current straps but would excite a broad Fourier mode spectrum. Instead we choose an antenna localised only radially at the edge of the plasma domain, which encloses the whole torus and excites only one mode ($m_{ant} = 3, n_{ant} = 15$). This model, although less realistic offers the first satisfactory results concerning the wave absorption for ICRH minority heating scenarios computed in W7X. Using this simplified antenna also allows us to take advantage of the 5-period symmetry of the device. Therefore the wave simulations were run only in one fifth of the full torus, limiting our toroidal angle domain between 0 and $2\pi/5$. Note that the antenna current density prescribed in LEMan should in principle be multiplied by a constant in order to account for the real value of the power coming from the antenna and coupled to the plasma. Therefore LEMan uses and computes only normalised quantities such as the power absorption results shown in the next section. LEMan computes in particular the power absorbed in the plasma inside each magnetic surface labeled s, the power coupled by the antenna in the same region and the Poynting vector, which read respectively [6] :

$$P_{pla}(s) = \frac{\omega}{8\pi}\int_{s'<s} \mathrm{d}V' \left(|\mathbf{B}|^2 - \mathbf{E}^*(\cdot\hat{\epsilon}\cdot\mathbf{E})\right) \tag{5}$$

$$P_{ant}(s) = \frac{i}{2}\int_{s'<s} \mathrm{d}V' \left(\mathbf{E}^*\cdot\mathbf{j}_{ant}\right) \tag{6}$$

$$S_{Poynt} = \frac{c}{8\pi}\int_{s} \mathrm{d}\mathbf{S}\left(\mathbf{E}^*\times\mathbf{B}\right) \tag{7}$$

where **B** and **E** are respectively the magnetic and electric wave fields. These quantities are used in LEMan in order to check the power balance $P_{pla}(s) = P_{ant}(s) + iS_{Poynt}$. A quantitative calculation of the power deposition can be performed using a particle code like VENUS-LEVIS [7].

In the following, the plasma is composed of ^{4}He with 5% of H minority. The central density is chosen as $n_{e0} = 8\times10^{19}\,m^{-3}$ and the profile takes the form $n_\alpha(s) = n_{\alpha 0}\left(0.99(1-s^4)^4 + 0.01\right)$, where α refers to the different electronic and ionic species in the plasma. This particular choice of density profile has a very low plasma density at the antenna, which is similar to the experimental setup where some gas is puffed near the antenna to tunnel the wave into the plasma The temperature profile is prescribed as $T_e = T_i = T_0(1-s)$ where $T_0 = 3\,keV$ is the central temperature. Coupling waves in the ion-cyclotron range of frequencies for efficient heating purposes requires a strong minority particle-wave absorption in the plasma centre. However, for a given wave frequency the resonance can be located only near the plasma edge and can even disappear because of the toroidal dependency of the magnetic field. Therefore a scan in the wave frequency allows us to qualitatively assess an optimal range of frequencies for which the power would be mostly deposited in the plasma core. The results of this scan are presented in the following section.

4. Power deposition

The wave calculation was done with 4 different frequencies : 32 MHz, 33.8 MHz, 36.6 MHz, 39.6 MHz. The plasma is composed of H minority in ^{4}He and we focus on fundamental minority heating only. These frequencies correspond respectively to the following resonant fields : 2.1T, 2.22T, 2.4T, 2.6T. The deposited power on the fast fraction of the minority species is displayed for each case in Figure 3 and shows some interesting consequences for the magnetic equilibrium characteristics of W7X. First, as mentioned before, the wave-particle resonance disappears in a certain range of toroidal angle for each frequency because of the high-mirror configuration we chose. Then, unlike in tokamak equilibrium where the resonance lies in a Doppler shifted vertical layer, the wave-particle interaction can display a more significant radial extent allowing a better power deposition around a certain toroidal position, e.g. $B_c = 2.1T, \varphi = \pi/5$.

On the other hand, in this particular case, the toroidal range for which the power is absorbed is rather limited. Higher frequencies, corresponding in our case to 2.4T and 2.6T, have a resonant layer geometry more localised to the curved boundary, and may besides lead to operational problems. Indeed, around the triangular shaped cross section, the resonance lies very close, if not on the wall which one must avoid for machine inner component safety.

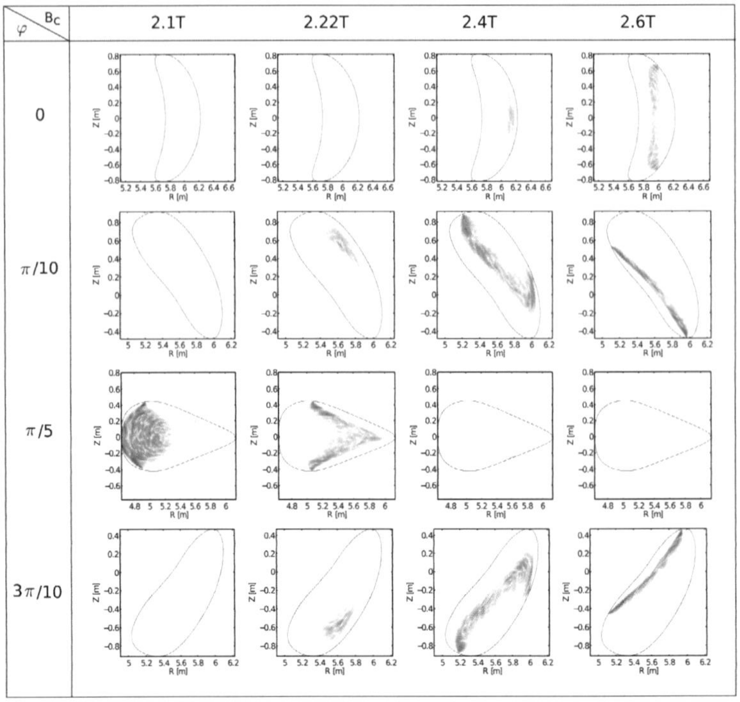

Figure 3. Power deposition in arbitrary units in the fast ion fraction of the minority species for the 4 frequencies of the scan at different toroidal angle

In Figure 4 we propose a more global diagnostic in order to describe the localisation of the power deposition around the torus. The idea is to investigate the radial and toroidal dependency of the power absorption. The radial domain is splitted into three sub-domains. Then, we integrate the absorbed power over the toroidal domain after integration over the poloidal variable and over each radial sub-domains separately. Precisely, if one calls the three-dimensional dependent deposited power density $P(s, \theta, \varphi)$, we compute the following power integrals :

$$\mathcal{P}_i(\varphi) = \frac{\displaystyle\int_{\Omega_i}\int_0^{2\pi} \sqrt{g}P(s,\theta,\varphi)\mathrm{d}s\mathrm{d}\theta}{\displaystyle\int_0^1\int_0^{2\pi}\int_0^{2\pi/5} \sqrt{g}P(s,\theta,\varphi)\mathrm{d}s\mathrm{d}\theta\mathrm{d}\varphi} \tag{8}$$

$$\text{with}\begin{cases} \Omega_1 \ \widehat{=}\ 0 \le s \le \dfrac{1}{4} \\ \Omega_2 \ \widehat{=}\ \dfrac{1}{4} \le s \le \dfrac{3}{4} \\ \Omega_3 \ \widehat{=}\ \dfrac{3}{4} \le s \le 1 \end{cases} \tag{9}$$

Here, Ω_i identifies a fractional volume within the plasma, i.e. Ω_1 and Ω_3 represent respectively the first and the last fourth of the plasma volume. We therefore have some information on the fraction of the power that goes to the edge of the plasma compared to that at the core.

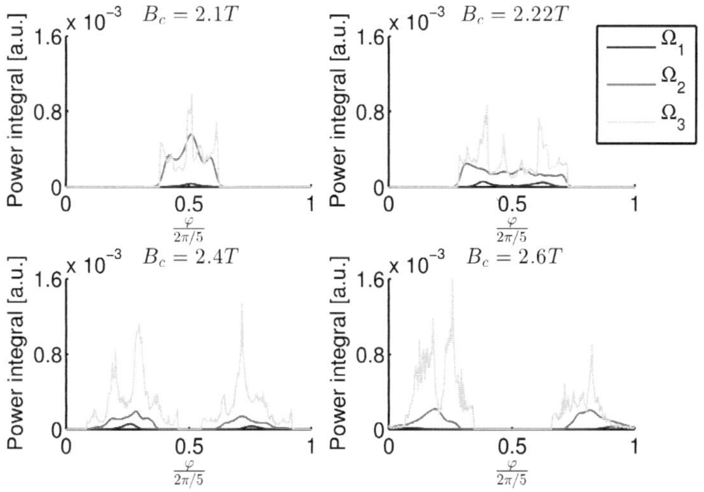

Figure 4. Power integrals computed for the three radial subdomains for the 4 frequencies of the scan as function of the normalised toroidal angle $\frac{\varphi}{2\pi/5}$.

Figure 4 shows the power integrals for each resonant field investigated and describes precisely the range of toroidal angle in which the wave transmits energy to the plasma. As the resonant field decreases this range becomes narrower. However it appears that the power deposition on the fast ion fraction of the minority species is most effective with an intermediate resonant field $B_c = 2.22T$. Indeed, it presents a toroidal range similar to the ones of higher resonant field.

In Figure 5 we plot the cumulative integration over the toroidal angle for the inner and outer radial domain, Ω_1 and Ω_3 respectively. Figure 5 shows that the $B_c = 2.22T$ case has the highest value of the power integral closest to the core flux surfaces, while limiting the relative amount of power going to the edge of the plasma. Finally it is clear that for higher resonant field, e.g. $B_c = 2.6T$, the wave-particle interaction disappears in almost half of the toroidal range and takes place mainly in the edge of the plasma with virtually no interaction in the core.

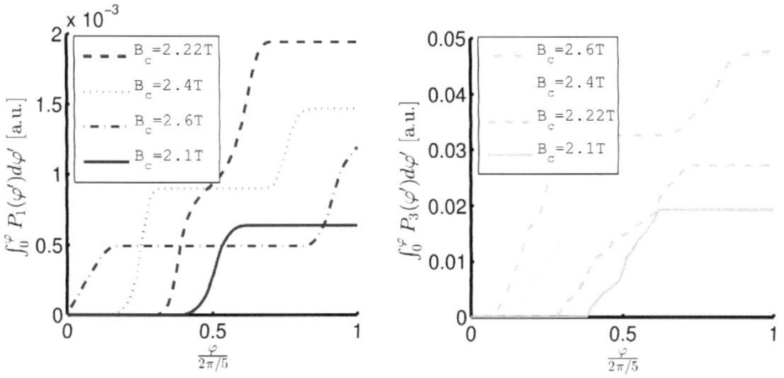

Figure 5. Cumulative integration over φ of the power integrals computed in the first and the last fourth of the plasma volume.

5. Conclusion

A 3D anisotropic high-mirror configuration W7X equilibrium was generated with ANIMEC. This equilibrium constitutes the basis for the application of by the 3D full-wave code LEMan in order to study the wave propagation in the ion-cyclotron range of frequency and the power deposition on fast minority ion fraction. A simplified antenna model was used with a single mode ($m_{ant} = 3, n_{ant} = 15$). This constitutes the first results concerning the wave absorption in an advanced stellarator system where it appears that some frequencies deposit power more efficiently on fast ions at the 0th order in Larmor radius. These preliminary calcutations indicate that a frequency of 33.8 MHz resonating with the minority species at $B_c = 2.22T$ appears to be close to optimal. Further simulations are planned with these 3D tools. For instance a more standard equilibrium, i.e. with a lower mirror ratio, will be investigated and compared to the study presented in this paper. Moreover it will be of interest to fully localise the antenna in one period therefore providing an excitation with many poloidal and toroidal modes describing a realistic antenna excitation spectrum and to study the wave propagation and absorption from that period to its neighbours.

References

[1] Cooper W A et. al. 2009 *Comput. Phys. Commun.* **180** 1524
[2] Popovich P et al. 2006 *Comput. Phys. Commun.* **175** 250
[3] Ongena J et al. 2014 *Phys. Plasmas* **21** 061514
[4] Geiger J et al. 2008 *35th EPS Conference on Plasma Phys. Hersonissos, 9-13 June 2008* **32D** 2.062
[5] Mellet N et al. 2010 *Comput. Phys. Commun.* **182** 570
[6] Mellet N 2009 *PhD Thesis n° 4398, EPFL Lausanne*
[7] Pfefferle D et al. 2014 *Comput. Phys. Commum.* **in press**

Journal of Physics: Conference Series **561** (2014) 012007 doi:10.1088/1742-6596/561/1/012007

Turbulent current drive

X. Garbet, D. Esteve, Y. Sarazin, G. Dif-Pradalier, P. Ghendrih, V. Grandgirard, G.Latu, A. Smolyakov*

CEA, IRFM, F-13108 St.Paul-lez-Durance cedex, France

* Department of Physics and Engineering Physics, U. of Saskatchewan, Canada

E-mail: xavier.garbet@cea.fr

Abstract. The Ohm's law is modified when turbulent processes are accounted for. Besides an hyper-resistivity, already well known, pinch terms appear in the electron momentum flux. Moreover it appears that turbulence is responsible for a source term in the Ohm's law, called here turbulent current drive. Two terms contribute to this source. The first term is a residual stress in the momentum flux, while the second contribution is an electro-motive force. A non zero average parallel wave number is needed to get a finite source term. Hence a symmetry breaking mechanism must be invoked, as for ion momentum transport. $E \times B$ shear flows and turbulence intensity gradients are shown to provide similar contributions. Moreover this source term has to compete with the collision friction term (resistivity). The effect is found to be significant for a large scale turbulence in spite of an unfavorable scaling with the ratio of the electron to ion mass. Turbulent current drive appears to be a weak effect in the plasma core, but could be substantial in the plasma edge where it may produce up to 10 % of the local current density.

1. Introduction

Turbulent acceleration was proposed as a possible mechanism for driving the intrinsic rotation observed in tokamaks [1, 2, 3]. This effect bears some similarity with turbulent heating, which was proposed as a potential heat source in magnetized plasmas [4, 5]. In practice, turbulent heating (resp. turbulent acceleration) corresponds to terms in the heat transport (resp. momentum) equation that do not depend on the temperature (current) or its gradient. When polarization terms are ignored, the sum over species of these sources vanishes, so that it corresponds in fact to heat and/or momentum transfer from some of the plasma constituents to others [5, 6, 7, 8, 9, 10, 11]. Net heating or acceleration is possible when radial fluxes at the edge are finite, due to finite Larmor radius effects and boundary conditions [10]. Thus a first question is whether turbulent acceleration, which appears as an effective electromotive force in the Ohm's law, is a significant effect. Another recent development comes from the calculation of various turbulent contributions to the ion momentum flux. Besides a turbulent viscosity, it has been shown that pinch terms contribute to the flux, and also a "residual stress" that is not proportional to the velocity or its gradient (see [12, 13] for overviews). Hence a second legitimate question is whether the residual stress contributes as a drive in the generalized Ohm's law.

The turbulent electro-motive force and the residual stress lead to a process that is called here "turbulent current drive". This modification of the Ohm's law was already mentioned in the past, and was dubbed "anomalous bootstrap current" [14], "turbulent dynamo" [15] or "source-like term in the Ohm's law" [16]. The terminology "anomalous bootstrap current" has become somewhat misleading since another effect was recently proposed, called "turbulence-driven bootstrap current" [17]. It is related to turbulent velocity scattering in the phase space and detrapping via radial transport, hence different from the processes investigated in the present work. As for ion momentum transport, a non zero average parallel wavenumber is needed to get finite turbulent electromotive force and residual stress - this is a consequence of general symmetry considerations [13]. Hence a symmetry breaking mechanism must be invoked. The final result depends sensitively on the efficiency of the mechanism at work. Moreover the source term competes with the Spitzer resistivity, so that the final result depends strongly on the collisionality. The purpose of this work is to estimate the amplitude of turbulent current drive for two symmetry breaking mechanisms, namely the effect of a shear flow and the inhomogeneity of the turbulence intensity profile. The mismatch between the mode frequency and the electron transit frequency introduces an unfavorable scaling with the ratio of the electron to the ion mass. Nevertheless the effect is significant for large scale fluctuations. Symmetry breaking due to $E \times B$ shear flow and gradient of turbulence intensity are shown to provide comparable contributions. Turbulence current drive is found to be quite small in the core, but not in the edge where the level of fluctuations is large and the gradient lengths are small.

The paper is organized as follows. In section 2, local conservation equations for heat and and momentum are derived from gyrokinetic equations. Explicit expressions of turbulent heating and momentum sources are detailed in section 3. Orders of magnitude are given and discussed. A conclusion follows.

2. Energy and momentum transport equations

2.1. Magnetic equilibrium and gyrokinetic equation

The spatial coordinates are (χ, θ, φ) where χ is the poloidal magnetic flux normalized to 2π, θ is the poloidal angle and φ the toroidal angle. The Jacobian of this metric is $1/\mathbf{B}\cdot\nabla\theta$ where the unperturbed magnetic field is

$$\mathbf{B} = I\nabla\varphi + \nabla\varphi \times \nabla\chi \tag{1}$$

We consider the gyrocenter distribution function $\bar{F}(\mathbf{z})$ where $\mathbf{z} = (\chi, \theta, \varphi, v_{G\parallel}, \mu)$. Here $v_{G\parallel}$ is the gyrocenter parallel velocity and μ is the magnetic moment. To derive the conservation laws, it is convenient to use the collisionless gyrokinetic equation written in its conservative form [18, 19]

$$\partial_t \bar{F} + \frac{1}{B_\parallel^*}\nabla_{\mathbf{z}} \cdot \left(\dot{\mathbf{z}} B_\parallel^* \bar{F}\right) = 0 \tag{2}$$

where $\dot{\mathbf{z}} = \mathrm{d}_t\mathbf{z}$ and $B_\parallel^* = B + \frac{m}{e} v_{G\parallel}\mathbf{b}\cdot(\nabla\times\mathbf{b})$ is the Jacobian of the gyrocenter transformation (m is the mass, e the charge and $\mathbf{b} = \mathbf{B}/B$ the unit vector along the field lines). For an electromagnetic turbulence, the gyrocenter equations of motion are

$$B_\parallel^* d_t\mathbf{x} = -B_\parallel^*[\bar{H}, \mathbf{x}] = v_{G\parallel}\mathbf{B}^* + \frac{1}{e}\mathbf{b}\times\nabla\Lambda \tag{3}$$

$$B_\parallel^* m d_t v_{G\parallel} = -B_\parallel^*[\bar{H}, v_{G\parallel}] = -\mathbf{B}^*\cdot\nabla\Lambda \tag{4}$$

where $\Lambda = \mu B + \bar{h}$,

$$\bar{h} = e(\bar{\phi} - v_{G\parallel}\bar{A}_\parallel), \tag{5}$$

is the gyroaverage perturbed hamiltonian, $\mathbf{B}^* = \mathbf{B} + \frac{m v_{G\parallel}}{e}\nabla\times\mathbf{b}$, and

$$\bar{H} = \frac{1}{2} m v_{G\parallel}^2 + \Lambda \tag{6}$$

is the hamiltonian, $\bar{\phi} = J \cdot \phi$ and $\bar{A}_\parallel = J \cdot A_\parallel$ are the perturbed gyro-averaged electric potential and parallel vector potential. The system is closed with a simplified Poisson equation

$$-\sum_{species} \nabla \cdot \left\{ \frac{\rho_{eq,s}}{B^2} \nabla_\perp \phi \right\} = e \sum_{species} n_G = \sum_{species} e \int \frac{2\pi}{m} B^*_\parallel \mathrm{d}\mu \mathrm{d}v_{G\parallel} J \cdot \bar{F} \tag{7}$$

and Ampère equation

$$-\nabla^2_\perp A_\parallel + \sum_{species} \frac{\omega_{ps}^2}{c^2} A_\parallel = \mu_0 \sum_{species} e \int \frac{2\pi}{m} B^*_\parallel \mathrm{d}\mu \mathrm{d}v_{G\parallel} v_{G\parallel} J \cdot \bar{F} \tag{8}$$

where $\rho_{eq,s}$ is the unperturbed mass density, and ω_{ps} the plasma pulsation of a species 's'. We will use in the following the long wavelength limit of the gyroaverage operator

$$J \simeq 1 + \frac{1}{2} \nabla \cdot \left(\frac{m\mu}{e^2 B} \nabla_\perp \right) \tag{9}$$

This operator is consistent with the left hand side of the Poisson equation Eq.(7), and is self-adjoint.

2.2. Momentum and energy transport equation

These equations have been derived for the set of equations given above in refs [3, 20]. The local energy transport equation reads for each species

$$\partial_t E_K + \partial_\chi Q = W_E \tag{10}$$

where

$$E_K = \int \mathrm{d}\tau^* \left(\frac{1}{2} m v_{G\parallel}^2 + \mu B \right) \bar{F} \tag{11}$$

is the kinetic energy, and $\mathrm{d}\tau^* = B^*_\parallel 2\pi \frac{d\mu}{m} dv_{G\parallel} \frac{d\theta d\varphi}{\mathbf{B} \cdot \nabla \theta}$ is the phase space volume element in between two magnetic surfaces χ and $\chi + d\chi$. The heat flux Q is defined as

$$Q = \int \mathrm{d}\tau^* \bar{H} v_G^\chi \bar{F} \tag{12}$$

and $v_G^\chi = \dot{\mathbf{z}} \cdot \nabla \chi$ is the radial projection of the gyrocenter velocity. The turbulent energy source is

$$W_E = -\int \mathrm{d}\tau^* \bar{h} \partial_t \bar{F} \tag{13}$$

In a similar way, the toroidal momentum conservation equation reads

$$\partial_t \mathcal{L}_\varphi + \partial_\chi \Pi^\chi_\varphi + \partial_\chi T^\chi_\varphi = \mathcal{J}^\chi \tag{14}$$

where

$$\begin{aligned} \mathcal{L}_\varphi &= m \int \mathrm{d}\tau^* u_\varphi \bar{F} \\ \Pi^\chi_\varphi &= m \int \mathrm{d}\tau^* \bar{F} u_\varphi v_G^\chi \\ T^\chi_\varphi &= \int^\chi d\chi \int \mathrm{d}\tau^* \bar{F} \partial_\varphi \bar{h} \\ \mathcal{J}^\chi &= e \int \mathrm{d}\tau^* v_G^\chi \bar{F} \end{aligned} \tag{15}$$

Here $u_\varphi = \frac{I}{B}v_{G\|}$ is the gyrocenter toroidal angular velocity at requested order. An explicit expression of the polarization term T_φ^χ is given in [20, 3]. In most cases, it can be ignored. The local angular momentum conservation equation can then be recast as

$$\partial_t \mathcal{L}_\varphi + \partial_\chi \Pi_\varphi^\chi = R_0 M_\| \tag{16}$$

where the local source of toroidal momentum source is

$$M_\| = -\int d\tau^* \bar{F}\frac{I}{BR_0}\nabla_\| \bar{h} \tag{17}$$

where the parallel gradient is defined as

$$\nabla_\| = \frac{1}{B}\mathbf{B}\cdot\nabla = \frac{I}{BR^2}\left(\partial_\varphi + \frac{1}{q}\partial_\theta\right) \tag{18}$$

and R is the major radius (R_0 is the major radius on axis).

3. Explicit expressions of fluxes and sources

The particle conservation equation and the equations Eqs.(10,16) can be recast as follows [3]

$$\partial_t N + \partial_\chi \Gamma_N^\chi = 0 \tag{19}$$

$$Nm\partial_t V_\| + \partial_\chi \Gamma_V^\chi = M_\| \tag{20}$$

$$\frac{3}{2}N\partial_t T + \partial_\chi \Gamma_T^\chi = W_T \tag{21}$$

where N is the density, $V_\|$ is the parallel fluid velocity and T the temperature. The quasilinear expressions of the fluxes and sources are [3]

$$\begin{pmatrix} W_T \\ M_\| \\ \Gamma_N^\chi \\ \Gamma_V^\chi \\ \Gamma_T^\chi \end{pmatrix} = \frac{T}{e}\sum_{\mathbf{n}}\int_{-\infty}^{+\infty} d\omega' \int d\tau^* F_{eq}\left|\frac{\bar{h}_\mathbf{n}}{T}\right|^2 \frac{\Delta\omega_\mathbf{n}}{(\omega'-\omega_\mathbf{n})^2 + \Delta\omega_\mathbf{n}^2}$$

$$n\left(\omega' - \mathbf{n}\cdot\boldsymbol{\Omega}_*\right)\delta\left(\omega' - \mathbf{n}\cdot\boldsymbol{\Omega}\right)\begin{pmatrix} e\left(\omega' - k_\| V_\|\right)/n \\ ek_\|/n \\ 1 \\ mv_{G\|} \\ \mathcal{E} - \frac{3}{2}T \end{pmatrix} \tag{22}$$

where F_{eq} is the unperturbed distribution function (typically a Maxwellian), $\mathcal{E}$ the particle kinetic energy, n the toroidal wavenumber. The triplets of integers $\mathbf{n}$ refer to the Fourier development of various fields when using angle variables. Here $\omega' = \omega - n\omega_E$, and $\mathbf{n}\cdot\boldsymbol{\Omega}$ is the resonant frequency calculated in the frame of reference where the radial electric field vanishes, i.e. $\mathbf{n}\cdot\boldsymbol{\Omega} = n\omega_D + k_\| v_{G\|}$ for passing particles, and $\mathbf{n}\cdot\boldsymbol{\Omega} = n\omega_{Dt}$ for trapped particles, where ω_{Dt} is the precession frequency. The pulsation $\mathbf{n}\cdot\boldsymbol{\Omega}_*$ is an energy dependent diamagnetic pulsation, which is detailed in the next section. From now on the calculations are performed in the frame of reference where ω_E locally vanishes. This does not forbid a finite velocity shear $\partial_\chi\omega_E$, which is a source of acceleration.

3.1. Practical expressions for fluxes and sources

Let us now consider one species with mass m_s and charge e_s (for electrons $e_s = -e$. For a Maxwellian equilibrium distribution function, sources and fluxes are given by the expressions Eq.(22), which can be recast as

$$\begin{pmatrix} \frac{W_T}{N_s e_s} \\ \frac{M_\parallel}{N_s e_s} \\ \frac{\Gamma_N^\chi}{N_s} \\ \frac{\Gamma_T^\chi}{N_s T_s} \\ \frac{\Gamma_V^\chi}{N_s m_s v_{Ts}} \end{pmatrix} = -\frac{T_s}{e_s}\sum_{\mathbf{n}} \int_{-\infty}^{+\infty} d\omega \int_0^{+\infty} du e^{-u} \int_{-\infty}^{+\infty} \frac{d\zeta}{\sqrt{2\pi}} e^{-\frac{\zeta^2}{2}}$$

$$n\left|\frac{\bar{h}_{\mathbf{n}}}{T}\right|^2 D_{\mathbf{n}\omega} \frac{\Delta\omega_{\mathbf{n}}}{(\omega-\omega_{\mathbf{n}})^2+\Delta\omega_{\mathbf{n}}^2} \delta\left(R_{\mathbf{n}\omega}\right) \begin{pmatrix} \frac{\omega}{n} \\ \frac{k_\parallel}{n} \\ 1 \\ u+\frac{\zeta^2}{2}-\frac{3}{2} \\ \varepsilon_t \zeta \end{pmatrix}$$

where the denominator $D_{\mathbf{n}\omega}$ is

$$D_{\mathbf{n}\omega} = \omega + n\frac{T_s}{e_s}\partial_\chi \Xi_s - \varepsilon_t k_\parallel v_{Ts} \mathcal{V}_{s\parallel} \tag{23}$$

with

$$\partial_\chi \Xi_s = \partial_\chi \ln N_s + \left(u + \frac{\zeta^2}{2} - \frac{3}{2}\right)\partial_\chi \ln T_s + \varepsilon_t \zeta \partial_\chi \mathcal{V}_{s\parallel} \tag{24}$$

The resonance term $R_{\mathbf{n}\omega}$ is

$$R_{\mathbf{n}\omega} = \omega - k_\parallel v_{Ts}\left(\mathcal{V}_{s\parallel} + \zeta\right) - n\omega_{Ds}\left(u + \left(\zeta + \mathcal{V}_{s\parallel}\right)^2\right) \tag{25}$$

for passing particles and

$$R_{\mathbf{n}\omega} = \omega - n\omega_{Ds} u \tag{26}$$

for trapped particles, where $\mathcal{V}_{s\parallel} = V_{\parallel s}/v_{Ts}$ is the parallel fluid velocity normalized to the species thermal velocity v_{Ts}. The variable $u = \mu B/T_s$ and $\zeta = v_{G\parallel}/v_{Ts}$ are respectively the normalized perpendicular energy and parallel velocity. For trapped (resp. passing) particles, $\varepsilon_t = 0$ (resp. $\varepsilon_t = 1$), i.e. trapped particles do not carry parallel momentum. The curvature frequency is defined as $\omega_{Ds} = \frac{q}{r}\frac{T_s}{e_s BR}$, the thermal velocity is $v_{Ts} = \sqrt{T_s/m_s}$, and the Larmor radius is $\rho_s = \frac{m_s v_{Ts}}{e_s B}$. The expression of the precession frequency that is used here is valid only for deeply trapped particles, and is given to illustrate the structure of fluxes. It should be replaced by its exact expression for accurate calculations. The structure of fluxes and sources given above is close to the one derived in [16], using similar techniques.

3.2. Orders of magnitude

Current is carried by passing electrons. We further simplify the problem by ignoring trapped electrons and keeping transit resonances only. In that case the angle variables are simply the cyclotron angle, and the poloidal and toroidal angles. The force acting on electrons is

$$M_{\|e} = \frac{N_e T_e}{R_0} \frac{1}{\sqrt{2\pi}} \sum_{\mathbf{k}} \exp\left\{ -\frac{1}{2} \left(\frac{\omega_{\mathbf{k}}}{\omega_{te}} \right)^2 \right\} \left| \frac{\bar{h}_{\mathbf{k}}}{T_e} \right|^2 k_\| R_0 \frac{\omega_{\mathbf{k}} - \omega_{*e}}{|\omega_{te}|} \tag{27}$$

where the components of $\mathbf{k}$ are the poloidal and toroidal wavenumbers (m, n), $\omega_{*e} = n\frac{T_e}{e} \partial_\chi \Xi_e|_{\zeta=\omega_{\mathbf{n}}/\omega_{te}}$, and $\omega_{te} = k_\| v_{Te}$ (n is the toroidal wavenumber).

Fluctuation measurements indicate that the spectral turbulence intensity $I_{\mathbf{k}} = |\bar{h}_{\mathbf{k}}/T_e|^2$ is maximum at low wavenumbers $k_\perp \rho_i = 0.1$, and decays as $k_\perp^{-3}$ for $k_\perp \rho_i \leq 1$, faster beyond $k_\perp \rho_i = 1$ [21] ($k_\perp$ is the perpendicular wavenumber and ρ_i the thermal ion gyroradius). For a frequency of the order of the diamagnetic frequency, the ratio $\frac{\omega_{\mathbf{k}}}{\omega_{te}}$ behaves as $\frac{qR_0}{sL_p} \frac{\rho_i}{w_k} \sqrt{\frac{m_e}{m_i}}$, where q is the safety factor, s the magnetic shear, w_k the mode width, L_p a gradient length (for simplicity all gradient lengths are assumed to be comparable). It appears that in spite of an unfavorable mass scaling, the dominant contribution comes from low wave number fluctuations, at long as quasi-linear theory is valid. An order of magnitude of the average wavenumber due to $E \times B$ shear flow is [22]

$$\bar{k}_\| R_0 = -2\frac{q}{s} k_\theta^2 \rho_i^2 \frac{1}{\omega_{\mathbf{k}}} \left(1 - \frac{\omega_{\mathbf{k}}}{n\omega_{De}} \right) \frac{dV_{E\theta}}{dr} \tag{28}$$

here $\gamma_E = \frac{dV_{E\theta}}{dr}$ is the flow shear rate. An order of magnitude of the flow shear rate at low parallel Mach number is $\gamma_E \simeq \rho_i \frac{v_{Ti}}{L_p^2}$. Hence the mean parallel wave number is $\bar{k}_\| R_0 \simeq \frac{q}{s} k_\theta \rho_i \frac{\rho_i}{L_p}$. Another mechanism comes from the gradient of turbulence intensity [23]. In that case $\bar{k}_\| R_0 \simeq s k_\theta \frac{w_k^2}{L_I}$, where $\frac{1}{L_I} = \frac{d\ln(I_{\mathbf{k}})}{dr}$. Hence both effects leads to the same scaling for the mean parallel wave numbers in the gyroBohm framework $w_k \equiv \frac{1}{k_\theta} \equiv \rho_i$, i.e. $\bar{k}_\| R_0 \equiv \frac{\rho_i}{L_p}$. Assuming isotropic spectra $w_k \simeq \frac{1}{k_\theta}$, it turns out that the effect of turbulence intensity gradient is stronger at low wave numbers. The turbulent force is then approximated by

$$M_{\|e} \simeq \frac{N_e T_e}{L_p} \frac{q\rho_i}{L_I} \sqrt{\frac{m_e}{m_i}} I_{turb} \tag{29}$$

where I_{turb} is the turbulence intensity at the spectrum peak. The effective electric field

$$E_{eff} \simeq \frac{M_{\|e}}{N_e e} \simeq \frac{T_e}{eR_0} \sqrt{\frac{m_e}{m_i}} \frac{qR_0}{L_p} \frac{\rho_i}{L_I} I_{turb} \tag{30}$$

can then be compared to the inductive field, of the order of $0.1V/m$. It appears that E_{eff} is usually small in the core plasma, i.e. of the order of $10^{-3}V/m$ or less. However, this is not true in the plasma edge, where the level of fluctuations is large, and the gradient scales are small. We choose the following parameters at the pedestal height $N_e = 1.10^{19}m^{-3}$, $T_e = 100eV$, $R_0 = 1m$, $a = 0.3m$, $L_p = L_I = 0.01m$, $B = 1T$, $q = 3$, $\sqrt{\frac{m_i}{m_e}} = 60$, $I_{turb} \simeq \left|\frac{\delta N}{N_e}\right|^2 \simeq 10^{-3}$. One then finds $\frac{T_e}{eR_0} \simeq 10^2\ V/m$, $\sqrt{\frac{m_e}{m_i}} \frac{qR_0}{L_p} \simeq 5$, and $\frac{\rho_i}{L_I} \simeq 0.072$. This leads to an effective electromotive field of the order of $3.610^{-2}V/m$, i.e. can be considered as significant. Another way to estimate the turbulence contribution to the current is to compare the driven current to the bootstrap current. The local bootstrap current in the edge scales as $J_{boot} \simeq 5\sqrt{\frac{R}{a}} q \frac{N_e T_e}{BL_p}$, while the turbulence driven current can be estimated by balancing the turbulent force $M_{\parallel e}$ with the collisional electron/ion friction force $\nu_{ei}\frac{m_e}{e}J_{turb}$ (ν_{ei} is the electron/ion collision frequency). An estimate of the turbulence driven current is

$$J_{turb} = \frac{e}{m_e \nu_{ei}} \frac{N_e T_e}{L_p} \frac{q\rho_i}{L_I} \sqrt{\frac{m_e}{m_i}} I_{turb} \tag{31}$$

so that the ratio of the turbulent to bootstrap current is

$$\frac{J_{turb}}{J_{boot}} = \frac{1}{5}\sqrt{\frac{a}{R_0}} \frac{v_{Te}}{L_I \nu_{ei}} I_{turb} \tag{32}$$

Using the parameters above, it is found that $J_{turb}/J_{boot} \simeq 0.11$, i.e. 10%.

Another turbulence contribution comes from the flux of momentum Γ_V^X, whose expression is given by Eq.(22). Using a radial coordinate close to a minor radius and neglecting the particle flux, the momentum flux is $\Gamma_V = N_e m_e \left\langle \tilde{v}_{Er} \tilde{V}_\parallel \right\rangle$. The structure of this flux is close to the one found for ions in the context of plasma "intrinsic rotation" (see overviews [12, 13] and references therein). It can be written as

$$\Gamma_V = -N_e m_e D_V \partial_r V_\parallel + N_e m_e v V_\parallel + \Pi_\parallel \tag{33}$$

where D_V is a viscosity, v is the pinch velocity and $\Pi_\parallel$ the residual stress. The electron viscosity D_V can also be interpreted as an hyper-resistivity since the electron parallel velocity is a good proxy of the current. This is not by far a new idea [24, 25]. The value of the turbulent viscosity is still subject to debate. It is probably in the range of $[0.1-1]m^2/s^{-1}$. The corresponding diffusion time is to be compared with the electron collision time. Its contribution to the Ohm's law is negligible for the current evolution at large scale, but could matter for some specific problems like MHD inertial layers. A second contribution comes from pinch terms. The expression of v is quite complex, as several effects contribute. However perpendicular compressibility often provides the dominant contribution, of the order of $v \simeq D_V/R$. If the viscosity is substantial, this effect cannot be ignored either for small scale dynamics. Finally a third contribution comes from the residual stress. An order of magnitude is

$$\Pi_\parallel \simeq N_e T_e \frac{L_p}{R_0} \sum_{\mathbf{k}} i k_\parallel R_0 \omega_{*e} \tau_{c\mathbf{k}} \left| \frac{\bar{h}_{\mathbf{k}}}{T_e} \right|^2 \tag{34}$$

where $\tau_{c\mathbf{k}} = 1/\Delta\omega_{\mathbf{k}}$ is a correlation time and $\Delta\omega_{\mathbf{k}} \simeq \omega_{te}$ for passing electrons. It appears that the divergence of this flux is of the same order as the source term Eq.(29).

4. Conclusion

In summary, turbulence processes may modify substantially the Ohm's law. It is already known that turbulence produces a significant hyper-resistivity. However it also appears that it is responsible for pinch terms in the momentum flux. Moreover turbulence produces a source term in the Ohm's law that is not proportional to the current or its gradient. Two terms contribute to this source. The first term is the residual stress in the momentum flux. The second one is an electromotive force, that competes with the electron-ion collision friction (Spitzer resistivity). A symmetry breaking mechanism is needed to provide a finite average parallel wave number that produces a non zero current drive. $E \times B$ shear flows and radial gradient of turbulence intensity give similar orders of magnitude. The effect is significant for a large scale turbulence in spite of an unfavorable scaling with the ratio of the electron to ion masses. Turbulent current drive appears to be a weak effect in the plasma core, but could be substantial in the plasma edge where the level of fluctuations is large and the gradient lengths are small.

Acknowledgements
This work was initiated at the Festival of Theory , Aix en Provence 2013. The authors thank P.K. Kaw, A. Das, P.H. Diamond, and Lu Wang, for fruitful discussions. This project has received funding from the European Union's Horizon 2020 research and innovation programme under grant agreement number 633053. The views and opinions expressed herein do not necessarily reflect those of the European Commission .

References

[1] Diamond P H, McDevitt C J, Gürcan Ö D, Hahm T S, and Naulin V 2008 *Physics of Plasmas* **15** 012303
[2] Wang L and Diamond P H 2013 *Phys. Rev. Lett.* **110** 265006
[3] Garbet X, Esteve D, Sarazin Y , Abiteboul J, Bourdelle C, Dif-Pradalier G, Ghendrih P, Grandgirard V, Latu G, and Smolyakov A 2013 *Phys. Plasmas* **20** 072502
[4] Ott E, McBride J B, Orens J H, and Boris J P 1972 *Phys. Rev. Lett.* **28** 88
[5] Manheimer W M, Ott E, Tang W M 1977 *Phys. Fluids* **20** 808
[6] Sugama H, Okamoto M, Horton W, Wakatani M 1996 *Phys. Plasmas* **3** 2379
[7] Waltz R E, Staebler G M, Dorland W, Hammett G W, Kotschenreuther M, Konings J A 1997 *Phys. Plasmas* **4** 2482
[8] Hinton F L and Waltz R E 2006 *Phys. Plasmas* **13** 102301
[9] Waltz R E and Staebler G M 2008 *Phys. Plasmas* **15** 014505
[10] Zhao L and Diamond P H 2012 *Phys. Plasmas* **19** 082309
[11] Barnes M, Parra F I, and Dorland W 2012 *Phys. Rev. Letters* **109** 185003
[12] Diamond P H, McDevitt C J, Gürcan Ö D *et al* 2009 Nucl. Fusion **49** 045002
[13] Peeters A G *et al* 2011 *Nucl. Fusion* **51** 094027
[14] Itoh S-I and Itoh K 1988 *Phys. Letters A* **125** 267
[15] Hinton F L, Waltz R E and Candy J 2004 *Phys. Plasmas* **11** 2433
[16] Gatto R, and Chavdarovski 2011 *Open Plasma Physics Journal* **4** 1
[17] McDevitt C J, Tang Xian-Zhu , and Guo Zehua 2013 *Phys. Rev. Letters* **111** 205002
[18] Brizard A J and Hahm T S 2007 *Rev. Mod. Phys.* **79** 421
[19] Garbet X , Idomura Y, Villard , and Watanabe T H 2010 *Nucl. Fusion* **50** 043002
[20] Abiteboul J, Garbet X, Grandgirard V, Allfrey S J, Ghendrih P, Latu G, Sarazin Y, and Strugarek A 2011 *Phys. Plasmas* **18** 082503
[21] Hennequin P *et al Nucl. Fusion* 2006 **46** S771-S779
[22] Garbet X, Sarazin Y, Ghendrih P *et al* 2002 *Phys. Plasmas* **9** 3893
[23] Gürcan Ö D, Diamond P H, Hennequin P *et al* (2010) *Phys. Plasmas* **17** 112309
[24] Schmidt J, and Yoshikawa S 1971 Phys. Rev. Lett. **26** 753
[25] Biskamp D 1984 Plasma Phys Contr Fusion **26** 311

Self organisation of plasma turbulence: impact on radial correlation lengths

Philippe Ghendrih, , Guilhem Dif-Pradalier, Claudia Norscini, Thomas Cartier-Michaud, Damien Estève, Xavier Garbet, Virginie Grandgirard, Guillaume Latu, Chantal Passeron, Yanick Sarazin

CEA, IRFM, F-13108 Saint-Paul-lez-Durance, France

E-mail: philippe.ghendrih@cea.fr

Abstract. The empirical scaling law used to support the projected ITER performance [1, 2] in analysed to determine the control parameters that must be considered in gyrokinetic simulations. The results of such an analysis are contradictory and do not appear to fit with present evidence from gyrokinetic simulations. Analysing the dependence of the correlation length on the ρ_* parameter, we show that local values of the correlation length are governed by the shearing effect of the corrugation patterns, but that coarse graining this value, in time, radially or poloidally does not allow one to recover this match. A comparison to the scaling law must then rely on the scaling properties of the Probability Density Function of the correlation length that is found to exhibit heavy tails with algebraic decay.

1. Introduction

The scaling law for the energy confinement time has been an area of very important research activity that both summarises the experimental effort and has allowed one to design devices with a prescribed fusion performance as in the case of ITER [3, 1, 2]. A key commitment of large scale computing with gyrokinetics models [4] is to back such a scaling law [5, 6, 7] and increase our confidence by combining understanding and precise predictions to be challenged by experimental investigation [8, 9, 10, 11].

As it stands the reference scaling law need clarification. The issue of the β dependence is often alluded to [12, 13, 14]. Here we concentrate on the leading dependence, namely the ρ_* dependence [14, 2], the only to exhibit a mass dependence. Some concern has already been pointed out regarding the isotope effect that appears to depart from that expected from the ρ_* effect[15, 16, 17]. More recent theoretical results on the change of transport from Bohm to gyro-Bohm when $1/\rho_*$ exceeds 300 [5, 6, 18] appear to be in contradiction with the more gyro-Bohm like feature of the empirical scaling law. Indeed, the present machine pool used to build the experimental evidence should exhibit a significantly weaker dependence on ρ_* than it actually does. This aspect is presently being revisited by scaling the heating power in the simulations with ρ_* [19]. Further effort should help in clarifying both the expected ρ_* and mass dependences in gyrokinetic flux-driven simulations [20, 21, 22, 23]. Also, a quite important issue is to bridge the key features of the scaling law with the meso-scale structures, such as avalanches

Table 1. Exponents of the energy confinement time scaling

τ_0	α_i	α_n	α_p	α_r	α_a	α_k	α_m	α_b
0.0562	0.93	0.41	0.69	1.97	0.58	0.78	0.19	0.15

[24, 25, 26, 27, 28] and corrugations [29, 30, 31, 32], that are believed to play a strong role in turbulent transport regulation [33].

The scope of our confinement analysis with global flux driven simulations is to determine the key control parameters and the physics that drive these dependences. In the framework of scaling laws, which hold over a couple of decades, we rely on the assumption that all details are not important and that confinement properties are based on a few robust features. Should this not be the case, or should the simple scaling that is observed be the result of competing mechanisms that tend to cancel out, would be an outstanding but quite problematic result for fusion. Whatever, stepping up our present predictive capability with global flux-driven simulations is a first priority and the scope of the present work.

The paper is organised as follows. In Section 2 we revisit the conversion of the scaling law for the energy confinement time from the so-called engineer parameters to dimensionless parameters. Our interest is to highlight the expressions that allow one to compute the exponents of the control parameters that are relevant to gyrokinetic simulation. The computation of the correlation length in a GYSELA [20] simulation evolving towards global balance between the heating source and the exhaust flux at the outer radius is described in Section 3. We then show in Section 4 that the correlation length is strongly related to local effects in time, radius and poloidal angle. The latter effect appears to stem from the interplay between corrugations and avalanches that self-consistently determine the transport level. The relevance of coarse graining action to extract from the broad Probability Density Function a relevant correlation length is discussed there. Finally, Section 5, dedicated to the Discussion and Conclusion, closes the paper.

2. Energy confinement scaling laws

The guideline for ITER design is the scaling law for the energy confinement τ_E built with an international data base of experimental results [2]. In its standard form it is expressed in terms of products of so-called engineer parameters at some power:

$$\tau_E = \tau_0 \; I_{MA}^{\alpha_i} \; n_{19}^{\alpha_n} \; P_{MW}^{-\alpha_p} \; R_{SI}^{\alpha_r} \; A^{-\alpha_a} \; \kappa_a^{\alpha_k} \; M^{\alpha_m} \; B_{SI}^{\alpha_b} \tag{1}$$

The values of the various exponents are given in table 1 and the subscripts indicate the normalisation of the various terms. Strangely enough, some parameters are already dimensionless such as the plasma elongation κ_a, the plasma aspect ratio A and the mass number M which is the mass of the ion species normalised by the nucleon mass. In this formulation, the various exponents, the α's, are defined in such a way that they are positive. The chosen parameters for this scaling law are the plasma current I, the plasma density n, the heating power P, the major radius R and the magnetic field amplitude B. This combination of parameters at various powers must yield by definition a time unit for τ_E, the so-called Kadomtsev constraint [34]. One the obtains the following relationship between the exponents:

$$5\;\alpha_b = 3\alpha_p + 4\alpha_r - \Big(8\alpha_n + \alpha_i + 5\Big) \tag{2}$$

This relation is satisfied by the coefficients of table 1 within less than 1 %. It is to be noted that the results appears as a difference between two large values that weigh more than 13 times their difference, a feature that is prone to error amplification. This done, the scaling law is expressed in terms of dimensionless control parameters, namely ρ_*, β and ν_*. We only address here the ρ_* dependence, this parameter being defined as:

$$\rho_* = \frac{\rho_p}{a} = \frac{(T/m_p)^{1/2}}{a\ \Omega_p} \tag{3}$$

where a is the plasma minor radius, $a = R/A$, and where ρ_p, m_p and Ω_p are the proton Larmor radius, mass and cyclotron frequency respectively. The temperature T is the volume-average thermal energy per particle. The ρ_* parameter is akin to the Reynolds number insofar that it accounts for the number of degrees of freedom in the system. However, this estimate only takes into account the effect of the temperature, of the magnetic field and that of the size of the device. The mass effect is not taken into account in this definition. When switching to dimensionless parameters one finds:

$$\Omega_p \tau_E \propto \nu_*^{\alpha_\nu}\ \rho_*^{-\alpha_\rho}\ \beta^{\alpha_\beta}\ q_{cyl}^{\alpha_{qcyl}}\ A^{\alpha_A}\ (q/q_{cyl})^{\alpha_q}\ \kappa_a^{\alpha_\kappa}\ M^{\alpha_M} \tag{4}$$

In this expression the dependence on q_{cyl} the cylindrical value of the safety factor essentially accounts for the plasma current dependence and directly impacts the dependence on the aspect ratio A. The dependence on (q/q_{cyl}) is related to shaping properties similarly to that on the elongation κ. The exponents for ν_* is of order zero while that of β is of order minus one, hence a rather severe degradation of the confinement with β, while α_ρ is found to be:

$$\alpha_\rho = \frac{6\Big(\alpha_i + \alpha_r\Big) - 2\ \alpha_n - 18\ \alpha_p}{5(1-\alpha_p)} \tag{5}$$

Given the data from table 1, one finds $\alpha_\rho \approx 2.68$ quite close to the value 3 of the gyro-Bohm scaling. Similarly one obtains:

$$\alpha_{qcyl} = \frac{\alpha_r - 2\alpha_n - 3\alpha_p - 4\alpha_i}{5(1-\alpha_p)} \tag{6}$$

$$\alpha_A = \frac{-\alpha_i + 3\alpha_r - 2\alpha_n - 5\alpha_p - 2\alpha_a}{2(1-\alpha_p)} \tag{7}$$

$$\alpha_M = \frac{\alpha_m}{1-\alpha_p} \tag{8}$$

which yields: $\alpha_{qcyl} \approx -2.99$, $\alpha_A \approx -0.73$, $\alpha_M \approx 0.61$. In such an analysis, small q_{cyl}, small A and large M are favourable. The largest dependence in the confinement scaling law is thus found to be approximately $(q_{cyl}\ \rho_*)^{-3}$, hence gyro-Bohm like regarding ρ_*.

In this framework, where transport is implicitly considered to be diffusive, one readily understands the exponent within the random-walk paradigm. Heat is then described to experience a random walk from the plasma core, where the heat source is located, to the heat sink at the plasma boundary. Within this $1D$ description the confinement time takes the form $\tau_E = \tau_{rw} N_*^2$, where N_* is the number of steps between the core and the edge, and τ_{rw} is the characteristic time of one step. For electrostatic turbulence in magnetised plasmas, the eddy size, hence the step of the random walk scales like ρ with characteristic velocity v_E the $E \times B$ electric drift velocity. As a consequence, one obtains $N_* = a/\rho$, and, given the drift ordering, $\tau_{rw} = \rho/v_E = a/c$, where c is the thermal velocity. Given $\rho_{s*} = \rho/a$ where $\rho_{s*} = \sqrt{M}\rho_*/Z$ is

Journal of Physics: Conference Series **561** (2014) 012008 doi:10.1088/1742-6596/561/1/012008

species dependent (hence the subscript s):

$$N_* = \frac{1}{\rho_{s*}} = \frac{a\,\Omega}{(T/(Mm_p))^{1/2}} = \frac{Z\sqrt{M}}{M}\frac{a\,\Omega_p}{(T/m_p)^{1/2}} == \frac{Z}{\sqrt{M}}\frac{1}{\rho_*} \qquad (9)$$

$$\tau_{rw} = \frac{1}{\Omega_p}\frac{M}{Z}\frac{1}{\rho_{s*}} = \frac{1}{\Omega_p}\frac{M}{Z}N_* \qquad (10)$$

One thus obtains $\Omega_p\tau_E = N_*^3 M/Z$. In this expression, the confinement time depends on the number of steps N_* which is an effective measure of the number of degrees of freedom. Following this understanding of the ρ_* dependence, the scaling of the confinement time should take the form $\Omega_p\tau_E \propto N_*^{\alpha_\rho} M^{\alpha_{sM}}$. The other dimensionless control parameters do not depend on the mass so that the new exponent for the mass dependence α_{sM} is then:

$$\alpha_{sM} = \frac{\alpha_m}{1-\alpha_p} + \frac{\alpha_\rho}{2} \approx 1.95 \qquad (11)$$

This introduces therefore a much stronger mass ratio M dependence than would be expected. In flux driven gyrokinetic simulations, one must also relate the injected power P to the dimensionless parameters. The power must then scale with ρ_* as well as with M. The ρ_* dependence stems from the increased volume to heat, scaling typically like ρ_*^{-3} balanced by the improved confinement at smaller ρ_*. The isotope effect is only governed by its impact on the confinement time. One thus obtains:

$$P \propto \rho_*^{-p_\rho} M^{-\alpha_M} \quad ; \quad p_\rho = 18/5 - \alpha_p \qquad (12)$$

The required power increases with $1/\rho_*$ with an exponent $p_\rho \approx 0.92$ [19] while the isotope effect allows one to reduce the power with increasing M to achieve a similar plasma performance. It is to be underlined that this increase is less that required to maintain a power density or a heat flux due to the confinement enhancement with $1/\rho_*$.

The latter result [19] is important when discussing the crossover from Bohm to gyro-Bohm as $1/\rho_*$ exceeds 300 as reported in gyrokinetic simulations [5, 6, 18]. Indeed, since the devices used in the scaling law database range from Compass with $1/\rho_* \approx 50$ to JET $1/\rho_* \approx 300$, one would then predict that the global energy confinement time to do no better than Bohm when only the size of the device is scanned. The gyrokinetic codes, used to shed light and build confidence in the empirical scaling law, must therefore be run in carefully determined scans. In the case of global and flux driven codes, the simulations can be addressed in the same framework as the experiments [19].

The scope of the following analysis is to address the issue of the ρ_* scaling, including the mass ratio dependence with the GYSELA code. However, given the computation cost to reach a confinement time with global, flux driven, gyrokinetic simulations, our approach is based on analysing the local features of turbulence during the transient towards plasma thermal equilibrium. Of particular interest, is to identify the characteristic step and duration of such a step in the transport process [35]. The underlying assumption is that turbulence has reached equilibrium locally so that standard measurements such as that of the correlation length L_c or correlation time τ_c are relevant of the statistical steady state. Alternatively, their evolution would also be indicative of the global feature of turbulence. Finally, assuming that the correlation length and time are a measure of the random walk step and characteristic time then allows one to link the local and global properties of turbulent transport. With respect to the scaling issue we then expect L_c/ρ_p to scale between ρ/ρ_p and $(\rho/\rho_p)^{1/2}\rho_*^{-1/2}$ while $\tau_c\, c_p\, M^{1/2}/a$ would scale from a constant to being proportional to $(\rho/\rho_p)^{1/2}\rho_*^{-1/2}$. Here $c_p = \rho_p\ \Omega_p$ and $\rho_p/\rho = \rho_*\ N_*$. The gyro-Bohm limit would correspond to $L_c/\rho_p \propto \rho/\rho_p$ and $\tau_c\ c_p\ M^{1/2}/a \propto 1$ while the Bohm limit would be met with either $L_c/\rho_p \propto (\rho/\rho_p)^{1/2}\rho_*^{-1/2}$ and $\tau_c\ c_p\ M^{1/2}/a \propto 1$ or $L_c/\rho_p \propto \rho/\rho_p$ and $\tau_c\ c_p\ M^{1/2}/a \propto (\rho/\rho_p)^{1/2}\rho_*^{-1/2}$. Such an approach of confinement properties matches that in experiments where the measurement of the turbulent correlations, hence local properties, complete the global measurements.

3. Correlation length of the electric potential fluctuations in a slowly evolving temperature profiles

In this Section we analyse the $1/\rho_* = 300$ GYSELA simulation described in [31] with fixed density and constant heating source. Profiles close to experimental ones from Tore Supra shot TS45511 have been used, in particular ν_* that weighs the normalised collision term figure 1, the safety factor q figure 1, the density profile figure 2, which is constant in time (given the assumption of adiabatic electrons), and the initial ion temperature profile figure 2. The safety factor, close to one towards the core increases monotonically to $r/a = 1$ where it reaches a relatively large value, $q_a \approx 4$. The circular cross section of Tore Supra are convenient to reduce the simulation cost. Still, in the present paper, the chosen ρ_*, $\rho_* = 1/300$, is larger than that of the Tore Supra shot, $\rho_* \approx 1/450$. The temperature profile is observed to evolve on long time

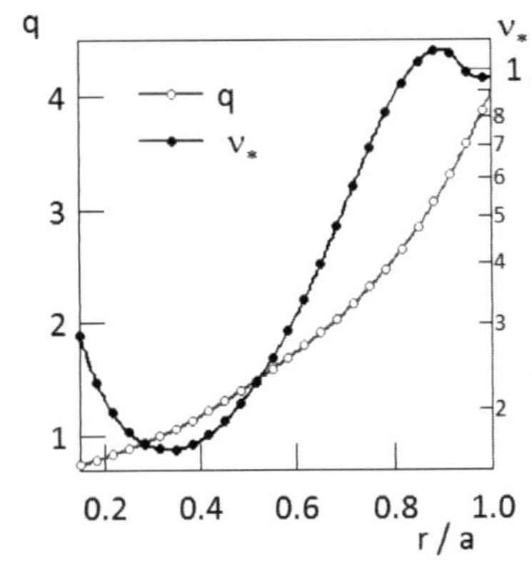

Figure 1. Safety factor q and collisionality ν_* profiles used in the simulation.

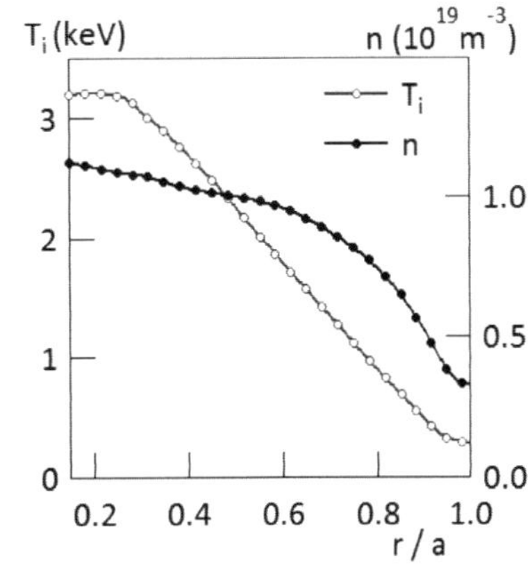

Figure 2. Density n and initial temperature T_i profiles used in the simulation.

scales, figure 3, with a broad region at constant temperature gradient, figure 4. In this paper we concentrate on the radial correlation length of the electric potential $\phi(r, \theta, \varphi = 0, t)$. In practise, the available data is given at an arbitrary value of the toroidal angle φ. As explained in [31], the poloidal average of ϕ is taken as a proxy for the flux averaged potential, an approximation that has been tested to be quite accurate. We thus determine the potential fluctuations as $\tilde{\phi}(r, \theta, \varphi = 0, t) = \phi(r, \theta, \varphi = 0, t) - < \phi(r, \theta, \varphi = 0, t) >_{\theta,[t]}$. Here the averaging operator $<>$ is applied over all poloidal angles θ and for shifting time windows indicated by the notation $[t]$. This allows one to remove the slowly varying features of the electrostatic potential. The time window is 129 points in time long, the delay between consecutive times being $0.8a/c_0$. This time window has been chosen as above the minimum time window providing correct values of L_c. Depending on the cases, we have allowed for an overlap of the times windows, typically of 50 %, but also 95 % and the reference case at 0 %. The time window is centred on the

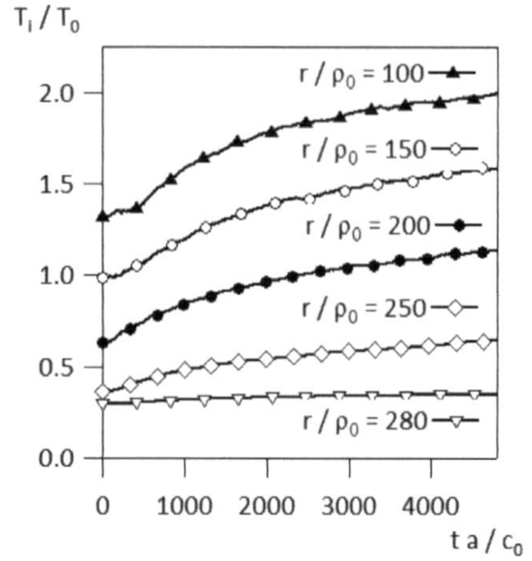

Figure 3. Time trace of the ion temperature at various radial positions.

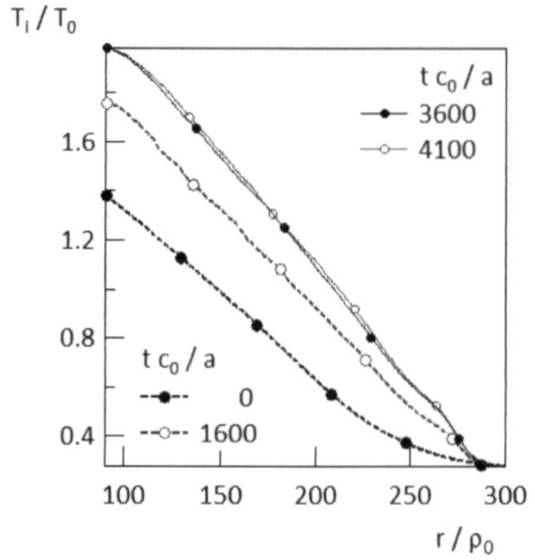

Figure 4. Temperature profiles in the gradient region displaying the gradual increase of the temperature gradient.

time of reference t of the expression of $\tilde{\phi}$. Due to the high frequency modulation in time of the electrostatic potential and the comparatively small variation in radius, the radial average value is also removed so that the field of interest is $\delta\phi = \tilde{\phi} - < \tilde{\phi} >_r$, see figure 5. In the case of adiabatic electrons, this field then corresponds to electron density fluctuations. The correlation function is defined as:

$$C_\phi(r,\theta,t,\delta r) = \frac{< \delta\phi(r,\theta,t)\ \delta\phi(r+\delta r,\theta,t) >_{[t]}}{\left(< \delta\phi(r,\theta,t)^2) >_{[t]}\ \ < \delta\phi(r+\delta r,\theta,t)^2) >_{[t]} \right)^{1/2}} \tag{13}$$

It decays close to exponentially from its maximum at 1 for $\delta r = 0$. The correlation length is defined here such that $C_\phi(r,\theta,t,\delta r = L_c) = 1/2$. The result of the calculation is then a

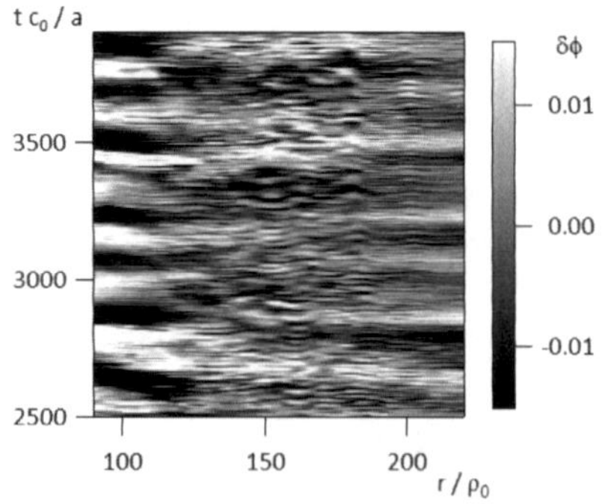

Figure 5. Fluctuations of the electric potential contour map akin to electron density fluctuations.

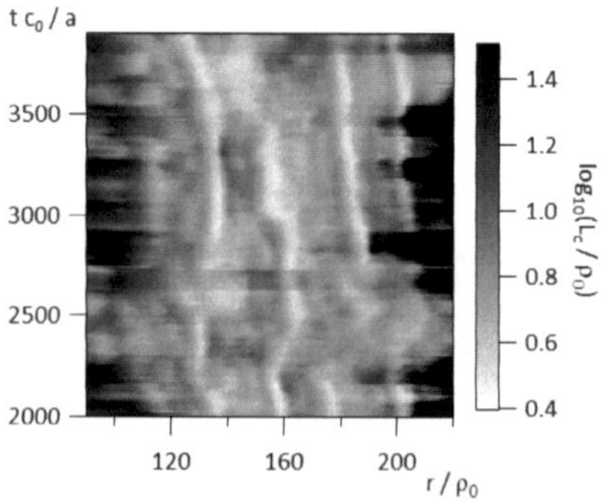

Figure 6. Contour map of the connection length L_c, computed as the local PDF average.

3D map of the correlation length depending on time, poloidal angle and radius. Taking data within various intervals of time, poloidal angle and even radius provides a way to build Probability Density Functions from which one can compute moments and in particular the mean value.

As a first example of the result of the calculation, the mean value of the correlation length in the low field side region $0 \leq \theta \leq \pi/4$ ($\theta = 0$ in the midplane low field side) is shown on figure 6. From the time trace of the radial profile of $\delta\Phi$, figure 5, one can identify three regions. Towards the plasma centre with $r \leq 120\ \rho_0$, a large amplitude and low frequency activity appears to dominate the fluctuation behaviour. On the correlation length counterpart, figure 6, it is characterised by large values of the correlation length $L_c \geq 10$. Towards the outer radius, $r \geq 200\ \rho_0$, in a radial region where turbulence appears to take the form of low frequency streamer like structures, an abrupt transition is observed. It corresponds in fact to a region where $L_c \geq 40$ which is the upper limit used in the calculation of the correlation function $C_\phi(r, \theta, t, \delta r)$. It is interesting to note that the boundary of these two regions appears to fluctuate in time with both high and meso frequencies [36]. In the correlation function contour plot, and between these two radial boundaries, one can readily identify a pattern of small L_c values that appears to be radially localised and evolve slowly in time. With "good eyes", one can identify a similar pattern in the $\delta\phi$ contour plot, figure 5.

Should L_c scale with ρ, hence with $\sqrt{T_i}$ one would expect a ratio of L_c of typically 1.6 between the inner and outer radius values of the region that is analysed. Such a trend is not readily observed in the simulation data.

The statistical analysis of L_c is first addressed with Probability Density Function (PDF). On

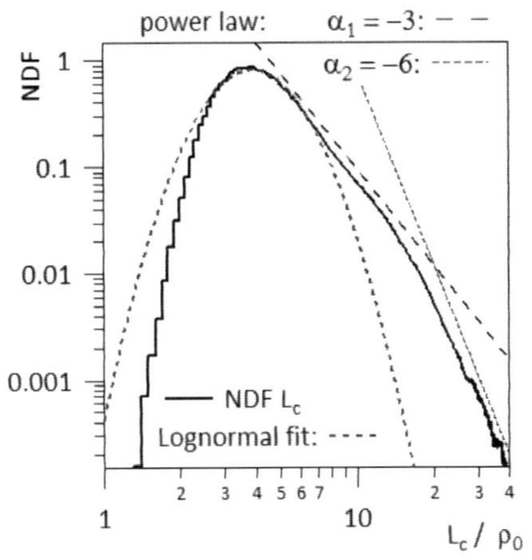

Figure 7. PDF of L_c (not normalised to unity integral), Note the loglog axis.

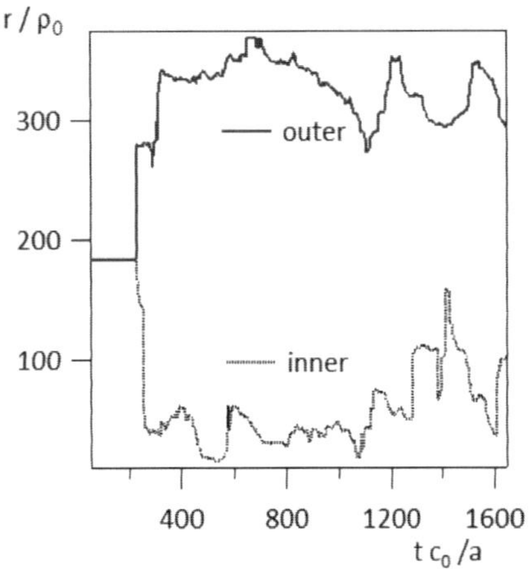

Figure 8. Inner and outer boundary of the region such that $L_c \leq 40\ \rho_0$.

figure 7 the PDF is plotted in a log-log scale which means that a PDF with parabolic shape corresponds to a lognormal distribution. The PDF for L_c is thus found to exhibit a heavy tail with respect to the lognormal PDF with a two slope algebraic tail. When computing this PDF, only the data points with $L_c \leq 40$ were taken into account leaving out the inner and outer regions characterised by $L_c > 40$. On figure 8, the time trace of the inner and outer boundaries of the region with $L_c \leq 15$ are plotted. One readily observes the sudden expansion towards the end of the linear phase, followed by fluctuating positions together with large amplitude and slower frequency modulation. Correlation and anti-correlation between the inner and outer boundary can be observed as well as different time scales for the boundary displacement. From figure 7 and figure 8, one concludes that L_c is a dynamical feature of the system, that it is specific of the chosen region within the plasma. The heavy tail structure also suggests that various events take place that are characterised by very different correlation lengths. It is then difficult to assign to this variability a single value as would have been required to readily fulfil the condition discussed in Section 2 on the scaling law.

4. Averages of the correlation length

4.1. Poloidal variation

One way to proceed towards a single characteristic value is to average the existing data despite the heavy tail PDF, figure 7. A standard procedure is the flux surface average. This allows one to reduce the space dependence to the single direction transverse to the flux surfaces. Confinement properties are then discussed in terms of profiles but it is then more problematic to seek agreement with poloidally localised measurements when poloidal dependences are expected, the ballooning structure of turbulent transport being a serious candidate for such a poloidal variation [18].

In a first step we investigate the poloidal variation of L_c assuming time homogeneity, and thus using all data points in time for the statistics. On figure 9, the poloidal dependence of the

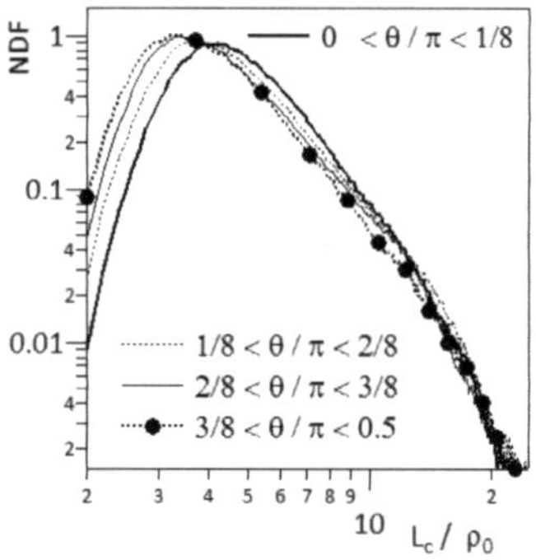

Figure 9. Poloidal dependence of the PDF, with the poloidal angle divided in 16 sectors, time coarse grained and in space such that $150 \leq r/\rho_* \leq 200$.

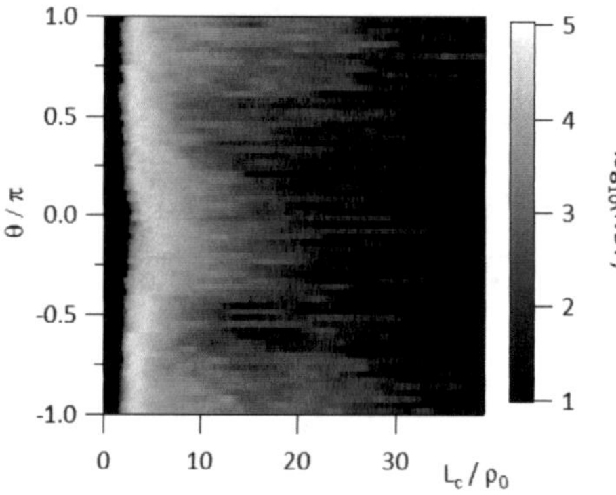

Figure 10. Poloidal distribution of the PDF, coarse grained in time, with 128 poloidal sectors and a single radial position.

PDF of L_c is shown in four different poloidal regions located on the low field side towards the top, $0 \leq \theta \leq \pi/2$. One finds that the PDF maximum is shifted towards the smaller values as θ is increasing [18]. For this plot a range of values of the radial position $150 \leq r/\rho_s \leq 200$ is used to increase the statistics. A complete poloidal dependence is illustrated on figure 10, in this case a single radial point $r/\rho = 150$ is used. These two PDFs are computed using all available points in time in the nonlinear regime. The poloidal dependence exhibits a near Top-Down symmetry, $\theta \to -\theta$. The PDF appears to be very broad with marked narrowing at $\theta \approx \pm\ \pi/4$. As already underlined, the PDF maximum decreases from the low-field midplane towards the high-field miplane. On the high-field side, the PDF appears to develop a double hump, the secondary structure building for $L_c \approx 25\ \rho_0$. From this analysis, one finds that there is no poloidal symmetry with in particular a difference between the low and high field side as expected from the known ballooned feature of ITG turbulence.

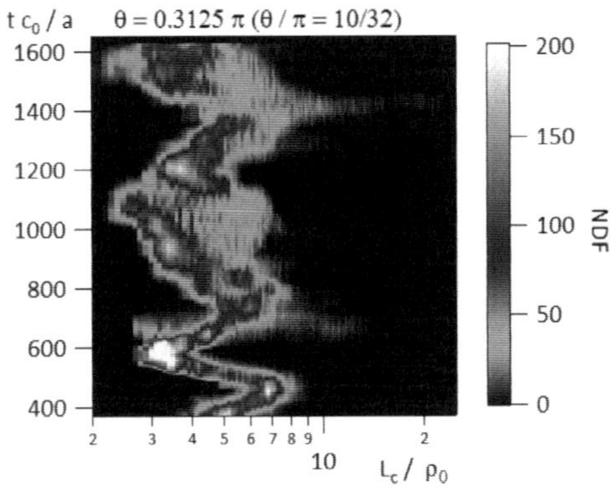

Figure 11. Contour plot of the PDF with its time evolution in the low field top sector.

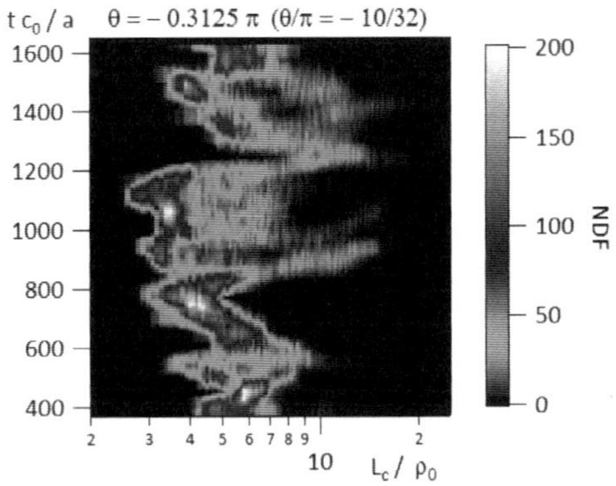

Figure 12. Contour plot of the PDF with its time evolution in the low field bottom sector.

We now address the time dependence of the PDF of L_c, see figure 11 for the low field region towards the top, and figure 12 for the low field towards the bottom. Two features are striking, first without time averaging one loses the top-bottom symmetry, second the evolution is marked by quiescent phases with narrow PDFs peaking towards the low values of the correlation length, interrupted by relaxation events with broad PDF's extending towards the large values of L_c. It thus appears that the top-down symmetry reported previously only holds when considering time averages. Furthermore, one finds that the time dependence is far from being homogeneous with qualitative and quantitative variations.

4.2. Comparing corrugation and correlation length patterns

The simulation that has been used is characterised by marked corrugation patterns [29, 30, 31, 32], see figure 13. The latter are identified as regions with large shear (both negative and positive shear of the $E \times B$ flow) and are readily identified on the contour plot figure 13 despite their variation in time, including rather high frequency behaviour, and radial position [36]. On the latter figure, one can identify two specific features; first the long lived shear patterns appear to go in pairs with a zero line separating them, second, short lived shear structure [37] appear to have a ballistic motion that is readily associated to avalanche transport [24, 25, 26, 27, 28]. The former pattern is correlated to the corrugations in the temperature gradient [29] and is in fact a double shear layer associated to a maximum or a minimum of the zonal flow, here defined as $< v_{E\theta} >_\theta$, namely the flux surface average of the poloidal component of the $E \times B$ drift velocity. Rather than defining a precise pattern of the corrugations, we define a series of points where the absolute value of the zonal flow shear goes through a local maximum in the radial direction. This data is superimposed on the 2D contour plot of the correlation length L_c plotted against radius and time, and for poloidally localised data $0 \leq \theta \leq \pi/4$, figure 14. On this figure one finds that the black dots of the shear extrema are organised in the form of corridors that bound regions of minimum correlation length, typically $L_c \leq 5$ which is the order of magnitude of the width of the local peak in zonal flow amplitude that drives the corrugation.

The corrugation pattern appears to wind around radially with time, figure 13. A time average will thus tend to smear out the correlation between the local shear regions and the minima in L_c. It will then be difficult to untangle the physics that finally yields a given average in time of

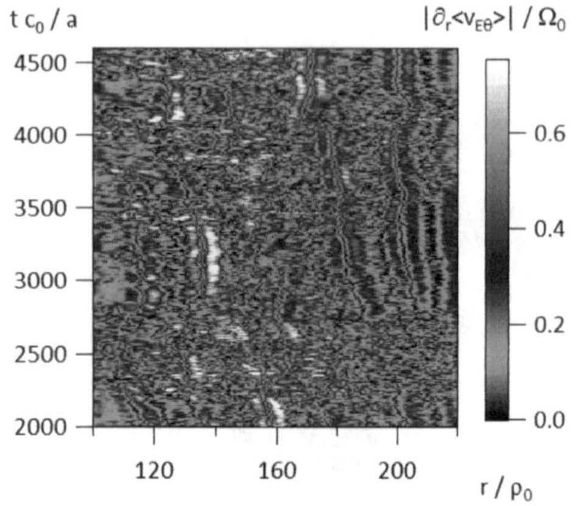

Figure 13. Contour plot of the absolute value of the zonal flow shear displaying the corrugation patterns.

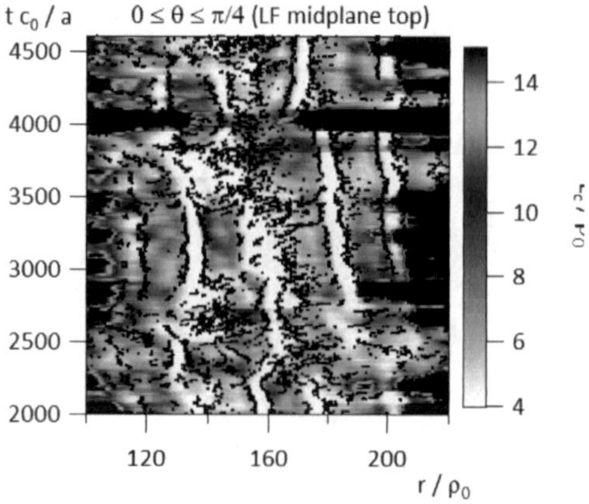

Figure 14. Contour plot of the correlation length and superimposed position (radius, time) of the local maximum of the zonal flow shear.

L_c. Furthermore, the noticeable agreement displayed on figure 14 for the low field side data of L_c vanishes and no meaningful matching is found between the shear layer pattern and that of averaged (coarse grained) L_c values. In agreement with the latter observation, one readily checks that the matching of the high field side values of L_c and the zonal flow shear is rather poor. One thus finds that the shearing effect does not act uniformly in the poloidal direction. Indeed

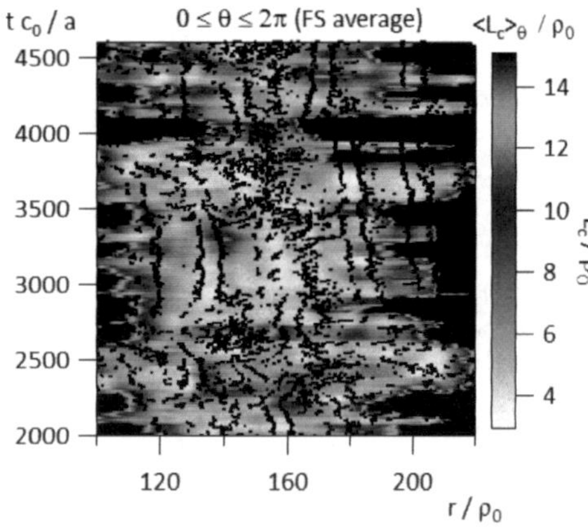

Figure 15. Contour plot of the correlation length coarse grained on a complete flux surface and superimposed position of the local maximum zonal flow shear.

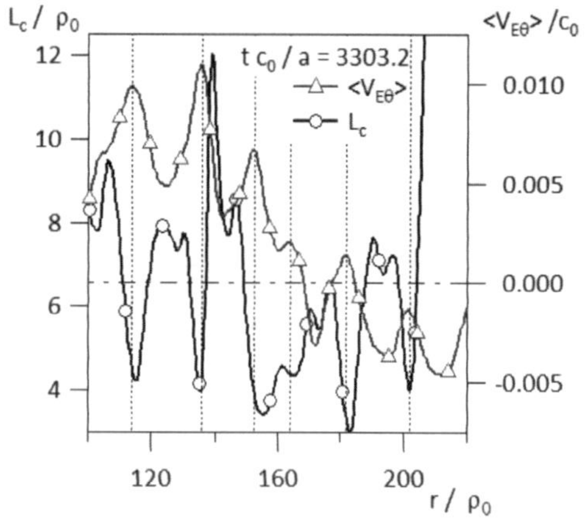

Figure 16. Profiles of the zonal flow shear and of the correlation length with matching between the maxima of the zonal flow and minima of L_c.

it appears to govern minima in the correlation length only in the low field side region. Consequently, the poloidal averaged value of L_c also falls short of reflecting the actual self-regulation of turbulent transport. Similarly, when considering the radial profiles of the correlation length, with minima aligned on the maxima of the zonal flow, figure 16, one finds that radial coarse

Journal of Physics: Conference Series **561** (2014) 012008 doi:10.1088/1742-6596/561/1/012008

graining will tend to mask this alignment property.

It thus seems that the self-regulation of turbulent transport by the zonal flows is a local feature in the poloidal and radial directions which also exhibits a clear evolution. Coarse graining in order to reduce the distribution of the correlation length to a single characteristic value will thus mask this key property. The investigation of the scaling law of the energy confinement on the basis of a correlation length analysis does not appear to be as straightforward as would be the case for homogeneous turbulent systems.

5. Discussion and Conclusions

Bridging the experimental effort on confinement performance, exemplified by the empirical scaling law of the energy confinement time, to the theoretical investigation is usually based on computing correlation lengths and times. Knowing the scaling of these quantities, one can then estimate that of diffusion coefficients and consequently that of the confinement time. Gyrokinetic codes have built-in self similarity so that one readily expects the radial correlation length to scale like $1/B$ according to the gyro-Bohm scaling. However, in view of the questions opened by the empirical scaling law, it is important to investigate this result in terms of the isotope (or mass) effect as well as with other dependencies.

In that spirit we have analysed a GYSELA simulation at $1/\rho_* \approx 300$ that has been run for a large fraction of the confinement time using as input data from a Tore Supra shot. The analysis of the correlation length has indicated a very rich distribution of values with a heavy tail and double algebraic law towards the large values. The detailed analysed has revealed that the correlation length exhibits patterns that are reminiscent of that of the corrugations with minimal values of the correlation length and of its rms value located at the position of the corrugations. Furthermore, in agreement with the ballooned structure of turbulent transport the correlation length exhibits a poloidal variation. Finally, one finds that the corrugations evolve slowly in time introducing consequently a time dependence of the radial pattern of the correlation length. An interesting and unexpected result of the analysis has been the poloidally localised matching of the corrugation and correlation length patterns. Consequently, one finds that any averaging process of the correlation length, in time, poloidal angle or radius, will remove the connection between the correlation length and the self-organised turbulent transport properties.

The result of the present analysis is thus twofold: on the one hand it is difficult to define a single correlation length for a given simulation region. The standard averaging procedures in radius, poloidal angle or time suppress the correlation between confinement properties and the correlation length and as such are not appropriate. The description of confinement in terms of a random walk must then be analysed on the bases of non-homogeneous models, including anomalous transport such as trapping or sticking. This becomes a considerable task of its own that is not yet available. Further effort is thus needed to scale the PDF of the correlation lengths and correlation times with the key dimensionless control parameters. Such an analysis must also involve the scaling properties of the meso scale features such as the corrugation width and their typical radial separation. On the other hand, one finds that the experimental analysis of the transport properties and scaling laws must face the same issues. Indeed, the experimental investigation of turbulence properties yields local measurements of the correlations. The issue of comparing local properties to global confinement properties is thus the same in the experiments and the simulations. Using the latter to define the appropriate framework for such comparison should help in clarifying some of the open issues such as the isotope effect. Finally, as underlined by the analysis of the scaling law of the confinement time in the first part of the paper, consistent changes of the simulation parameters are mandatory to avoid introducing spurious correlations within the scans that would open the way to misleading interpretation of the results.

Acknowledgments
The authors acknowledge the fruitful and lively discussions held during the Festival de Théorie in Aix-en-Provence. They are most indebted to Pat Diamond for his strong support and very active scientific participation to this event. This work is carried out in the framework of the ANR-GYPSI and ANR-ESPOIR projects (ANR-10-BLAN-941 and ANR-09-BLAN-0035-01).

References

[1] I. P. E. G. on Confinement, Transport, I. P. E. G. on Confinement Modelling, Database, and I. P. B. Editors, "Chapter2: Plasma confinement and transport," *Nuclear Fusion*, vol. 39, no. 12, p. 2175, 1999.

[2] M. Shimada, D. Campbell, V. Mukhovatov, M. Fujiwara, N. Kirneva, K. Lackner, M. Nagami, V. Pustovitov, N. Uckan, J. Wesley, N. Asakura, A. Costley, A. Donn, E. Doyle, A. Fasoli, C. Gormezano, Y. Gribov, O. Gruber, T. Hender, W. Houlberg, S. Ide, Y. Kamada, A. Leonard, B. Lipschultz, A. Loarte, K. Miyamoto, V. Mukhovatov, T. Osborne, A. Polevoi, and A. Sips, "Chapter 1: Overview and summary," *Nuclear Fusion*, vol. 47, no. 6, p. S1, 2007.

[3] F. W. Perkins, C. W. Barnes, D. W. Johnson, S. D. Scott, M. C. Zarnstorff, M. G. Bell, R. E. Bell, C. E. Bush, B. Grek, K. W. Hill, D. K. Mansfield, H. Park, A. T. Ramsey, J. Schivell, B. C. Stratton, and E. Synakowski, "Nondimensional transport scaling in the tokamak fusion test reactor: Is tokamak transport bohm or gyro?bohm?," *Physics of Fluids B: Plasma Physics (1989-1993)*, vol. 5, no. 2, pp. 477–498, 1993.

[4] J. Candy and R. E. Waltz, "Anomalous transport scaling in the diii-d tokamak matched by supercomputer simulation," *Phys. Rev. Lett.*, vol. 91, p. 045001, Jul 2003.

[5] Z. Lin, S. Ethier, T. S. Hahm, and W. M. Tang, "Size scaling of turbulent transport in magnetically confined plasmas," *Phys. Rev. Lett.*, vol. 88, p. 195004, Apr 2002.

[6] B. F. McMillan, X. Lapillonne, S. Brunner, L. Villard, S. Jolliet, A. Bottino, T. Görler, and F. Jenko, "System size effects on gyrokinetic turbulence," *Phys. Rev. Lett.*, vol. 105, p. 155001, Oct 2010.

[7] T. Vernay, S. Brunner, L. Villard, B. F. McMillan, S. Jolliet, T. M. Tran, and A. Bottino, "Synergy between ion temperature gradient turbulence and neoclassical processes in global gyrokinetic particle-in-cell simulations," *Physics of Plasmas (1994-present)*, vol. 19, no. 4, pp. –, 2012.

[8] D. Schissel, J. DeBoo, K. Burrell, J. Ferron, R. Groebner, H. S. John, R. Stambaugh, B. Tubbing, K. Thomsen, J. Cordey, M. Keilhacker, D. Stork, P. Stott, A. Tanga, and J. Team, "H-mode energy confinement scaling from the diii-d and jet tokamaks," *Nuclear Fusion*, vol. 31, no. 1, p. 73, 1991.

[9] J. Christiansen, J. Cordey, K. Thomsen, A. Tanga, J. DeBoo, D. Schissel, T. Taylor, O. Kardaun, F. Wagner, F. Ryter, S. Kaye, Y. Miura, N. Suzuki, M. Mori, T. Matsuda, H. Tamai, T. Takizuka, S.-I. Itoh, and K. Itoh, "Global energy confinement h-mode database for iter," *Nuclear Fusion*, vol. 32, no. 2, p. 291, 1992.

[10] C. C. Petty, T. C. Luce, K. H. Burrell, S. C. Chiu, J. S. deGrassie, C. B. Forest, P. Gohil, C. M. Greenfield, R. J. Groebner, R. W. Harvey, R. I. Pinsker, R. Prater, R. E. Waltz, R. A. James, and D. Wrblewski, "Nondimensional transport scaling in diii?d: Bohm versus gyro?bohm resolved," *Physics of Plasmas (1994-present)*, vol. 2, no. 6, pp. 2342–2348, 1995.

[11] C. C. Petty and T. C. Luce, "Scaling of heat transport with collisionality," *Physics of Plasmas (1994-present)*, vol. 6, no. 3, pp. 909–921, 1999.

[12] D. C. McDonald, J. G. Cordey, C. C. Petty, M. Beurskens, R. Budny, I. Coffey, M. de Baar, C. Giroud, E. Joffrin, P. Lomas, A. Meigs, J. Ongena, G. Saibene, R. Sartori, I. Voitsekhovitch, and J. E. contributors, "The beta scaling of energy confinement in elmy h-modes in jet," *Plasma Physics and Controlled Fusion*, vol. 46, no. 5A, p. A215, 2004.

[13] C. C. Petty, T. C. Luce, D. C. McDonald, J. Mandrekas, M. R. Wade, J. Candy, J. G. Cordey, V. Drozdov, T. E. Evans, J. R. Ferron, R. J. Groebner, A. W. Hyatt, G. L. Jackson, R. J. La Haye, T. H. Osborne, and R. E. Waltz, "Beta scaling of transport on the diii-d tokamak: Is transport electrostatic or electromagnetic?," *Physics of Plasmas (1994-present)*, vol. 11, no. 5, pp. 2514–2522, 2004.

[14] D. McDonald, J. Cordey, K. Thomsen, O. Kardaun, J. Snipes, M. Greenwald, L. Sugiyama, F. Ryter, A. Kus, J. Stober, J. DeBoo, C. Petty, G. Bracco, M. Romanelli, Z. Cui, Y. Liu, Y. Miura, K. Shinohara, K. Tsuzuki, Y. Kamada, T. Takizuka, H. Urano, M. Valovic, R. Akers, C. Brickley, A. Sykes, M. Walsh, S. Kaye, C. Bush, D. Hogewei, Y. Martin, A. Cote, G. Pacher, J. Ongena, F. Imbeaux, G. Hoang, S. Lebedev, A. Chudnovskiy, and V. Leonov, "Recent progress on the development and analysis of the itpa global h-mode confinement database," *Nuclear Fusion*, vol. 47, no. 3, p. 147, 2007.

[15] S. D. Scott, M. C. Zarnstorff, C. W. Barnes, R. Bell, N. L. Bretz, C. Bush, Z. Chang, D. Ernst, R. J. Fonck,

L. Johnson, E. Mazzucato, R. Nazikian, S. Paul, J. Schivell, E. J. Synakowski, H. Adler, M. Bell, R. Budny, E. Fredrickson, B. Grek, A. Janos, D. Johnson, D. McCune, H. Park, A. Ramsey, M. H. Redi, G. Taylor, M. Thompson, and R. Wieland, "Isotopic scaling of confinement in deuteriumtritium plasmas," *Physics of Plasmas (1994-present)*, vol. 2, no. 6, pp. 2299–2307, 1995.

[16] J. Cordey, B. Balet, D. Bartlett, R. Budny, J. Christiansen, G. Conway, L.-G. Eriksson, G. Fishpool, C. Gowers, J. de Haas, P. Harbour, L. Horton, A. Howman, J. Jacquinot, W. Kerner, C. Lowry, R. Monk, P. Nielsen, E. Righi, F. Rimini, G. Saibene, R. Sartori, B. Schunke, A. Sips, R. Smith, M. Stamp, D. Start, K. Thomsen, B. Tubbing, and M. von Hellermann, "Plasma confinement in jet hmode plasmas with h, d, dt and t isotopes," *Nuclear Fusion*, vol. 39, no. 3, p. 301, 1999.

[17] T. Hahm, L. Wang, W. Wang, E. Yoon, and F. Duthoit, "Isotopic dependence of residual zonal flows," *Nuclear Fusion*, vol. 53, no. 7, p. 072002, 2013.

[18] Y. Sarazin, V. Grandgirard, J. Abiteboul, S. Allfrey, X. Garbet, P. Ghendrih, G. Latu, A. Strugarek, G. Dif-Pradalier, P. Diamond, S. Ku, C. Chang, B. McMillan, T. Tran, L. Villard, S. Jolliet, A. Bottino, and P. Angelino, "Predictions on heat transport and plasma rotation from global gyrokinetic simulations," *Nuclear Fusion*, vol. 51, no. 10, p. 103023, 2011.

[19] M. Nakata and Y. Idomura, "Plasma size and collisionality scaling of ion-temperature-gradient-driven turbulence," *Nuclear Fusion*, vol. 53, no. 11, p. 113039, 2013.

[20] V. Grandgirard, Y. Sarazin, P. Angelino, A. Bottino, N. Crouseilles, G. Darmet, G. Dif-Pradalier, X. Garbet, P. Ghendrih, S. Jolliet, G. Latu, E. Sonnendrcker, and L. Villard, "Global full- f gyrokinetic simulations of plasma turbulence," *Plasma Physics and Controlled Fusion*, vol. 49, no. 12B, p. B173, 2007.

[21] Y. Idomura, H. Urano, N. Aiba, and S. Tokuda, "Study of ion turbulent transport and profile formations using global gyrokinetic full- f vlasov simulation," *Nuclear Fusion*, vol. 49, no. 6, p. 065029, 2009.

[22] S. Ku, C. Chang, and P. Diamond, "Full-f gyrokinetic particle simulation of centrally heated global itg turbulence from magnetic axis to edge pedestal top in a realistic tokamak geometry," *Nuclear Fusion*, vol. 49, no. 11, p. 115021, 2009.

[23] J. Heikkinen, S. Janhunen, T. Kiviniemi, and F. Ogando, "Full f gyrokinetic method for particle simulation of tokamak transport," *Journal of Computational Physics*, vol. 227, no. 11, pp. 5582 – 5609, 2008.

[24] Y. Sarazin, V. Grandgirard, J. Abiteboul, S. Allfrey, X. Garbet, P. Ghendrih, G. Latu, A. Strugarek, and G. Dif-Pradalier, "Large scale dynamics in flux driven gyrokinetic turbulence," *Nuclear Fusion*, vol. 50, no. 5, p. 054004, 2010.

[25] Y. Sarazin and P. Ghendrih, "Intermittent particle transport in two-dimensional edge turbulence," *Physics of Plasmas (1994-present)*, vol. 5, no. 12, pp. 4214–4228, 1998.

[26] B. F. McMillan, S. Jolliet, T. M. Tran, L. Villard, A. Bottino, and P. Angelino, "Avalanche like bursts in global gyrokinetic simulations," *Physics of Plasmas (1994-present)*, vol. 16, no. 2, p. 022310, 2009.

[27] L. Villard, A. Bottino, S. Brunner, A. Casati, J. Chowdhury, T. Dannert, R. Ganesh, X. Garbet, T. Grler, V. Grandgirard, R. Hatzky, Y. Idomura, F. Jenko, S. Jolliet, S. K. Aghdam, X. Lapillonne, G. Latu, B. F. McMillan, F. Merz, Y. Sarazin, T. M. Tran, and T. Vernay, "Gyrokinetic simulations of turbulent transport: size scaling and chaotic behaviour," *Plasma Physics and Controlled Fusion*, vol. 52, no. 12, p. 124038, 2010.

[28] S. Jolliet and Y. Idomura, "Plasma size scaling of avalanche-like heat transport in tokamaks," *Nuclear Fusion*, vol. 52, no. 2, p. 023026, 2012.

[29] G. Dif-Pradalier, P. H. Diamond, V. Grandgirard, Y. Sarazin, J. Abiteboul, X. Garbet, P. Ghendrih, A. Strugarek, S. Ku, and C. S. Chang, "On the validity of the local diffusive paradigm in turbulent plasma transport," *Phys. Rev. E*, vol. 82, p. 025401, Aug 2010.

[30] L. Villard, P. Angelino, A. Bottino, S. Brunner, S. Jolliet, B. F. McMillan, T. M. Tran, and T. Vernay, "Global gyrokinetic ion temperature gradient turbulence simulations of iter," *Plasma Physics and Controlled Fusion*, vol. 55, no. 7, p. 074017, 2013.

[31] P. Ghendrih, "Phase space structures in gyrokinetic simulations of fusion plasma turbulence." to be published in *Eur. Phys. J. D.*

[32] Norscini, "Turbulent transport close to marginal instability: role of the source driving the system out of equilibrium." 2014.

[33] P. H. Diamond, S.-I. Itoh, K. Itoh, and T. S. Hahm, "Zonal flows in plasmaa review," *Plasma Physics and Controlled Fusion*, vol. 47, no. 5, p. R35, 2005.

[34] J. Connor and J. Taylor, "Scaling laws for plasma confinement," *Nuclear Fusion*, vol. 17, no. 5, p. 1047, 1977.

[35] J.-P. Bouchaud and A. Georges, "Anomalous diffusion in disordered media: Statistical mechanisms, models and physical applications," *Physics Reports*, vol. 195, no. 45, pp. 127 – 293, 1990.

[36] E. Floriani, G. Ciraolo, P. Ghendrih, R. Lima, and Y. Sarazin, "Self-regulation of turbulence bursts and transport barriers," *Plasma Physics and Controlled Fusion*, vol. 55, no. 9, p. 095012, 2013.

[37] J. Abiteboul, P. Ghendrih, V. Grandgirard, T. Cartier-Michaud, G. Dif-Pradalier, X. Garbet, G. Latu, C. Passeron, Y. Sarazin, A. Strugarek, O. Thomine, and D. Zarzoso, "Turbulent momentum transport in core tokamak plasmas and penetration of scrape-off layer flows," *Plasma Physics and Controlled Fusion*, vol. 55, no. 7, p. 074001, 2013.

Parallel Kelvin-Helmholtz instability in edge plasma

H. Guillard[1], M. Bilanceri[1,4] C. Colin[2] P. Ghendrih[3], G. Giorgiani[1], B. Nkonga[1], F. Schwander[2], E. Serre[2] and P. Tamain[3]

[1] Inria Sophia Antipolis Mediterranee and Nice Sophia Antipolis Univ., LJAD, UMR 7351, France

[2] M2P2-UMR 7340, Aix-Marseille Univ., France

[3] CEA - IRFM, Saint Paul Lez Durance., France

E-mail: `Herve.Guillard@inria.fr`

Abstract. In the scrape-off layer (SOL) of tokamaks, the flow acceleration due to the presence of limiter or divertor plates rises the plasma velocity in a sonic regime. These high velocities imply the presence of a strong shear between the SOL and the core of the plasma that can possibly trigger some parallel shear flow instability. The existence of these instabilities, denoted as parallel Kelvin-Helmholtz instability in some works [1, 2] have been investigated theoretically in [3] using a minimal model of electrostatic turbulence composed of a mass density and parallel velocity equations. This work showed that the edge plasma around limiters might indeed be unstable to this type of parallel shear flow instabilities. In this work, we perform 3D simulations of the same simple mathematical model to validate an original finite volume numerical method aimed to the numerical study of edge plasma. This method combines the use of triangular unstructured meshes in the poloidal section and structured meshes in the toroidal direction and is particularly suited to the representation of the real complex geometry of the vacuum chamber of a tokamak.

The numerical results confirm that in agreement with the theoretical expectations as well as with other numerical methods, the sheared flows in the SOL are subject to parallel Kelvin-Helmholtz instabilities. However, the growth rate of these instabilities is low and these computations require both a sufficient spatial resolution and a long simulation time. This makes the simulation of parallel Kelvin-Helmholtz instabilities a demanding benchmark.

1. Introduction

Poloidal up-down asymmetries of turbulence are known to exist in tokamaks as Tore-Supra. These asymmetries correlate with the edge geometry and the location of the limiters : Increased turbulence level being observed in the close vicinity of limiters. An explanation on these poloidal asymmetries relies on the triggering of parallel shear flow instabilities between the rotating core plasma and the accelerated parallel flow caused by the limiters : In the scrape-off layer (SOL), the flow acceleration due to the presence of limiter or divertor plates rises the plasma velocity in a sonic regime. These high velocities imply the presence of a strong shear between the SOL and the core of the plasma that can possibly trigger some unstable behavior denoted as parallel shear flow instability in [1, 2]. In this contribution, we investigate this possible explanation with the help of 3D numerical simulations and indeed show that instabilities can develop due to the presence of limiters. The outline of this paper is as follows : in section 2, we present

[4] Now with General Electric Oil & Gas Florence Area, Italy

the mathematical model used for this study. Section 3 is devoted to a brief description of the numerical method that we have used. We conclude by presenting our results in section 4.

2. Mathematical model

The possible presence of a parallel instability created by the strong shear between the SOL velocity and the core plasma is here investigated using the same minimal electrostatic model than in [3]. In this model, the electron are adiabatic and ions are assumed to be cold, moreover the temperature variation across the SOL is not taken into account. The magnetic field is considered steady and given in a toroidal coordinate system (r, θ, ϕ) by

$$\vec{B} = B_t\vec{e}_\phi + B_p\vec{e}_\theta \text{ with } q = \frac{rB_t}{R_0B_p} = 6 \tag{1}$$

The electric field is electrostatic $\Rightarrow \vec{E} = -\nabla\Phi$ and the velocity is assumed to have the following form

$$\vec{u} = u_{||}\vec{b} + \vec{u}_{drift} \tag{2}$$

where the perpendicular velocity retains only the electric drift contribution $\vec{u}_{drift} = \dfrac{\vec{B} \times \nabla\Phi}{B^2}$

Given an electron reference temperature T_e and length scale L_{ref}, the variables are made dimensionless as follows

$$\tilde{\vec{x}} = \frac{\vec{x}}{L_{ref}}, \quad \tilde{u}_{||} = \frac{u_{||}}{c_s}, \quad \tilde{t} = t\frac{c_s}{L_{ref}}, \tag{3}$$

$$\tilde{\Phi}, = \Phi\frac{e}{T_e} \quad \tilde{\vec{B}} = \frac{\vec{B}}{B_0} \tag{4}$$

where $c_s = \sqrt{\dfrac{T_e}{m_i}}$ is the sound speed. With this normalization, the model takes the form

$$\begin{cases} \dfrac{\partial n}{\partial t} + \nabla \cdot n\vec{u} = \nabla \cdot (\mu_n D \nabla n) \\ \dfrac{\partial n u_{||}}{\partial t} + \nabla \cdot (n u_{||}\vec{u}) + \nabla_{||} p = -n\vec{u} \cdot \dfrac{D\vec{b}}{Dt} + \nabla \cdot (\mu_u D \nabla n u_{||}) \\ \vec{u} = u_{||}\vec{b} + \rho^\star \dfrac{\vec{B} \times \nabla\Phi}{B^2}, \quad \Phi = \ln\left(\dfrac{n}{n_0}\right) \\ p = n, \quad D = \mathrm{I} - \vec{b} \otimes \vec{b}^T \end{cases} \tag{5}$$

and is completely defined by the following three parameters representing respectively the dimensionless mass diffusion, parallel momentum diffusion and normalized Larmor radius.

$$\tilde{\mu}_n = \frac{\mu_n}{L_{ref}c_s}, \quad \tilde{\mu}_u = \frac{\mu_u}{L_{ref}c_s}, \quad \rho^\star = \frac{m_i c_s}{eB_0L_{ref}} \tag{6}$$

The computational domain is the full torus in ϕ and an annulus of the poloidal cross section limited by a limiter of extend $2\pi/10$ in θ and 0.25 in r. The boundary conditions are given by

- Bohm boundary conditions on the limiter target plates : $\boldsymbol{u_{\|}/c_s \geq 1}$
- slip boundary condition on the exterior wall
- interior wall
  - Imposed parallel Mach number $\boldsymbol{Ma_{\|} = Ma_{\|}(R_{min})}$
  - Imposed mass flux $\boldsymbol{(\mu_n D \nabla n).\vec{e}_r = given}$

3. The numerical scheme

3.1. Spatial discretization

The numerical scheme is described in details in [4]. It uses an unstructured finite element triangular mesh in the poloidal plane and a structured mesh in the toroidal direction. The use of Finite Element type mesh makes easy the representation of complex geometries as well as the use of possible adaptive mesh refinement. Although these possibilities are not exploited here, these features may be extremely useful as the need of more realism in plasma simulations will grow. The figure 1 displays an enlarged view of the mesh near the limiters target plates. The

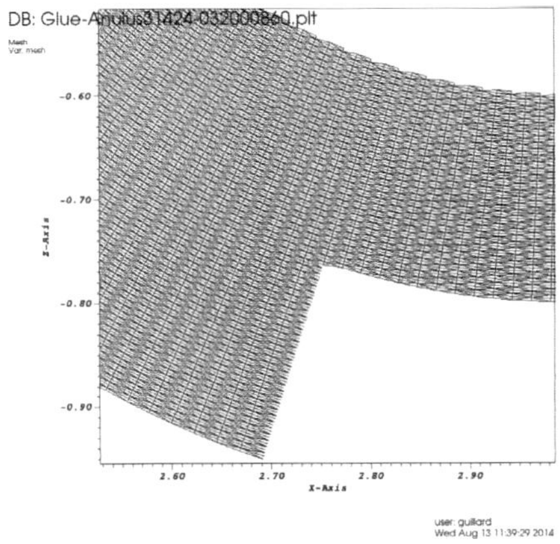

Figure 1. Zoom on the poloidal mesh near the limiter,

discretization method uses a finite volume method using an exact Riemann solver to compute the convective fluxes at the interfaces between cells. This type of numerical method do not separate explicitly the parallel and perpendicular directions. This can be a shortcoming of the method in case of strongly anisotropic flow as this has to be compensated by an increased resolution and the use of large meshes. On the other hand, the present numerical method is genuinely conservative, is very robust with no addition of explicit artificial diffusivity and incorporates in an automatic way the curvature terms. Since, the system (5) is not written in cartesian coordinates, we briefly comment on the way to treat vector equations in our scheme : Assume we want to solve a (vector) conservative equation.

$$\frac{\partial \vec{u}}{\partial t} + \nabla . F = \vec{0} \tag{7}$$

After integration on the cell $\boldsymbol{i}$ this equation becomes :

$$Vol_i \frac{d\vec{u}_i}{dt} + \sum_{j \in \kappa(i)} F_{ij} = \vec{0} \tag{8}$$

where $\boldsymbol{F_{ij}}$ are the numerical fluxes and $\boldsymbol{\kappa(i)}$ denotes the set of the neighbors of $\boldsymbol{i}$. Note that the spatial integration is a mathematical operation that does not depend on the system of coordinate used. This step of the method is thus independent of this system of coordinates and can be performed in any convenient system (cartesian or not). However, in curvilinear system of coordinates, the basis functions depend on the space position and their projection does not commute with space integration. In our numerical scheme, we take this into account by projecting the sum of the fluxes on a representative basis vector of the cell $\boldsymbol{i}$ (that is thus different from cells to cells).

$$\left(Vol_i \frac{d\vec{u}_i}{dt} + \sum_{j\in\kappa(i)} F_{ij} \right) \cdot \vec{e_i} = 0 \Rightarrow \tag{9}$$

or with $u_i = \vec{u}.\vec{e_i}$

$$Vol_i \frac{du_i}{dt} + \left(\sum F_{ij}\right) \cdot \vec{e_i} = 0 \tag{10}$$

This procedure can be applied to any general curvilinear coordinate system. In this work, we have applied it with $\vec{e_i} = \vec{b}$ to obtain the parallel velocity. Note that the curvature terms are implicitly discretized during this projection step. The method is made $\boldsymbol{2^{nd}}$-order in space and time with the help of the MUSCL method in space and RK2 in time.

3.2. Parallel efficiency

The method uses domain partitioning for parallel computations with the MPI communication library to transmit information between processors. In the present study, we have used a mesh of 31424 x 128 = 4.02 M of nodes. This corresponds to a spatial resolution of $\sim 5mm$ in the radial and of $\mathbf{1.2^o}$ in the poloidal directions This large number of nodes and in particular the large number of poloidal planes have been first used to ensure a sufficient resolution and accuracy of the numerical scheme but also to test the parallel implementation of the method and its performance when a large number of processors is used. Since, the method is explicit in time, we expect its scalability performances to be good. Figure 2 shows that this is indeed the case and that the speed-up obtained with up to 1024 processors scales linearly with the number of processors when the mesh is large enough.

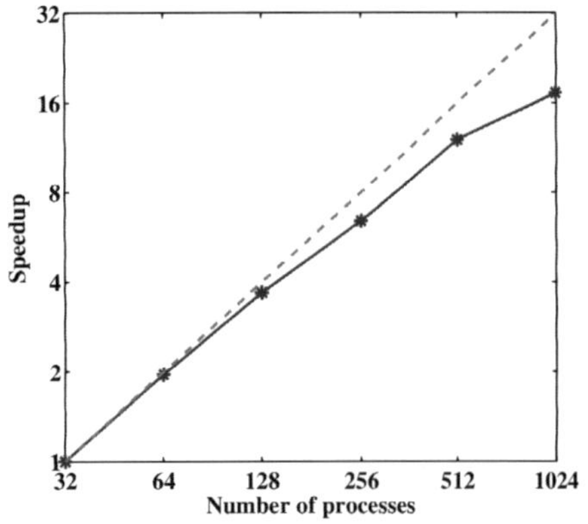

Figure 2. Scalability on a mesh of 4.02 M nodes

4. Results

The local linear stability theory developed in [3] predict instability growth if the background axisymmetric steady state verifies :

$$(\frac{dM_{//}}{dR})^2 - (\frac{d\ln n}{dR})^2 > 0 \tag{11}$$

For the present simulations, the values of the physical parameters defining the problems were: $B_0 = 1.3T, L_{ref} = 0.25m$. An uniform temperature set to $100eV$ corresponding to a sound speed of

$$c_s = \sqrt{\frac{T}{m_i}} = \sqrt{\frac{100 \cdot 1.6022 \cdot 10^{-19}}{1.6727 \cdot 10^{-27}}} \simeq 100000m/s \tag{12}$$

was used and the momentum diffusivity coefficient was $\mu_u = 1m^2/s$ With this choice, the dimensionless parameters governing the flows are equal to : $\tilde{\mu}_n \simeq 0.410^{-3}$ $\tilde{\mu}_u = 10^{-1}\tilde{\mu}_n$ and $\rho^\star = 3.10^{-3}$. The core flow in the tokamak is defined by a velocity corresponding to $Ma_i = 0.75$.

For these values, the criterion (11) is satisfied on a large part of the domain upstream of the target plates. To check this theory, we have thus perturbed the 2D axisymmetric steady state displayed in figure 3

Figure 3. The figure displays the momentum perturbation i.e the difference between the parallel momentum and its corresponding value in the 2D axisymmetric steady state. At time $t = 0$, these perturbations are of order 10^{-6} and cannot be observed on this figure.

by small parallel momentum fluctuations

$$\delta\Gamma = \Gamma_0(R, Z)(1 + \sum A_m \cos(m\phi)) \tag{13}$$

where the A_m are of order 10^{-6}.

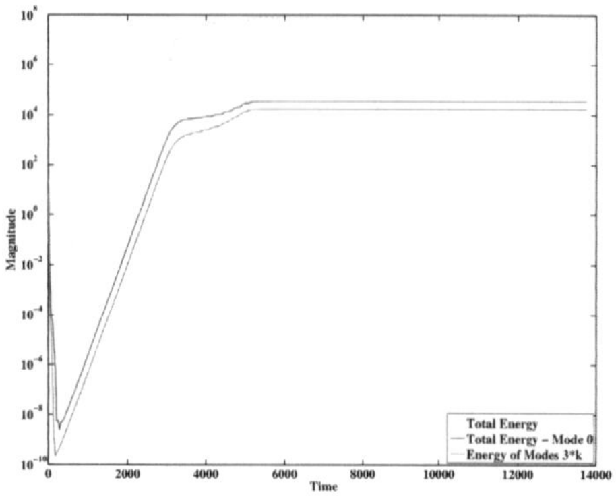

Figure 4. Growth of the perturbation

As shown in figure 4 after a phase where the perturbation decreases, the flow re-organizes and enters in a quasi-periodic behavior with regular creation/destruction of perturbations. The figure 5 displays this saturated quasi - steady state characterized by continuous creation/convection and destruction of perturbation.

5. Concluding remarks

We have performed 3D simulations of a simple mathematical model to study the possible triggering of parallel shear flow instabilities in a tokamak in presence of limiters. These numerical results confirm that in agreement with the theoretical expectations, the sheared flows in the SOL can be subject to parallel Kelvin-Helmholtz instabilities. These results have also been confirmed using a different numerical method as reported in [5]. The growth rate of these instabilities is however low and it takes a very long time for an unstable mode to arise. Due to the large mesh used in this study, we believe that the numerical resolution is sufficient and that these simulation have been able to capture the parallel Kelvin-Helmholtz instability phenomenon. This must however be checked by performing simulations with increased resolution. In the future, it would be useful to consider also simpler benchmarks such as linear growth rates of drift or ballooning modes. This will be a next step in the development of this numerical method.

Acknowledgments

Part of this work has been supported by the French ANR (Agence Nationale de la Recherche) through the project ESPOIR : ANR-09-BLAN-0035-01.

[1] Garbet X, Fenzi C, Capes H, Devynck P and Antar G 1999 *Phys. Plasmas* **6** 3955â3965
[2] Catto P, Rosenbluth M and Liu C 1973 *Phys. Fluids* **16** 1719
[3] Schwander F, Chiavassa G, Ciraolo G, Ghendrih P, Isoardi L, Paredes A, Sarazin Y, Serre E and Tamain P 2011 *Journal of Nuclear Materials* **415** S601 – S604 URL `http://www.sciencedirect.com/science/article/pii/S0022311510006677`
[4] Bonnement A, Fajraoui T, Guillard H, Martin M, Mouton A, Nkonga B and Sangam A 2011 *ESAIM-Proc* **32** 163–176
[5] Serre E, Bigot B, Colin C, Paredes A, Schwander F, Bufferand H, GCiraolo, Ghendrih P, Tamain P, Bilanceri M, Guillard H and Nkonga B *15th European Fusion Theory Conference, 23-26 September 2013, Merton College, Oxford, UK*

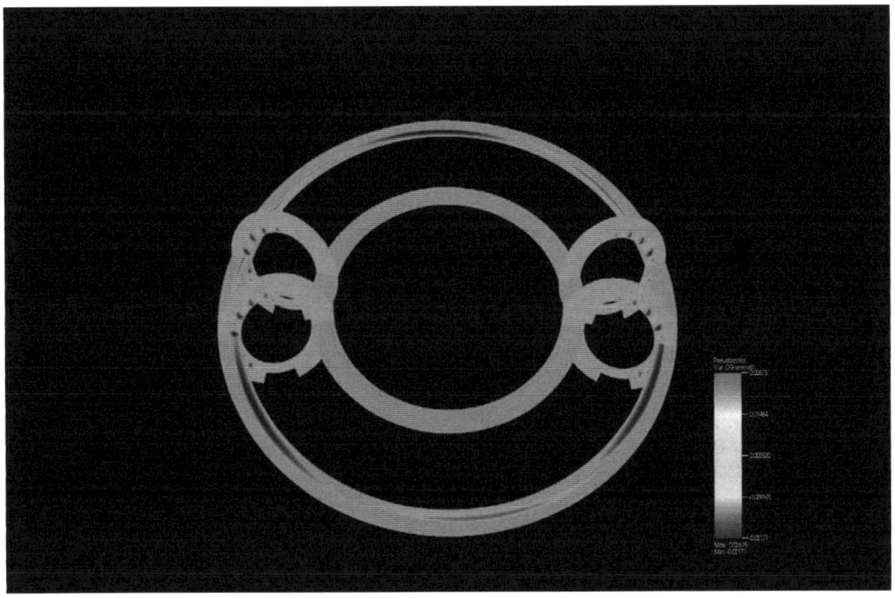

Figure 5. Saturated quasi - steady state : The figure displays the difference between the parallel momentum and its corresponding value in the 2D axisymmetric steady state.

An iterative method for including Doppler shift in global wave solvers using FEM decomposition

T. Hellsten, T. Johnson, P. Vallejos

Dept. of Fusion Plasma Physics, EES KTH, SE-100 44 Stockholm, Sweden

E-mail: thel@kth.se

Abstract. A method for calculating the wave field for spatial dispersive media is proposed suitable for FEM. The method is based on operator splitting by separating the induced current and wave field calculations, and solving the system by means of iterations. In order to take into account several coexisting waves with different poloidal mode numbers when calculating the induced current the wave field is decomposed into wavelets, for which the current is calculated assuming the plasma to be weakly non-uniform.

1. Introduction

Modelling of the wave field for RF-heating is a challenging task because of the spatial dispersive nature of magnetised plasmas and the co-existence of different waves. The spatial dispersion requires that the different waves have their own response. The issue is often handled by Fourier decomposing the wave equation [1]. Hybrid methods using Fourier decomposition in the toroidal and the poloidal directions and FEM (finite element methods) decomposition across the flux surfaces are commonly used. The Fourier decomposition leads to dense matrixes that become time consuming to invert. FEM methods have the advantages of producing local decompositions which, in general, are faster to invert, but for which it is more difficult to include spatial dispersive effects. Lately a method to include or correct for spatial dispersive effects was developed applicable for strong single pass damping for which the local wave field is described by a single wave [2]. Another method is to use operator splitting and solve an inhomogeneous wave equation by means of iterations with the induced current as a source term calculated from the wave field [3, 4]. Instead of calculating the current by tracing the particle orbits we use here a higher order quasi-homogeneous approximation consistent with energy conservation. In order to take into account the non-local response from several coexisting waves to include the Doppler shift the wave field is locally decomposed into poloidal wavelets, for which the quasi-homogeneous approximation is applicable.

2. Wave-field

The wave equation for a spatial dispersive medium can be written as

$$\nabla\times\nabla\times\mathbf{E}-\left(\frac{\omega}{c}\right)^2\mathbf{E}=i\omega\mu_0\mathbf{J}_{ant}+i\omega\mu_0\mathbf{J}_{ind}\,. \tag{1}$$

Here we assume the induced current

$$\mathbf{J}_{ind}\left(\mathbf{r},t\right)=\mathbf{J}_{ind}\left(\mathbf{E}\right)\,, \tag{2}$$

to be a linear function of the wave field **E**. The conductivity tensor, $\sigma_{jl}(\omega,\mathbf{k})$, for a fluctuating electric field can straightforwardly be defined in Fourier space (ω, **k**), for which the induced current can be calculated

$$J_{ind,j}(\omega,\mathbf{k})=\sigma_{jl}(\omega,\mathbf{k})E_l. \tag{3}$$

Transforming the response into the real space gives

$$J_{ind,j}(t,\mathbf{x})=\hat{\sigma}_{jl}(t,\mathbf{x})*E_l(t,\mathbf{x}), \tag{4}$$

where * denotes the convolution between the electric field and $\hat{\sigma}_{jl}(t,\mathbf{x})$, the inverse Fourier transform of $\sigma_{jl}(\omega,\mathbf{k})$. The convolution describes a non-local dependence of the induced current on the electric field; the local induced current depends on the wave-particle interactions the particles experienced earlier. Multiplying Eq. (3) with the electric field **E** and transforming it to the real space gives

$$P(t,\mathbf{x})=E_l(t,\mathbf{x})*\left(\hat{\sigma}_{jl}(t,\mathbf{x})*E_l(t,\mathbf{x})\right), \tag{5}$$

which represents the time derivative of the free energy density available for dissipation. This quantity differs from the work done locally by the wave field obtained by multiplying Eq. (4) with **E** in real space. If one instead calculates the work from the power theorem, one obtains a relation between the total transferred energies from the waves to the particles in the whole Fourier space to the whole real space from $t = -\infty$ to $t = +\infty$. Note that the work done by the wave field on the particles is not the same as heating, which is dissipation of the wave field into thermal energy of the particles. Dissipation requires decorrelation of the changes in energy due to wave-particle interactions e.g. by collisions. In absence of decorrelation, the energy the particles receives from wave-particle interaction is transferred to a coherent motion of the particles, which independent of the slope of the distribution function may result in energy transferred from the particles to the wave or vice versa.

An often used approximation of the conductivity tensor in an inhomogenous medium is that for a local homogenous medium with **k** as the local wave vector:

$$J_{ind,j}(\mathbf{x},t)=\sigma_{jl}(\mathbf{x};\omega,\mathbf{k})E_l(\mathbf{x},t;\omega,\mathbf{k}). \tag{6}$$

The local wave vector **k,** which may be chosen in different ways, should be consistent with the local dispersion relation. Assuming that the wave field can be decomposed in orthogonal modes one would then obtain an energy relation of the form

$$\frac{1}{2\mu_0}\frac{\partial}{\partial t}\left(\tilde{\mathbf{B}}^*\cdot\mathbf{B}\right)+\frac{\varepsilon_0}{2}\frac{\partial}{\partial t}\sum_{\mathbf{k}}\left(\mathbf{E}_{\mathbf{k}}^*\cdot\mathbf{K}_{\mathbf{k}}^H\cdot\mathbf{E}_{\mathbf{k}}\right)+\frac{1}{\mu_0}\nabla\cdot\left(\mathbf{E}^*\times\tilde{\mathbf{B}}\right)-i\omega\varepsilon_0\sum_{\mathbf{k}}\mathbf{E}_{\mathbf{k}}^*\mathbf{K}_{\mathbf{k}}^A\mathbf{E}_{\mathbf{k}}=-\mathbf{E}^*\cdot\mathbf{J}_{ant}, \tag{7}$$

where $\mathbf{E}=\sum_{\mathbf{k}}\mathbf{E}_{\mathbf{k}}$, $\tilde{\mathbf{B}}$ is the magnetic induction related to the wave field **E**, $\mathbf{K}^{\mathrm{H}}$ and $\mathbf{K}^{\mathrm{A}}$ are the Hermitian and antiHermitian part of the dielectric tensor, respectively. The conductivity and dielectric tensors are related through $i\omega\varepsilon_0\left(K_{ij}-\delta_{ij}\right)=\sigma_{ij}$. The two first terms in Eq. (7) represent the change in the local wave energy density, the third term the divergence of the Poynting flux, the fourth term the dissipation of wave energy and the left hand side the work done by the antenna, the emitted wave energy. In this approximation the total energy flow is carried by the Poynting flux, thus the energy flux carried by particles is not included. If one instead starts with the energy conservation law including the kinetic energy flux,

$$\begin{aligned}&\frac{1}{2\mu_0}\frac{\partial}{\partial t}\left(\tilde{\mathbf{B}}^*\cdot\tilde{\mathbf{B}}\right)+\frac{\varepsilon_0}{2}\frac{\partial}{\partial t}\sum_{\mathbf{k}}\left(\mathbf{E}_{\mathbf{k}}^*\cdot\mathbf{K}_{\mathbf{k}}^H\cdot\mathbf{E}_{\mathbf{k}}\right)+\frac{1}{\mu_0}\nabla\cdot\left(\mathbf{E}^*\times\tilde{\mathbf{B}}\right)-\omega\varepsilon_0\sum_{\mathbf{k}}\nabla\cdot\left(\mathbf{E}_{\mathbf{k}}^*\cdot\frac{\partial\mathbf{K}_{\mathbf{k}}^H}{\partial\mathbf{k}}\cdot\mathbf{E}_{\mathbf{k}}\right)\\&+i\omega\varepsilon_0\sum_{\mathbf{k}}\mathbf{E}_{\mathbf{k}}^*\mathbf{K}^A\mathbf{E}_{\mathbf{k}}=-\mathbf{E}^*\cdot\mathbf{J}_{ant},\end{aligned} \tag{8}$$

and assumes that the electric field of the modes $\mathbf{E_k}$ can be decomposed into slowly varying amplitude functions and rapidly oscillating phases with a slowly varying wave vectors $\mathbf{k_k}$ of the form $\mathbf{E_k}(\mathbf{x},t)=\mathbf{E}_{\mathbf{k},0}(\mathbf{x},t)\exp i\left(\int \mathbf{k_k}\cdot d\mathbf{x}-\omega t\right)$, the induced current for a single mode, $\mathbf{J}^{s}_{ind,\mathbf{k}}(\mathbf{x},t)$, then becomes

$$\mathbf{J}^{s}_{ind,\mathbf{k}}(\mathbf{x},t)=\exp i\left(\int \mathbf{k}\cdot d\mathbf{x}-\omega t\right)\times\left\{\boldsymbol{\sigma}_\mathbf{k}\mathbf{E}_{\mathbf{k},0}(\mathbf{x},t)+i\frac{\partial\boldsymbol{\sigma}_\mathbf{k}}{\partial\mathbf{k}}\cdot\frac{\partial\mathbf{E}_{\mathbf{k},0}}{\partial\mathbf{x}}\right.$$
$$\left. i\frac{1}{2}\left(\frac{\partial\omega_\mathbf{k}}{\partial\mathbf{k}}\frac{\partial\omega_\mathbf{k}}{\partial\mathbf{k}}:\frac{\partial^2\boldsymbol{\sigma}_\mathbf{k}}{\partial\mathbf{x}\partial\mathbf{k}}-\frac{\partial\omega_\mathbf{k}}{\partial\mathbf{k}}\frac{\partial\omega_\mathbf{k}}{\partial\mathbf{x}}:\frac{\partial^2\boldsymbol{\sigma}_\mathbf{k}}{\partial\mathbf{k}\partial\mathbf{k}}\right)\mathbf{E}_{\mathbf{k},0}(\mathbf{x},t)\left(\frac{\partial\omega_\mathbf{k}}{\partial\mathbf{k}}\cdot\frac{\partial\omega_\mathbf{k}}{\partial\mathbf{k}}\right)^{-1}\right\} \tag{9a}$$

Or written in component form in a Cartesian coordinate system

$$J^{s}_{ind,\mathbf{k}}(\mathbf{x},t)=\exp i\left(\int \mathbf{k_k}\cdot d\mathbf{x}-\omega t\right)\times\left\{\sigma_{\mathbf{k},ij}E_{\mathbf{k},0,j}(\mathbf{x},t)+i\frac{\partial\sigma_{\mathbf{k},ij}}{\partial k_l}\cdot\frac{\partial E_{\mathbf{k},0,j}}{\partial x_l}+\right.$$
$$\left. i\frac{1}{2}\frac{\partial\omega_\mathbf{k}}{\partial k_s}\left(\frac{\partial\omega_\mathbf{k}}{\partial k_l}\frac{\partial^2\sigma_{\mathbf{k},ij}}{\partial x_l\partial k_s}-\frac{\partial\omega_\mathbf{k}}{\partial x_l}\frac{\partial^2\sigma_{\mathbf{k},ij}}{\partial k_s\partial k_l}\right)\left(\frac{\partial\omega_\mathbf{k}}{\partial k_u}\frac{\partial\omega_\mathbf{k}}{\partial k_u}\right)^{-1}E_{\mathbf{k},0,j}(\mathbf{x},t)\right\}, \tag{9b}$$

where $\omega_\mathbf{k}=\omega_\mathbf{k}(\mathbf{k},\omega,\mathbf{x})$ is the local dispersion relation for the mode $\mathbf{k}$.

A major problem when solving the wave equation for a spatial dispersive medium for which $\mathbf{K_k}$ differ for the different modes, $\mathbf{k}$, is that the composition of the modes is not known in advance. An alternative to calculate the wave field by first substituting the expression for the induced current into the wave equation and then solving the equation is to solve the wave equation (1) and the equation for the induced current (2) by means of iterations, so called operator splitting. Green and Berry proposed to solve the system by calculating the induced current by integrating the acceleration due to the wave field along the unperturbed orbits [3]. Here the idea is to separate the wave field locally into wavelets and use a quasi-homogeneous approximation based on Eq. (9) for the induced current in weakly inhomogeneous plasma.

When solving the wave equation by means of iterations it is an advantage, if the initial guess is close to the solution. An initial guess of the wave field can be obtained by using an approximation of the induced current with $\mathbf{J}_{0,ind}(\mathbf{x},t)$, which does not depend on how the waves are separated

$$\nabla\times\nabla\times\mathbf{E}^{(n+1)}-\left(\frac{\omega}{c}\right)^2\mathbf{K}_0\mathbf{E}^{(n+1)}=i\omega\mu_0\mathbf{J}_{ant}+i\omega\mu_0\left(\mathbf{J}^{(n)}_{ind}-\mathbf{J}^{(n)}_{0,ind}\right), \tag{10}$$

where the induced current $\mathbf{J}^{(n)}_{ind}$ is calculated from Eq. (9) with $\mathbf{E}_{\mathbf{k},0}=\mathbf{E}^{(n)}_{\mathbf{k},0}$ and $\mathbf{J}^{(n)}_{0,ind}=\sigma_0\mathbf{E}^{(n)}_{\mathbf{k},0}$, where the upper index (n) represent the n:th iteration. Here we are interested to correct for the upshift of $k_\parallel=n_\phi B_\phi/RB+m_\theta B_\theta/rB$ when the fast wave has different poloidal mode numbers, where R denotes the major radius, r the minor radius θ the poloidal angle and ϕ the toroidal angle with m_θ and n_ϕ as the polodal and toroidal mode numbers, respectively. As the zeroth order approximation σ_0 we use the conductivity tensor in homogeneous plasma with $k_\parallel=n_\phi/R$.

The boundary condition, in addition to the standard boundary conditions on continuity across the boundary between the plasma and the vacuum regions, is that there should be not net emission or damping at the plasma boundary i. e.

$$\mathbf{E}^*\cdot\mathbf{J}_{ind}=0. \tag{11}$$

3. The induced current by wave-particle interactions

Calculating the induced current at a point $\mathbf{x}_0$ and at a time t_0 by wave-particle interactions from the Vlasov equation gives

$$\mathbf{J}_{ind}\left(\mathbf{x}_0,t_0,\mathbf{E}\right)=-\int d^3 v Ze\mathbf{v}\left(\delta\mathbf{v}\left(\mathbf{x}_0,t_0,\mathbf{v},\mathbf{E}\right)\cdot\frac{\partial f_0}{\partial\mathbf{v}}\right), \tag{12}$$

where f_0 is the unperturbed distribution function at $\mathbf{x}_0$ and $\delta\mathbf{v}(\mathbf{x}_0, t_0, \mathbf{v}, \mathbf{E})$ the change in velocity caused by the wave field $\mathbf{E}$ including interactions all along the orbit for $t \leq t_0$. How far back in time the interactions have to be taken into account depends on the decorrelation time, which is determined by the decorrelation of the particle phase relative the wave phase. When the change in the particle energy caused by the wave field is averaged over the difference between the wave and particle phase the linear contribution vanishes. The second order contribution is given by

$$\Delta W\left(z_0,t_0,\mathbf{v},\mathbf{E}\right)=\sum_{\mathbf{k}_1}\sum_{\mathbf{k}_2}\int d^3 v\left\{\left(m\mathbf{v}^*\cdot\delta\mathbf{v}_{\mathbf{k}_1}^*\right)\left(\delta\mathbf{v}_{\mathbf{k}_2}\cdot\frac{\partial f_0}{\partial\mathbf{v}}\right)+\frac{m}{2}\delta\mathbf{v}_{\mathbf{k}_1}^*\cdot\delta\mathbf{v}_{\mathbf{k}_2}f_0\right\}, \tag{13}$$

where the sum includes interactions $\mathbf{k}_1$ and $\mathbf{k}_2$ before $t = t_0$ and $\delta\mathbf{v}_{\mathbf{k}}=\delta\mathbf{v}_{\mathbf{k}}\left(z_0,t_0,\mathbf{v},\mathbf{E}_{\mathbf{k}}\right)$. That the change in $\delta\mathbf{v}$ arises from interactions all along the orbit before the considered time, is consistent with that the relation between the current and field appear as a convolution of the electric field and the conductivity tensor as in Eq. (4) and that the time derivative of the free energy density due to the wave field is given by a double convolution as in Eq. (5).

The aim of this section is to calculate the change in $\delta\mathbf{v}$. Wave-particle interactions can be separated in resonant and non-resonant interactions. Resonant high frequency interactions in inhomogeneous plasma differ conceptually from that in homogeneous plasmas. In homogeneous plasma the resonant particles are continuously accelerated. In inhomogeneous plasma the particle orbits are characterised by different nearly periodic cycles. If one defines resonances as large constructive changes for nearly periodic cycles, the definition of a resonance will then depend on the decorrelation time. When the decorrelation time is shorter than the bounce time, the resonances are characterised by acceleration over several wave periods for which the wave phase and particle phase coincides. The rapid variation of the phase difference between the wave and the particle position for high frequency waves results in a localisation of the effective resonant interaction into small regions in real space. The Doppler shifted cyclotron and Landau resonances become then the resonances. The net increase in velocity can for these be calculated with the stationary phase method, which for a single wave when gyro averaged gives

$$\begin{aligned}\delta v_\perp &= \frac{Ze}{m}\left(1-\frac{k_\parallel v_\parallel}{\omega}\right)\left(E_+ \mathrm{J}_{n-1}(k_\perp\rho)+E_- \mathrm{J}_{n+1}(k_\perp\rho)+\frac{v_\parallel}{v_\perp}E_\parallel \mathrm{J}_n(k_\perp\rho)\right)\exp i\left(\Theta+\pi/4\right)\sqrt{\frac{2\pi}{|\ddot{\vartheta}|}}\\ \delta v_\parallel &= \frac{Ze}{m\omega}\left(v_\perp k_\parallel E_+ \mathrm{J}_{n-1}(k_\perp\rho)+v_\perp k_\parallel E_- \mathrm{J}_{n+1}(k_\perp\rho)+v_\parallel k_\parallel E_\parallel \mathrm{J}_n(k_\perp\rho)\right)\exp i\left(\Theta+\pi/4\right)\sqrt{\frac{2\pi}{|\ddot{\vartheta}|}},\end{aligned} \tag{14}$$

where Θ is the difference between the gyro phase and wave phase at the stationary phase point where $\dot{\vartheta}=\omega-n\omega_c\left(\mathbf{x}\right)-k_\parallel\cdot v_\parallel=0$ for which $\ddot{\vartheta}\approx n\omega_c r v_\parallel\sin\theta/qR^2$, ω_c denotes the cyclotron frequency, n the harmonic number, $k_\parallel$ the parallel wave number and $v_\parallel$ the parallel velocity. A local Cartesian coordinate system (x, y, z) is used, where the z-direction is parallel with the magnetic field and rotated such that $k_y = 0$. $E_+ = 0.5(E_x + iE_y)$ and $E_- = 0.5(E_x - iE_y)$ are the electric field components rotating co and counter to the ions. Note for $n = 0$, i.e. for the Landau resonances $\delta v_\perp = 0$ because $\omega-k_\parallel\cdot v_\parallel=0$. In the limit of small $k_\perp\rho$ we have for these resonances

$$\delta\upsilon_{\parallel} = \left(-\frac{\mu}{m} k_{\parallel} \tilde{B}_{\parallel} + \frac{Ze}{m} E_{\parallel} \right) \exp i(\Theta + \pi/4) \sqrt{\frac{2\pi}{|\ddot{\vartheta}|}}, \tag{15}$$

where μ is the magnetic moment. The first term is the acceleration by the TTMP and the second term the Landau damping. A more detailed analysis of how to add and calculate the coherent interactions can be found in Ref. [5].

Although the wave-particle interactions are global interactions the net changes in velocity can be described by local quantities. In order to estimate the effective localisation in space of the wave-particle interaction we estimate the interaction time with twice the time it takes for the phase difference ϑ to increase with $\pi/2$. Assuming $\upsilon_{\parallel}$ to be constant gives $2\delta t = 2\sqrt{\pi q R^2 / r n \omega_c \upsilon_{\parallel} \sin\theta}$. The corresponding distance over which the interaction takes place is given by $2\delta l = 2\upsilon_{\parallel}\delta t = 2\sqrt{\pi q R^2 \upsilon_{\parallel} / r n \omega_c \sin\theta}$, where q is the safety factor. The cyclotron resonances with $n \neq 0$ are rather well localised because of the high cyclotron frequencies. The interaction time $2\delta t$ can be compared with the effective interaction time obtained from the stationary phase method $\delta t_{res} = \sqrt{2\pi/|\ddot{\vartheta}|}$, which differs with $\sqrt{2}$ from the estimation. For Landau resonances, those with n = 0, the characteristic time and distance of the resonant variation are determined by the variation of $k_{\parallel}$ and $\upsilon_{\parallel}$ along the orbit. To calculate the change in velocity due to resonant interactions, it is sufficient that the wave field is well described by a planar wave near the resonant region, which is typically shorter than the distance the particle makes under a decorrelation time.

In order to include interactions at earlier times, within a decorrelation time, we assume the wave field can be decomposed locally into different coherent planar waves with a single frequency for which the local phases of the waves are determine by the phase of the antenna current. A particle moving along a magnetic field line may interact with several planar waves. For simplicity we assume the interactions to be well localised and to be defined by the stationary phase points. A single particle will then only interact with one wave at the same location described by ($t_{\mathbf{k}}$, $z_{\mathbf{k}}$) provided $\upsilon_{\parallel} \neq 0$ and the parallel wave numbers $k_{\parallel}$ differ for the different waves. Between the interactions the particle phase may be decorrelated or partly decorrelated. The change in the velocity at (t_0, z_0) of a particle having a velocity $\mathbf{v}_0$ depends then on all previous interactions taking into account the decorrelation

$$\delta\mathbf{v}(z_0, t_0, \mathbf{v}_0, \mathbf{E}) = \sum_{\mathbf{k}} \delta\mathbf{v}_{\mathbf{k}} \exp i(\Theta_{\mathbf{k}} + \pi/4)\delta_{\mathbf{k}}, \tag{16}$$

where $\delta_{\mathbf{k}}$ is the coherence between the interaction at the resonance at $z_{\mathbf{k}}$ and that remaining when the particle reaches the point z_0 at the time $t = t_0$.

In order to be able to add up the interactions we need to calculate the phase difference $\Theta_{\mathbf{k}}$ between the different interactions. The difference between the angular wave frequency and the harmonic of the gyro frequency at the point z_0 and the resonance at $z_{\mathbf{k}}$ in the guiding centre frame of the particle is given by $\Delta\vartheta_{\mathbf{k}} = \int_{t_{\mathbf{k}}}^{t_0} (\omega - n\omega_c - k_{\parallel,\mathbf{k}} \cdot \upsilon_{\parallel}) dt$. Expanding the harmonic cyclotron frequency around the point z_0 yields $z_k = (\omega - n\omega_{c0} - k_{\parallel,\mathbf{k}} \cdot \upsilon_{\parallel}) / (n\partial\omega_c/\partial z) + z_0$. Assuming $\upsilon_{\parallel}$ = const. yields

$$\Delta\vartheta_{\mathbf{k}} = \left[\omega - \upsilon_{\parallel} k_{\mathbf{k},\parallel} - n\omega_{c0} - 0.5n\frac{\partial\omega_c}{\partial z}(z - z_0) \right] \frac{(z - z_0)}{\upsilon_{\parallel}}. \tag{17}$$

The difference in gyro phase at the stationary phase points for a particle interacting with two waves $\mathbf{k}_1$ and $\mathbf{k}_2$ becomes

$$\Theta_{\mathbf{k}_1} - \Theta_{\mathbf{k}_2} = 0.5 v_{\parallel} \left(k_1 - k_2\right)^2 \left(n \frac{\partial \omega_c}{\partial z}\right)^{-1}, \tag{18}$$

where k_1 and k_2 denote the parallel wave number for the waves 1 and 2, respectively, typically $\left|\Theta_{\mathbf{k}_1} - \Theta_{\mathbf{k}_2}\right| \ll \pi/2$ for thermal particles.

When treating the wave-particle interactions we have to distinguish between coherent interactions and decorrelated interactions. The interactions have to be regarded as coherent when the interactions for a single mode are represented as a superposition of several interactions related to different basis functions describing one mode. If the time between the interactions is not sufficient for decorrelation, coherent interactions may also take place by interactions with different modes e. g. toroidal Fourier modes, mode converted waves or reflected waves because their phases are all determined by the phase of the antenna current. For fully coherent interactions we have for the first term in Eq. (13)

$$\begin{aligned} &\sum_{\mathbf{k}_1} \sum_{\mathbf{k}_2} \int d^3 v \left(m\mathbf{v}^* \cdot \delta \mathbf{v}_{\mathbf{k}_1}^*\right)\left(\delta \mathbf{v}_{\mathbf{k}_2} \cdot \frac{\partial f_0}{\partial \mathbf{v}}\right) = \\ &\sum_{\mathbf{k}_1} \sum_{\mathbf{k}_2} \int d^3 v \left|m\mathbf{v}^* \cdot \delta \mathbf{v}_{\mathbf{k}_1}^*\right| \left|\delta \mathbf{v}_{\mathbf{k}_2} \cdot \frac{\partial f_0}{\partial \mathbf{v}}\right| \cos\left(-\Theta_{\mathbf{k}_1} - \alpha_{\mathbf{k}_1}\left(z_{\mathbf{k}_1}\right) + \Theta_{\mathbf{k}_2} + \alpha_{\mathbf{k}_2}\left(z_{\mathbf{k}_2}\right)\right), \end{aligned} \tag{19}$$

where $\mathbf{E}_{\mathbf{k}} = \left|\mathbf{E}_{\mathbf{k}}\right| \exp i\alpha_{\mathbf{k}}\left(z_{\mathbf{k}}\right)$. When averaged over all phases ϑ for which the interactions with the modes $\mathbf{k}_1$ and $\mathbf{k}_2$ are given by the phases ϑ and $\vartheta - \Theta_{\mathbf{k}_1} + \Theta_{\mathbf{k}_2}$ where ϑ is random number we have

$$\Delta W_{\mathbf{k}_1\mathbf{k}_2} \propto \frac{\left|\mathbf{E}_{\mathbf{k}_1}\right|^2}{\left|\ddot{\vartheta}_{\mathbf{k}_1}\right|} + \frac{\left|\mathbf{E}_{\mathbf{k}_2}\right|^2}{\left|\ddot{\vartheta}_{\mathbf{k}_2}\right|} + \frac{2\left|\mathbf{E}_{\mathbf{k}_1}\right|\left|\mathbf{E}_{\mathbf{k}_2}\right|}{\sqrt{\left|\ddot{\vartheta}_{\mathbf{k}_1}\right|\left|\ddot{\vartheta}_{\mathbf{k}_2}\right|}} \cos\left(-\Theta_{\mathbf{k}_1} - \alpha_{\mathbf{k}_1}\left(z_{\mathbf{k}_1}\right) + \Theta_{\mathbf{k}_2} + \alpha_{\mathbf{k}_2}\left(z_{\mathbf{k}_2}\right)\right). \tag{20}$$

Which of the two interactions is the first interaction for particles with low $v_{\parallel}$, thermal particles, depends only on the wave numbers $\mathbf{k}_1$ and $\mathbf{k}_2$ and the location of the resonance, but not on the parallel velocity. For the first interactions the incoming phases are random, the net energy taken from the first mode constitutes of those from the first interactions only. The induced current caused at the first resonance is given by Eq. (9). The second interactions are affected by the accelerations at the first interactions because the changes in energy depend on the velocity the particles have before the interactions. The net exchange of energy at the second interactions includes therefore both the cross term and the interactions with the second mode alone. The absorption described by the cross term depends on the phase difference $\left(-\Theta_{\mathbf{k}_1} - \alpha_{\mathbf{k}_1}\left(z_{\mathbf{k}_1}\right) + \Theta_{\mathbf{k}_2} + \alpha_{\mathbf{k}_2}\left(z_{\mathbf{k}_2}\right)\right)$. Assuming $\ddot{\vartheta}_{\mathbf{k}_1} \approx \ddot{\vartheta}_{\mathbf{k}_2}$ and $\left|\Theta_{k_1} - \Theta_{k_2}\right| \ll \pi/2$ the induced current at the second resonance can be approximated by

$$\mathbf{J}_{ind,\mathbf{k}_2}\left(\mathbf{x},t\right) \approx \mathbf{J}_{ind}^{s}\left(\mathbf{E}_{\mathbf{k}_2}\right) + \frac{2\left|\mathbf{E}_{\mathbf{k}_1}\left(\mathbf{x}_{\mathbf{k}_1}\right)\right|}{\left|\mathbf{E}_{\mathbf{k}_2}\left(\mathbf{x}_{\mathbf{k}_2}\right)\right|} \mathbf{J}_{ind}^{s}\left(\mathbf{E}_{\mathbf{k}_2}\right) \cos\left(-\alpha_{\mathbf{k}_1}\left(z_{\mathbf{k}_1}\right) + \alpha_{\mathbf{k}_2}\left(z_{\mathbf{k}_2}\right)\right). \tag{21}$$

A more practical approximation in the presence of several modes is to neglect the order of the interactions and assume for all modes $\mathbf{k}_2$

$$\mathbf{J}_{ind,\mathbf{k}_2}\left(\mathbf{x},t\right) \approx \sum_{\mathbf{k}_1} \frac{\left|\mathbf{E}_{\mathbf{k}_1}\left(\mathbf{x}_{\mathbf{k}_1}\right)\right|}{\left|\mathbf{E}_{\mathbf{k}_2}\left(\mathbf{x}_{\mathbf{k}_2}\right)\right|} \mathbf{J}_{ind}^{s}\left(\mathbf{E}_{\mathbf{k}_2}\right) \cos\left(-\alpha_{\mathbf{k}_1}\left(z_{\mathbf{k}_1}\right) + \alpha_{\mathbf{k}_2}\left(z_{\mathbf{k}_2}\right)\right). \tag{21b}$$

Coherent interactions can also be approximated by representing the different interactions by a single $k_{\parallel}$ with the electric field component added up linearly including the phase differences.

For completely decorrelated interactions the phase difference between the interactions are random, and the cross terms vanish after averaging. Then only the contributions from the wave resonating at z_0 remain which give $\delta_\mathbf{k} = 0$ if $z_\mathbf{k} \neq z_0$ and $\delta_\mathbf{k} = 1$ if $z_\mathbf{k} = z_0$. When the modes are added incoherently, which is a typical assumption for different toroidal modes, one obtains

$$\Delta W \propto \frac{\left|\mathbf{E}_{\mathbf{k}_1}\right|^2}{\left|\ddot{\vartheta}_{\mathbf{k}_1}\right|} + \frac{\left|\mathbf{E}_{\mathbf{k}_2}\right|^2}{\left|\ddot{\vartheta}_{\mathbf{k}_2}\right|}. \tag{22}$$

4. Calculation of induced current by decomposing the wave field into wavelets

Because of the semi-local character of high frequency wave-particle interactions a local decomposition of the wave field into planar waves is sufficient for calculating the response. The localisation is of the order of a gyro radius for the gyro motion and $2\delta l$ for the effective region of wave-particle interaction. The continuous Morlet wavelets, which have been used for analyzing time sequences, are well suited for such a decomposition, for which the transform is defined by

$$S_{m,n} = \int_{-\infty}^{+\infty} s(x')\psi^*_{m,n}(x')dx', \tag{23}$$

where * denotes the complex conjugate and $\psi_{\mathrm{m,n}}$ is related to the mother wavelet ψ by

$$\psi_{\mathrm{m,n}}(x) = a^{-m/2}\psi\left(\frac{x-nb}{a^m}\right), \tag{24}$$

where $\psi(\mathrm{x}) = \left(e^{-i\kappa_0 x} - e^{-\frac{\kappa_0^2}{2}}\right)e^{-\frac{x^2}{2}}$. The inverse transform is given by

$$s(\mathrm{x}) = k_\psi \sum_{m=-\infty}^{\infty}\sum_{n=-\infty}^{\infty} \psi_{m,n}(x) S_{m,n}, \tag{25}$$

where k_ψ is a normalisation constant, $\psi_{m,n}$ is called a voice. The effective number of oscillations of the wavelet can be defined arbitrarily by the parameter κ_0. The 1D-wavelet expresses the wave field in terms of translations (space) by nb and dilatations (wave numbers) by $a^{-\mathrm{m}}$. Contrary to the Fourier expansion, which is an expansion in one variable k, wavelet transformation is in an expansion in two variables corresponding to x and k. The most common form is $a = 2^{1/j}$, where j is an integer. Because of the finite width of the wavelets a reduced set voices can be used compared to a Fourier expansion.

For our purpose of resolving waves, we rewrite the wavelets in a more convenient form by making the substitution $a^{-m}\kappa_0 = 2\pi k$, which gives

$$\psi_{k,\mathrm{n}}(\mathrm{x}) = \sqrt{\frac{2\pi k}{\kappa_0}}\left(e^{-2\pi i k(x-nb)} - e^{-\frac{\kappa_0^2}{2}}\right)e^{-\frac{2\pi^2 k^2(x-nb)^2}{\kappa_0^2}}. \tag{26}$$

Note that k is a countable set of numbers, which may not necessary be integers. The set of k has to be chosen together with the points $x_\mathrm{i} = nb$, which typically are the grid points in the FEM decomposition. A local mode, for which the wave vector changes slowly, will be composed of neighbouring voices i.e. neighbouring k-values. The derivative $\partial E_{0,l}/\partial x_s$ in Eq. (6b) is calculated by taking the derivative on the envelope function of the wavelet given by $\exp(-2\pi^2 k^2 (x-nb)^2/\kappa_0^2)$, weighted with the coefficient $S_{\mathrm{k,n}}$. The responses from them have then to be added coherently, whereas the response from different modes could be added coherently or by assuming them to be fully decorrelated.

Modelling the upshift with FEM codes for the fast magnetosonic wave can be done by using 1D wavelets to decompose the wave in the poloidal direction on each flux surface.

5. Conclusions and discussions

Modelling of the wave field of spatial dispersive plasma is a challenging task because the co-existence of different waves requiring different responses. Here a method has been proposed suitable for FEM codes. The method is based on solving the wave equation by operator splitting, for which the induced current is calculated from an approximate solution and the inhomogeneous wave equation is solved by means of iterations. In order to separate the response for the different modes and include the resonant response for an inhomogeneous spatial dispersive medium the wave field is separated into wavelets. When calculating the response it is important to separate coherent interactions from decorrelated interactions.

Calculating the wave field by making operator splitting and solving an inhomogeneous wave equation by means of iterations put severe restrictions on the iteration method, since for each current distribution there exists a wave field solving Eq. (10). For the wave field to be a physical solution the local wave vector has to be consistent with the local dispersion function and the wave field satisfying energy conservation. To obtain converged results requires advanced iteration schemes such as the minimum polynomial extrapolation system, so called Anderson acceleration method [6], which has been found by Green et al to give good results [4].

References

[1] E. F. Jaeger et al, Physics of Plasmas **15** (2008) 072513
[2] T. Hellsten et al, Nuclear Fusion **53** (2013) 09
[3] O. Meneghini et al, Physics of Plasmas **16** (2009) 080705
[4] D.L. Green and L.A. Berry, Computer Physics Communication **185** (2014) 736
[5] P.U. Lamalle (1997) Plasma Physics and Controlled Fusion **39** 1409
[6] H. F Walker and P. Ni, SIAM J Numer. Anal., (2011) **49** 1715

Non-linear Simulations of MHD Instabilities in Tokamaks Including Eddy Current Effects and Perspectives for the Extension to Halo Currents

M. Hoelzl[1], G.T.A. Huijsmans[2], P. Merkel[1], C. Atanasiu[3], K. Lackner[1], E. Nardon[4], K. Aleynikova[2], F. Liu[2], E. Strumberger[1], R. McAdams[5,6], I. Chapman[6], A. Fil[4]

[1] Max-Planck-Institute for Plasmaphysics, Boltzmannstr. 2, 85748 Garching, Germany
[2] ITER Organisation, Route de Vinon sur Verdon, St-Paul-lez-Durance, France
[3] Association EURATOM-MEdC Romania, P.O. Box MG-36, Magurele, Bucharest, Romania
[4] CEA, IRFM, CEA Cadarache, F-13108 St Paul-lez-Durance, France
[5] York Plasma Institute, University of York, York, YO10 5DD, UK
[6] Culham Science Centre, Abingdon, OX14 3DB, UK

E-mail: `mhoelzl@ipp.mpg.de`

Abstract. The dynamics of large scale plasma instabilities can be strongly influenced by the mutual interaction with currents flowing in conducting vessel structures. Especially eddy currents caused by time-varying magnetic perturbations and halo currents flowing directly from the plasma into the walls are important.

The relevance of a resistive wall model is directly evident for Resistive Wall Modes (RWMs) or Vertical Displacement Events (VDEs). However, also the linear and non-linear properties of most other large-scale instabilities may be influenced significantly by the interaction with currents in conducting structures near the plasma. The understanding of halo currents arising during disruptions and VDEs, which are a serious concern for ITER as they may lead to strong asymmetric forces on vessel structures, could also benefit strongly from these non-linear modeling capabilities.

Modeling the plasma dynamics and its interaction with wall currents requires solving the magneto-hydrodynamic (MHD) equations in realistic toroidal X-point geometry consistently coupled with a model for the vacuum region and the resistive conducting structures. With this in mind, the non-linear finite element MHD code JOREK [1, 2] has been coupled [3] with the resistive wall code STARWALL [4], which allows us to include the effects of eddy currents in 3D conducting structures in non-linear MHD simulations.

This article summarizes the capabilities of the coupled JOREK-STARWALL system and presents benchmark results as well as first applications to non-linear simulations of RWMs, VDEs, disruptions triggered by massive gas injection, and Quiescent H-Mode. As an outlook, the perspectives for extending the model to halo currents are described.

1. Introduction

Large scale plasma instabilities can be strongly influenced by the interaction with currents in conducting structures. The resulting forces on wall structures are an important topic for ITER as well. In order to address those questions involving non-linear plasma dynamics and its interaction with wall currents requires a code that solves the magneto-hydrodynamic (MHD) equations in

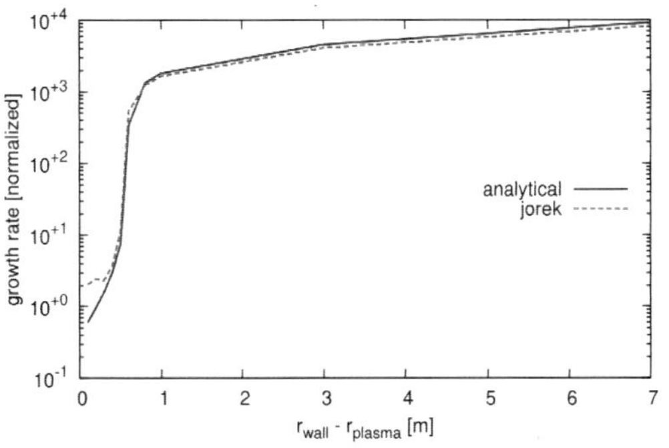

Figure 1 *Comparison of linear growth rates between analytical theory and JOREK-STARWALL results for an RWM case in a circular limiter plasma with minor radius* $a = 1$ m, *major radius* $R = 10$ m *and at a wall resistivity of* $\eta_{wall} = 2.5 \cdot 10^{-5}\ \Omega$ m.

realistic toroidal X-point geometry consistently coupled with a model for the vacuum region and the resistive conducting structures. With this in mind, the non-linear finite element MHD code JOREK [1, 2] has been coupled [3] with the resistive wall code STARWALL [4], which allows us to include the effects of eddy currents in 3D conducting structures in non-linear MHD simulations.

This article summarizes the capabilities of the coupled JOREK-STARWALL system (Section 2) and presents benchmark results (Section 3) as well as first physics applications to non-linear simulations (Section 4) of resistive wall modes (RWMs), vertical displacement events (VDEs), disruptions triggered by massive gas injection (MGI), and quiescent H-Mode (QH mode). This publication only provides a brief overview for the physics investigations and refers to separate papers going into more detail on each subject. As an outlook, the perspectives for extending the model to halo currents are described (Section 5).

2. The Model

2.1. The non-linear MHD code JOREK

The non-linear MHD code JOREK [1] solves a set of reduced MHD equations [5, 6] in realistic toroidal X-point geometry. Two-fluid extensions (diamagnetic drift, separate electron and ion temperatures) and a model for deuterium neutrals exist. A full MHD model is available [7], but needs to be extended by X-point boundary conditions and two-fluid terms.

2.2. The resistive wall code STARWALL

The standalone STARWALL code is a resistive wall code [4, 8] capable of linear stability analysis of MHD modes in the presence of 3D resistive walls. Since kinetic energy is neglected, its validity is restricted to cases where plasma inertia does not play an important role. Resistive walls are discretized by triangles using the thin-wall approximation. The validity of this approximation is discussed for instance in Reference [9] and Appendix A.

Parts of STARWALL are currently implemented into the linear MHD code CASTOR [10, 11, 12] along with additional extensions to create the new code CASTOR3D [13] for linear studies including coupling of different toroidal modes by 3D resistive walls, kinetic energy, plasma resistivity and plasma rotation. A modified version of STARWALL (sometimes called STARWALL_J) has been coupled to JOREK to allow non-linear simulations with 3D resistive wall effects. This coupling is described in the following Section.

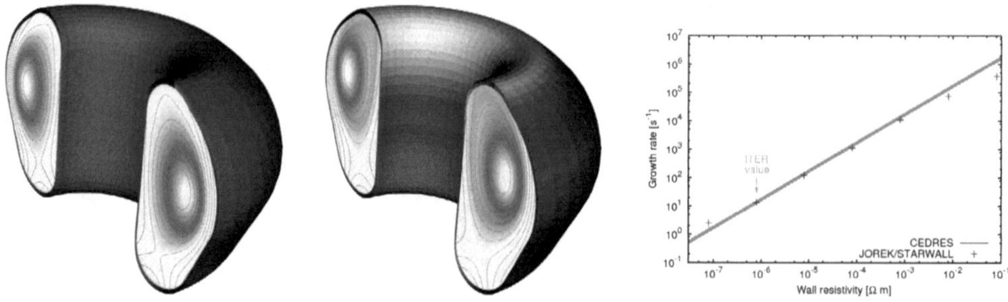

Figure 2 *ITER-like X-point plasma before (left) and during (middle) a VDE with an axisymmetric wall. Colors correspond to the plasma current density (multiplied by R due to normalization) and wall current density, respectively. The right figure shows the benchmark of linear growth rates for different wall resistivities against the CEDRES++ code with excellent agreement. Small deviations are observed at large wall resistivities since CEDRES++ neglects plasma inertia and at extremely small wall resisistivities below the realistic ITER values due to resolution limitations.*

2.3. The JOREK-STARWALL Coupling

The fully implicit JOREK-STARWALL coupling via a natural boundary condition has already been described previously in detail [3]. This Section only contains a brief summary.

The boundary of the JOREK computational domain is used as "interface" for the coupling of JOREK and STARWALL. STARWALL solves the field equations outside the interface (vacuum region and resistive wall structures) by a variational ansatz with the poloidal magnetic flux ψ as boundary condition which allows us to obtain an expression for the time evolution of wall currents. STARWALL computes a set of matrices representing this time evolution equation, which are used inside JOREK to keep track of the wall currents.

By considering all possible "unit perturbations", STARWALL also calculates matrices which allow to express the tangential magnetic field at the JOREK boundary in terms of wall currents and poloidal magnetic flux at the interface. This correlation is used inside JOREK to compute the natural boundary condition – a surface integral in the current definition equation originating from partial integration.

The implementation in JOREK is performed such that the full implicitness of the code is preserved without introducing additional unknowns. To achieve this, the unknown wall currents at time step $n+1$ are expressed in terms of the (as well unknown) JOREK variables at the same time step. The coupling is currently implemented only for the reduced MHD model, where the tangential magnetic field perturbation has no toroidal component. STARWALL and the coupling need to be extended in the future by additional matrices to add this additional component. As the STARWALL representation of wall currents allows only divergence-free surface currents, halo currents can presently not be included. Consequently, simulations are limited to eddy currents.

JOREK-STARWALL has already been benchmarked successfully against analytical theory and linear codes as briefly shown in Section 3. First physics results obtained with JOREK-STARWALL are presented in Section 4. An outlook to the inclusion of halo currents, which will be especially important for disruption simulations, is given in Section 5. In separate projects, extensions to impurity massive gas injection and runaway electrons are currently started such that we will be able to address the most important questions related to disruption physics in the future.

Journal of Physics: Conference Series **561** (2014) 012011 doi:10.1088/1742-6596/561/1/012011

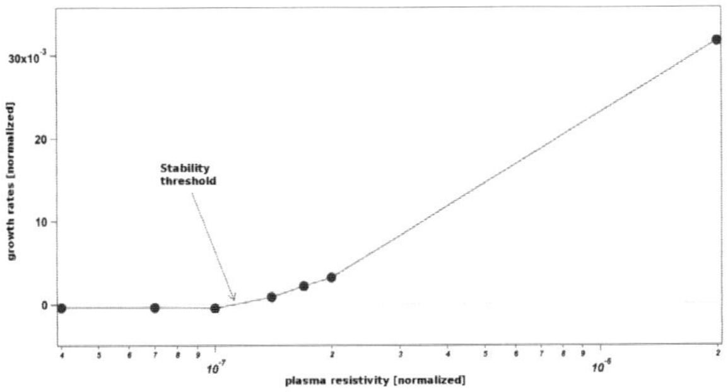

Figure 3 *Simulation of an RWM in ITER geometry, however at artificially increased values of resistivity and viscosity due to computational limitations. The growth rate of the observed instability shows a strong dependence on the plasma resistivity indicating a resistive instability. It is fully stabilized below a threshold in the plasma resistivity. Below this value, stabilizing dissipative effects (like viscosity) become dominant over the growth rate of the resistive instability.*

3. Benchmarks

This section briefly describes benchmarks carried out for the validation of JOREK-STARWALL. Linear growth rates are compared for resistive wall modes against analytical theory, and for a vertical displacement event against the CEDRES++ code [14]. Additional tests have been carried out earlier [3] by reproducing free-boundary equilibria and comparing tearing mode simulations against a linear code.

3.1. Resistive Wall Modes (RWMs)

The growth rates of a resistive wall mode (RWM) are compared [15, 16] to analytical theory [17] for a circular equilibrium. As seen in Figure 1, linear growth rates show very good agreement except for some deviations at small plasma-wall distances. Such cases where the resistive wall decreases the growth rates of the instability to the order of 1 s^{-1} or even below are hard to resolve since the natural boundary conditions tend to induce osciallatory behaviour in these cases. We will identify the reasons and improve this in the near future. This problem disappears for larger plasma-wall distances or higher wall resistivities, where growth rates stay higher.

3.2. Vertical Displacement Events (VDEs)

For an ITER-like VDE testcase with an axisymmetric wall, linear growth rates obtained with JOREK-STARWALL are compared in Figure 2 to the CEDRES++ code and show excellent agreement at realistic ITER wall resistivities [18]. For this benchmark, the plasma resistivity needs to be chosen between an upper limit above which the equilibrium changes too quickly, since we did not use a current source term, and a lower limit below which eddy currents in the scrape off layer would become dominant compared to the resistive wall currents.

4. Current Physics Applications

This Section briefly presents first physics applications of JOREK-STARWALL to RWMs, VDEs, QH-Mode, and disruptions triggered by massive gas injection.

4.1. Resistive Wall Modes (RWMs)

Resistive wall modes were simulated in an ITER scenario with reversed q-profile [16]. Results with the realistic location of the conducting wall in ITER were compared to an artificial case with closer wall, and the no-wall case. The comparison of those cases, and additional wall resistivity

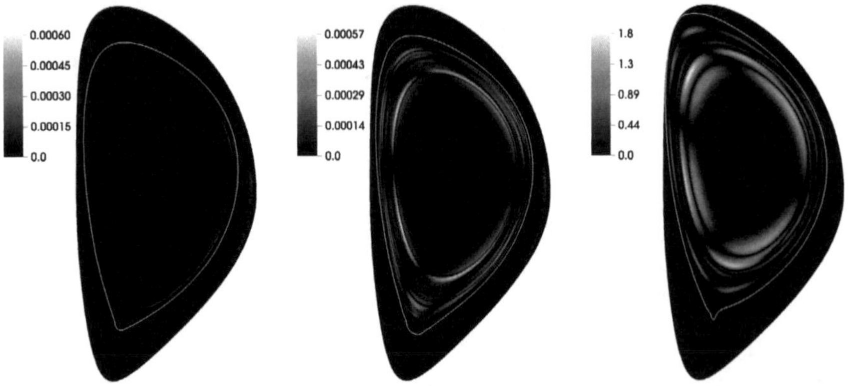

Figure 4 *The absolute value of the $n = 1$ component of the current density is plotted (in normalized units) close to the beginning of the simulation where the instability is almost axisymmetric (left, $t = 40\ t_A$), around the onset of the $n = 1$ perturbation where the axis has already moved upwards by about 1 cm due to the vertical instability (middle, $t = 1200\ t_A$) and at a later stage where the vertical axis displacement is about 32 cm (right, $t = 9600\ t_A$). The grey contour line corresponds to the pedestal temperature indicating the location of the separatrix. Clearly, the plasma has changed its shape in the course of the instability and the evolving $n = 1$ mode has become much less radially localized.*

scans allows us to quantify the stabilizing effect of the walls. Figure 3 shows a threshold observed in the plasma resistivity, below which the mode is fully stabilized. Consequently, this mode is a resistive mode. The existence of the threshold and the resistivity value it is observed at depend on the considered equilibrium, wall configuration, and plasma parameters.

4.2. Vertical Displacement Events (VDEs)

A vertical displacement event is an axisymmetric instability where the whole plasma moves up- or downwards with an approximately exponential evolution of the displacement in time. The time scale is governed by geometry and conductivity of conducting structures. A non-axisymmetric component can arise during the non-linear evolution of the VDE (external kink) when the plasma begins to shrink and the q-profile changes. Figure 4 shows a simulation of a VDE in ITER geometry (however with increased wall and plasma resistivities) which develops such a non-axisymmetric feature [19]. The vertical upward movement of the initially axisymmetric plasma is triggered by a small perturbation of coil currents. A small $n = 1$ perturbation at the separatrix can be observed (left Figure). A bit later, a 2/1 mode sets in at the $q = 2$ surface while some perturbations between this surface and the separatrix are still visible (middle). The 2/1 mode continues to grow and gets radially less localized while the vertical movement of the plasma continues (right).

4.3. Quiescent H-Mode (QH-Mode)

Quiescent H-Mode has been obtained in DIII-D [20], ASDEX Upgrade [21] and other devices. It is characterized by a saturated kink/peeling mode [22] and does not feature periodic ELM crashes. The influence of the vacuum vessel wall on the destabilization and saturation of edge modes for DIII-D QH-mode plasma has been studied with JOREK-STARWALL. Figure 5 shows the flux perturbation and the wall current stream function in the saturated state. When comparing to a fixed boundary simulation (i.e., ideal wall at the JOREK boundary; see Fig. 5 left) the saturated amplitude of the kink/peeling mode is larger in the "free boundary" simulation

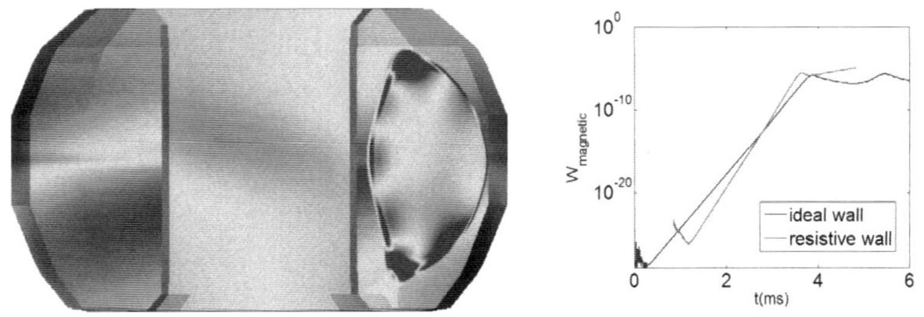

Figure 5 *The perturbed flux and wall potential on the resistive wall are plotted for a saturated kink/peeling mode for a simulation of DIII-D shot #153440 (left) with JOREK-STARWALL. The time evolution of the $n = 1$ magnetic energy (right) shows that the mode goes into a saturated state. Both, the initial growth rate as well as the saturation level are higher with the resistive wall boundary conditions of JOREK-STARWALL compared to ideal wall boundary conditions at the boundary of the JOREK computational domain.*

(resistive wall outside the JOREK domain; see Fig. 5 left). The initially rotating mode gets locked to the wall in the saturated phase of those simulations. To reproduce experimental observations in more detail, toroidal rotation needs to be taken into account (partly done already in Reference [23]), and also effects from neoclassical poloidal and diamagnetic rotation are important (simulations in preparation).

4.4. Disruption Events

JOREK simulations of disruption mitigation by Massive Gas Injection (MGI) are in progress [24]. The focus at the moment is on the cooling phase (when the gas enters the plasma and generates a cold front progressing from the edge to the core) and the triggering and development of the thermal quench (TQ). The MGI is modeled as a poloidally localized (and due to the low toroidal resolution of our first simulations toroidally weakly localized) source of neutrals which are assumed to move only by diffusive processes. Simulations with JOREK standalone (without STARWALL) for a JET plasma showed that an m=2, n=1 mode is triggered when the cold front reaches the q=2 surface, and that the TQ ensues when an m=3, n=2 mode is destabilized [24]. These simulations have been repeated with STARWALL (for the n=1 harmonic only). Even though the behavior is qualitatively similar, these simulations show quantitative differences and will allow to study, among other things, the dynamics of the locked mode (LM). Indeed, in JET disruptions, a LM is systematically observed. The LM signal is one of the main elements that can be exploited for disruption prediction [25], hence it is important to understand LM physics and its relation with disruptions. Figure 6 shows synthetic LM signals constructed by post-processing the above-mentioned JOREK simulations for JET with and without STARWALL. The left (resp. right) figure shows the amplitude (resp. phase) of the LM signal, which is defined as the n=1 component of the radial magnetic field at the machine midplane. The TQ sets in at about 7 ms, when the signals appear to be noisy. One can see that the LM amplitude peaks slightly before the TQ, with a similar peak value with and without STARWALL, on the order of 4 mT. Experimentally, the LM amplitude also typically presents a peak, although usually it is slightly after the TQ and its magnitude is smaller, typically 0.5 . . . 2 mT [De Vries EPS 2014]. The LM phase shows a different behavior with and without STARWALL, probably due to the modified torque exerted by the resistive wall on the plasma. These first MGI simulations do not pretend to be quantitatively correct, for several reasons: only $n = 0 \ldots 2$ harmonics are treated, the atomic physics is simplified, the neutral source is constant in time instead of having a realistic shape, etc. Present work is dedicated to improving all these points in order to simulate

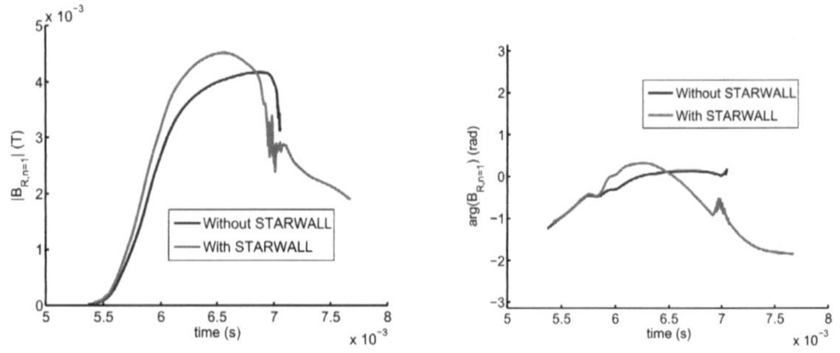

Figure 6 *Amplitude (left) and phase (right) of the synthetic locked mode signal produced as a post-processing of JOREK simulations of MGI in JET. In our simulations, the massive gas injection (currently assuming a neutral source at zero momentum) probably is the main cause for the locking of the MHD activity, since the electro-magnetic force responsible for mode locking vanishes in the ideal wall case.*

as realistically as possible a discharge with a D_2 MGI in JET, and to compare the results with experimental measurements in order to validate the model. JOREK-STARWALL will also be essential to study current quench and VDE phases.

5. Perspectives for an Extension to Halo Currents

Halo currents arising during a disruption event are a major concern for ITER due to the strong rotating asymmetric forces acting on vessel structures they may cause. Non-linear modeling will be important to predict these forces and also the effectiveness of disruption mitigation techniques aiming to reduce them. The JOREK-STARWALL code, however, currently assumes the wall surface currents to be divergence free, which allows us to include only eddy current effects. This Section provides a brief outlook to a future extension of JOREK-STARWALL to halo currents which effectively are current sources or sinks for the wall currents.

In the present STARWALL code, wall currents are given by $\mathbf{j}_\mathrm{w} = \hat{n} \times \nabla \phi_\mathrm{w}$ where the current stream functions ϕ_w are discretized by values at each triangle node and $\hat{n}$ denotes the normal vector at each triangle. In order to describe halo currents, which effectively enter as sources and sinks for the wall currents, this expression has to be generalized [26] to

$$\mathbf{j}_\mathrm{w} = \hat{n} \times \nabla \phi_\mathrm{w} + \nabla \varphi_\mathrm{w},$$

by adding a second non-divergence free surface current component, where φ_w takes the role of the electric potential on the wall.

In the derivation of the STARWALL equations, the starting point is a Lagrangian describing the energy associated to the vacuum fields and wall currents which involves terms depending on virtual surface currents (representing the plasma side) and the poloidal magnetic flux at the boundary of the JOREK domain as well as terms depending on the wall currents. A set of time evolution equations can be obtained by applying a variational principle and eliminating the virtual surface currents from the system. The coefficients of the equations can be written in terms of matrices (see Ref. [3]) and are used in JOREK to evolve the wall currents in time. The generalization of the wall current expression will extend the system of equations, i.e., introduce additional matrices and increase the size of some existing matrices. To make further required extensions clear, let us briefly summarize some technical details of the JOREK-STARWALL coupling:

(i) The JOREK computational grid is constructed and the information describing the boundary of the computational domain is written out, i.e., the boundary nodes and elements.

(ii) STARWALL calculates the "vacuum + wall response" matrices based on this JOREK boundary information and the desired wall geometry. Another JOREK simulation with a different computational grid requires a new STARWALL run.

(iii) The JOREK time evolution is carried out in free boundary mode using the STARWALL response matrices to evolve wall currents and to calculate the tangential magnetic field at the boundary of the JOREK domain. The tangential field is required for a boundary integral in the current definition equation arising from partial integration which acts as a natural boundary condition [3].

When introducing halo currents, consistency needs to be guaranteed between plasma and wall for currents flowing between both (current conservation). Additionally, the electric potential must be consistent between plasma and walls. In practice, STARWALL will provide a matrix correlation between currents flowing into the wall and electric potential. This correlation will be used as a boundary condition in JOREK. Although the reduced MHD model of JOREK only has a variable for the toroidal current density, the full current density can be reconstructed (Appendix B) such that halo currents will be compatible with, both, reduced and full MHD models.

A different longer term option would be to implement 3D conducting walls with finite thickness directly into the JOREK domain. A planned future implementation of toroidal finite elements in JOREK (instead of the present toroidal Fourier expansion) will clear the way for this option. STARWALL would then be used for free boundary conditions and conducting structures further away from the plasma. This approach will be tested soon by first implementing an axisymmetric conducting wall inside the JOREK domain.

6. Summary

We have presented some benchmarks of the coupled non-linear MHD code JOREK and the resistive wall code STARWALL against linear codes and analytical theory. Some first physics applications to RWMs, VDEs, QH-mode and disruptions were also shown.

An outlook is given to a future extension of JOREK-STARWALL to include halo currents. This will, together with separate other ongoing physics development projects in JOREK, allow for non-linear simulations of the most important processes relevant for disruption physics in the future.

Acknowledgements

This work was carried out using the HELIOS supercomputer system at Computational Simulation Centre of International Fusion Energy Research Centre (IFERC-CSC), Aomori, Japan, under the Broader Approach collaboration between Euratom and Japan, implemented by Fusion for Energy and JAEA.

References

[1] Huysmans G and Czarny O 2007 *Nuclear Fusion* **47** 659
[2] Czarny O and Huysmans G 2008 *Journal of Computational Physics* **227** 7423 – 7445
[3] Hoelzl M, Merkel P, Huysmans G, Nardon E, Strumberger E, McAdams R, Chapman I, Günter S and Lackner K 2012 *Journal of Physics: Conference Series* **401** 012010
[4] Merkel P and Sempf M 2006 *Proceedings of the 21st IAEA Fusion Energy Conference* (Chengdu, China) TH/P3-8 URL www-naweb.iaea.org/napc/physics/FEC/FEC2006/papers/th_p3-8.pdf
[5] Strauss H 1997 *Journal of Plasma Physics* **57** 83–87
[6] Franck E, Hoelzl M, Lessig A and Sonnendrücker E *Mathematical Models and Methods in Applied Sciences* In preparation
[7] Haverkort J 2013 *Magnetohydrodynamic Wavesand Instabilities in Rotating Tokamak Plasmas* Ph.D. thesis Eindhoven University of Technology, The Netherlands URL http://oai.cwi.nl/oai/asset/21317/21317B.pdf

[8] Merkel P and Strumberger E MHD stability studies with the STARWALL code in preparation
[9] Gimblett C 1986 *Nuclear Fusion* **26** 617–625 ISSN 0029-5515
[10] Huysmans G, Goedbloed J and Kerner W 1993 *Phys. Fluids B* **5** 1545
[11] Kerner W, Goedbloed J, Huysmans G, Poedts S and Schwarz E 1998 *Journal of Computational Physics* **142** 271 – 303
[12] Strumberger E, Merkel P, Tichmann C, and Günter S 2011 *Proceedings of the 38th EPS Conference on Plasma Physics* (Strasbourg, France) p5.082
[13] Strumberger E, Günter S, Merkel P and Tichmann C *Journal of Physics: Conference Series* Submitted
[14] Hertout P, Boulbe C, Nardon E, Blum J, Brémond S, Bucalossi J, Faugeras B, Grandgirard V and Moreau P 2011" *Fusion Engineering and Design* **86** 1045–1048 Proceedings of the 26th Symposium of Fusion Technology (SOFT-26)
[15] McAdams R, Chapman I, Wilson H, Hölzl M, Huysmans G, Liu Y and Merkel P 2013 (55th APS Conference)
[16] McAdams R 2014 *Non-linear Magnetohydrodynamic Instabilities in Advanced Tokamak Plasmas* Ph.D. thesis University of York
[17] Liu Y, Albanese R, Portone A, Rubinacci G and Villone F 2008 *Physics of Plasmas* **15** 072516
[18] Hoelzl M, Krebs I, Lessig A, Lackner K, Günter S, Merkel P, Wenninger R, Huysmans G, Nardon E, Orain F and ASDEX Upgrade Team 2013 (15th European Fusion Theory Conference (EFTC), Oxford, UK)
[19] Aleynikova K and Huijsmans G 2014 Private communication
[20] Burrell K H, Austin M E, Brennan D P, DeBoo J C, Doyle E J, Gohil P, Greenfield C M, Groebner R J, Lao L L, Luce T C, Makowski M A, McKee G R, Moyer R A, Osborne T H, Porkolab M, Rhodes T L, Rost J C, Schaffer M J, Stallard B W, Strait E J, Wade M R, Wang G, Watkins J G, West W P and Zeng L 2002 *Plasma Physics and Controlled Fusion* **44** A253–A263 ISSN 07413335
[21] Suttrop W, Hynönen V, Kurki-Suonio T, Lang P, Maraschek M, Neu R, Stäbler A, Conway G, Hacquin S, Kempenaars M, Lomas P, Nave M, Pitts R, Zastrow K D, Team t A U and Workprogramme c t t J E 2005 *Nuclear Fusion* **45** 721–730 ISSN 0029-5515
[22] Snyder P, Burrell K, Wilson H, Chu M, Fenstermacher M, Leonard A, Moyer R, Osborne T, Umansky M, West W and Xu X 2007 *Nuclear Fusion* **47** 961
[23] Liu F, Huijsmans G, Loarte A, Garofalo A M, Solomon W M, Snyder P B and Hoelzl M 2014 *Proceedings of the 41st EPS Conference on Plasma Physics* (Berlin, Germany) o5.135
[24] Fil A, Nardon E, Beyer P, Becoulet M, Dif-Pradalier G, Grandgirard V, Guirlet R, Hoelzl M, Huijsmans G, Latu G, Lehnen M, Loarte A, Orain F, Pamela S, Passeron C, Reux C, Saint-Laurent F, Tamain P and JET EFDA Contributors 2014 *Proceedings of the 41st EPS Conference on Plasma Physics* (Berlin, Germany) p1.045
[25] de Vries P, Pautasso G, Nardon E, Cahyna P, Gerasimov S, Maraschek M, Lehnen M, Huijsmans G, Hender T and JET EFDA contributors 2014 *Proceedings of the 41st EPS Conference on Plasma Physics* (Berlin, Germany) o5.133
[26] Atanasiu C and Zakharov L 2013 *Physics of Plasmas* **20** 092506

Appendix A. Validity of the thin wall approximation

STARWALL uses the thin wall approximation for 3D resistive wall structures. Conductors are discretized by triangles. The resistivity acting on the surface currents under this assumption is $\eta_{\text{surf}} = \eta/d$ where η is the specific resistivity of the wall material. We will discuss the validity of the thin wall approximation in the following, showing that it is a reasonable model in most situations. For the discussion, we will consider a typical time scale τ of the perturbation, which can either be the inverse of its rotation frequency or the inverse of an instability growth rate.

A magnetic field varying on the characteristic time scale τ penetrates into a conductor limited to the skin depth $\delta_{\text{skin}} = \sqrt{2\eta\tau/(2\pi\mu_r\mu_0)}$, where μ_0 denotes the vacuum permeability, and μ_r the relative permeability. ITER wall structures are mainly built from austenitic steel, and even martensitic components will be deep in the saturated regime due to the large toroidal field, such that $\mu_r = 1$ can be taken in estimates. For the steel walls ITER will have ($\eta \approx 10^{-6}\ \Omega$ m) we obtain $\delta_{\text{skin}} \approx 0.5\text{m} \cdot \sqrt{\tau[\text{s}]}$. The characteristic time for a magnetic perturbation to become homogeneous across the $d = 6$cm thick ITER walls is given by $\tau_{\text{skin}} = (0.06\text{m})^2/(0.5\text{m}^2/\text{s}) = 7\text{ms}$.

An external kink mode which can fully be stabilized by an ideally conducting wall, but is unstable in the presence of a resistive wall is called a resistive wall mode. Its growth time τ is approximately given by the characteristic L/R time τ_{wall} of the wall structures, which for the ITER conducting vessel is of the order of 200ms. For such a mode with $\tau \approx \tau_{\text{wall}} \gg \tau_{\text{skin}}$, the field is homogeneous across the wall and the thin wall approximation is well justified.

The opposite limit of an instability that is weakly coupled to the walls and can not be stabilized by ideally conducting walls corresponds to $\tau \ll \tau_{\text{skin}} \ll \tau_{\text{wall}}$. In this case, the magnetic perturbation can only penetrate into a very thin surface layer of the conductors. In principle, an effective – increased – resistivity needs to be considered for the wall current evolution: $\eta_{\text{surf,eff}} = \eta/\delta_{\text{skin}}$ instead of $\eta_{\text{surf}} = \eta/d$. However, the wall completely shields the magnetic field associated to such a fast mode and effectively acts as an ideally conducting wall – however with little influence on the mode growth. So a correction for the resistivity is not necessary to describe the dynamics of the mode, the thin wall approximation is well justified.

The remaining case would be an instability with $\tau \approx \tau_{\text{skin}}$ where the magnetic perturbation associated to the mode has partially penetrated through the wall. This situation corresponds to an increased wall resistivity $\eta_{\text{surf,eff}}$, which now really has to be taken into account in order to keep the thin wall approximation valid. The non-linear behaviour of an instability might not be obtained fully accurately with the thin-wall approximation in this case. Tokamak instabilities with such growth rates are, however, rather untypical since the wall coupling must be very marginal.

The geometrical simplification by the thin wall only plays a role if the mode wall distance is smaller than the wall thickness. In such an unusual case (e.g., large passive stabilizing loop present in ASDEX Upgrade), conductors can be discretized by several layers of thin walls.

Appendix B. Halo currents with reduced MHD

In the JOREK reduced MHD model, only a toroidal current component exists as a variable. Since the boundary of the JOREK computational domain is axi-symmetric, no currents could enter the walls preventing halo current modeling. However, the full current vector can be reconstructed assuming low pressure at the JOREK boundary: The momentum conservation equation then turns into $\mathbf{j} \times \mathbf{B} = 0$, i.e., the plasma current is parallel to the magnetic field. The current component flowing into the wall is then given by $j_\perp = j\mathbf{b} \cdot \hat{\mathbf{n}}$, where j denotes the toroidal current component in JOREK, $\mathbf{b} = \mathbf{B}/B$, and $\hat{\mathbf{n}}$ is the unit vector perpendicular to the conducting wall.

Joint Varenna-Lausanne International Workshop 2014 IOP Publishing
Journal of Physics: Conference Series **561** (2014) 012012 doi:10.1088/1742-6596/561/1/012012

On collisional impurity transport in nonaxisymmetric plasmas

A Mollén[1], **M Landreman**[2], **H M Smith**[3]

[1] *Department of Applied Physics, Chalmers University of Technology, Göteborg, Sweden*
[2] *Institute for Research in Electronics and Applied Physics, University of Maryland, College Park, Maryland 20742, USA*
[3] *Max-Planck-Institut für Plasmaphysik, 17491 Greifswald, Germany*

E-mail: albertm@chalmers.se

Abstract. The presence of impurity species in magnetic confinement fusion devices leads to radiation losses and plasma dilution. Thus it is important to analyze impurity dynamics, and search for means to control them. In stellarator plasmas the neoclassical ambipolar radial electric field often points radially inwards (referred to as the ion root regime), causing impurities to accumulate in the core. This can limit the performance of nonaxisymmetric devices.

In the present work we analyze neoclassical impurity transport in stellarator plasmas using a recently developed continuum drift-kinetic solver, the SFINCS code (the Stellarator Fokker-Planck Iterative Neoclassical Conservative Solver). The study is performed for a case close to the edge of W7-X using the standard configuration magnetic geometry. We investigate the sensitivity of impurity transport to impurity charge, main species density and temperature gradients, as well as ion temperature.

At the studied radial location we find that the neoclassical impurity peaking factor can be very large, particularly for high-Z impurities. The ambipolar radial electric field is in the ion root regime, and impurity accumulation can thus be expected. The accumulation is strengthened by the large main species density and temperature gradients. Moreover we find that the size of the bootstrap current is affected by the value of the plasma effective charge, suggesting that employing a realistic ion composition can be important when calculating the bootstrap current.

1. Introduction

The stellarator concept has an advantage over the tokamak concept, as it has the potential of steady state plasma operation with no current drive [1, 2]. However, stellarators exhibit some problems due to their inherent 3D nature. In 3D the existence of magnetic flux surfaces is not guaranteed as in axisymmetric configurations. The 3D magnetic equilibrium implies that particles can be trapped in helical magnetic wells where they can escape the plasma even in the absence of collisions. As a consequence the neoclassical transport is typically considerably larger in stellarators than in tokamaks. Furthermore, since collisionless trajectories are not necessarily confined, each species can have a different radial transport rate, resulting in a radial electric field to restore ambipolarity. This electric field is referred to as the ambipolar radial electric field, and can be determined without knowledge of the turbulent transport since the radial turbulent fluxes are automatically ambipolar to leading order in ρ_i/L, with ρ_i being the ion Larmor radius and L the smallest macroscopic scale length of interest [3].

In stellarators, the neoclassical ambipolar radial electric field often points radially inwards (referred to as the ion root regime), and this causes impurities to accumulate in the core [4–6]. This accumulation can be particularly strong for impurities of high charge Z. Since the presence of impurities leads to plasma dilution and radiation losses, where high-Z impurities are especially detrimental, it is important to study their behavior in fusion plasmas. This is an area which to date has been more extensively studied in tokamaks than in stellarators, related to the fact that the complicated stellarator geometry makes calculations demanding. In tokamak plasmas the transport is often dominated by turbulence, but since the neoclassical transport is typically much larger in stellarators it can be the main transport channel.

In the present work we perform a study of neoclassical impurity transport in stellarators using a recently developed neoclassical code SFINCS (the Stellarator Fokker-Planck Iterative Neoclassical Conservative Solver) described in [7]. This continuum code solves the 4D drift-kinetic equation, retaining coupling in four of the independent phase space variables (two spatial and two velocity), but neglecting radial coupling. The code can be used for an arbitrary number of plasma species, but in our study we will employ electrons (e), main hydrogen ions (i) and one impurity species (z).

The paper is organized as follows. In Sec. 2 we describe experimental parameters, introduce the impurity peaking factor and explain how to estimate it with SFINCS. The results are then discussed and summarized in Sec. 3.

2. Impurity density peaking

Flux-surface quantities only depend on the flux-surface label, for which we will use the normalized toroidal flux $\psi_N \equiv \psi/\psi_a$, where $2\pi\psi$ is the toroidal flux and $2\pi\psi_a$ the toroidal flux of the last closed flux-surface. This label is related to the effective radius r according to $\psi_N = (r/a)^2$, where a is the outermost effective minor radius. The density and temperature gradient scale lengths of species s are defined according to $L_{ns} = -\left[\partial\left(\ln n_s\right)/\partial r\right]^{-1}$ and $L_{Ts} = -\left[\partial\left(\ln T_s\right)/\partial r\right]^{-1}$. We now define the impurity peaking factor a/L_{nz}^0 (also referred to as the zero-flux impurity density gradient) as a/L_{nz} for which the flux-surface averaged impurity flux vanishes

$$\langle \boldsymbol{\Gamma}_z \cdot \nabla r \rangle \equiv \left\langle \int d^3v f_{z1} \mathbf{v}_{dz} \cdot \nabla r \right\rangle = 0. \tag{1}$$

Here $\langle X \rangle = \left(\int_0^{2\pi} d\theta \int_0^{2\pi} d\zeta \frac{X}{B^2} \right) / \left(\int_0^{2\pi} d\theta \int_0^{2\pi} d\zeta \frac{1}{B^2} \right)$ in Boozer coordinates, $f_{z1} = f_z - f_{Mz}$ is the departure from the Maxwellian part of the distribution function, and $\mathbf{v}_{dz}$ is the drift velocity.

The neoclassical impurity peaking factor is found with SFINCS through an iterative procedure, varying $\partial n_z/\partial r$ and the radial electric field $E_r = -\partial\Phi/\partial r$ until $\langle \boldsymbol{\Gamma}_z \cdot \nabla r \rangle = 0$ and $\sum_s Z_s e \langle \boldsymbol{\Gamma}_s \cdot \nabla r \rangle = 0$ simultaneously. The radial electric field for which the radial current vanishes is the ambipolar field. We will study a plasma edge scenario originally used in [8] but also considered in [7], with an extra single impurity species introduced into the hydrogen plasma. The magnetic equilibrium is the W7-X standard configuration at $r/a = 0.88$, with $n_e = 6.6 \times 10^{19}$ m^{-3} and $T_e = T_i = 1$ keV. The pressure gradient is large, $a/L_{ne} \approx 11.9$ and $a/L_{Te} = a/L_{Ti} \approx 5.8$, and we note that $\rho_i/L_n \approx 3 \times 10^{-2}$. The collisionality (specified on p. 10 in [7]) is below the Pfirsch-Schlüter (high collisionality) regime. As our baseline scenario we will study C^{6+} with $Z_{\text{eff}} = 1.05$, but we also analyze Ar^{18+}, Ni^{28+} and W^{40+}. The resolution used in the SFINCS runs is $N_\theta = 19$, $N_\zeta = 71$ grid points in the poloidal and toroidal direction (per identical segment of the stellarator, where W7-X has a five-fold symmetry in the toroidal direction), $N_x = 8$ grid points in energy and $N_\xi = 60$ Legendre polynomials to represent the distribution function (here $\xi = v_\parallel/v$), $N_L = 4$ Legendre polynomials and $N_{x,\text{Pot}} = 40$ grid points in energy to represent the Rosenbluth potentials, and $x_{\text{Max}} = 5.0$ as the maximum normalized speed for the Rosenbluth potential grid.

Figure 1 shows how the neoclassical impurity peaking factor and the ambipolar electric field vary with impurity charge, electron density gradient, ion temperature gradient, and ion temperature. We note that in the scalings where the pressure gradient is changed one should ideally generate a new magnetic equilibrium for each point to match the new pressure profile from force balance. This has however not been made due to the significant extra workload such a procedure requires; instead the baseline magnetic equilibrium is used for all the points. The transport of each plasma species in stellarators is, in general, proportional to a linear combination of the thermodynamic forces, where ZeE_r/T_z is included as a term in one of the forces of the impurity transport equation. It is thus not surprising that we find that the peaking factor is heavily dependent on the charge of the impurity, and a very strong peaking is found for high-Z impurities. The ambipolar electric field (which is in the radially inward ion root regime) is not significantly affected by changing the impurity charge, but the reason for this is that the impurity content is kept low to maintain $Z_{\text{eff}} = 1.05$ and consequently the ambipolar electric field is close to its value in a corresponding pure plasma. Furthermore we see that both the peaking factor and the strength of the radial electric field experience approximately linear increase with main species gradients. Increasing the ion (and impurity) temperature leads to an increase in the strength of the electric field, but the impurity peaking factor is reduced.

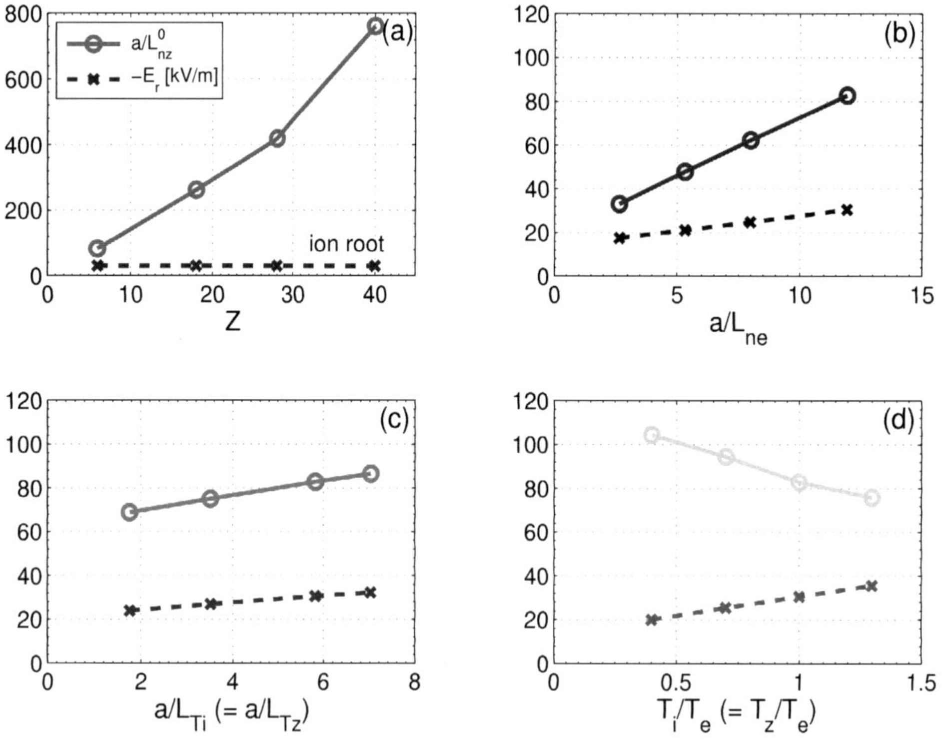

Figure 1. Impurity peaking factor (——○) and ambipolar radial electric field (- - - -×) as functions of impurity charge (a), electron density gradient (b), ion temperature gradient (c), and ion temperature (d) with electron temperature fixed. Note that in (d) $\partial T_i/\partial r$ $(= \partial T_z/\partial r)$ is fixed which implies that a/L_{Ti} $(= a/L_{Tz})$ is varying.

We can try to obtain a further insight for the results in Fig. 1 by using a simplified picture. In a typical ion root scenario, the radial electric field is approximately determined by the condition that the radial ion flux vanishes, since the electron particle flux is typically much smaller and we look at a case when the impurity particle flux is zero. The radial ion flux is

$$\langle \Gamma_i \cdot \nabla r \rangle \propto \frac{1}{n_i}\frac{\partial n_i}{\partial r} + \frac{e}{T_i}\frac{\partial \Phi}{\partial r} + \beta \frac{1}{T_i}\frac{\partial T_i}{\partial r}, \quad (2)$$

where β is a parameter which in general depends on the radial electric field. Assuming $n_e \approx n_i$ which follows from the low impurity content, and setting Eq. (2) to zero gives an approximation to the ambipolar radial electric field

$$E_r = -\frac{\partial \Phi}{\partial r} = \frac{T_i}{ae}\left[\frac{a}{n_i}\frac{\partial n_i}{\partial r} + \beta \frac{a}{T_i}\frac{\partial T_i}{\partial r}\right] = -\frac{T_i}{ae}\left[\frac{a}{L_{ne}} + \beta \frac{a}{L_{Ti}}\right]. \quad (3)$$

Neglecting the E_r dependence in β, Equation (3) is in qualitative agreement with Figs. 1 (a)-(c), since E_r is approximately independent of Z while depending linearly (with an offset) on a/L_{ne} and a/L_{Ti}. Also Fig. 1 (d) is consistent with Eq. (3), as E_r depends linearly on T_i.

Let us now consider the impurity peaking factor. Since the peaking factor is very large and with the assumption $Z \gg 1$, in the simplest model, the only thermodynamic drive terms that matter for the impurities are $(1/n_z)(\partial n_z/\partial r)$ and $(Ze/T_z)(\partial \Phi/\partial r)$. The peaking factor is thus approximately determined from

$$0 = \langle \Gamma_z \cdot \nabla r \rangle \propto a\left[\frac{1}{n_z}\frac{\partial n_z}{\partial r} + \frac{Ze}{T_z}\frac{\partial \Phi}{\partial r}\right]. \quad (4)$$

In a more complete model, there would be other drive terms in Eq. (4) from all the other inhomogeneous terms in the system of kinetic equations. Nevertheless substituting Eq. (3) into Eq. (4) we obtain the impurity peaking factor as

$$\frac{a}{L^0_{nz}} = Z\left[\frac{a}{L_{ne}} + \beta \frac{a}{L_{Ti}}\right]. \quad (5)$$

This equation approximately reproduces the linear dependence of a/L^0_{nz} on Z shown in Fig. 1 (a). Similarly to Eq. (3) for E_r, the linear dependence of a/L^0_{nz} on a/L_{ne} and a/L_{Ti} is reproduced in Figs. 1 (b) and (c) with a small offset. The variation of a/L^0_{nz} with T_i in Fig. 1 (d) is likely connected to the fact that we keep $\partial T_i/\partial r$ $(= \partial T_z/\partial r)$ fixed rather than a/L_{Ti} $(= a/L_{Tz})$ when performing the scan, this implies that a/L_{Ti} is decreasing with increasing T_i/T_e.

In Fig. 2 it is shown how the main species radial flux and the bootstrap current vary with impurity charge, electron density gradient, ion temperature gradient, and ion temperature. Increasing the gradients unsurprisingly leads to a larger radially outward flux. The size of the bootstrap current is increased by increasing the ion temperature gradient, but slightly reduced by increasing the main species density gradients. Increasing the ion (and impurity) temperature leads to an increased outward flux, and has a non-monotonic effect on the bootstrap current.

Finally we investigate how the impurity peaking factor, the ambipolar radial electric field, the main species radial flux and the bootstrap current are affected by the plasma effective charge Z_{eff}. The analysis is made by changing the impurity concentration, and the results are illustrated in Fig. 3. As long as the impurity concentration is low, the main species density gradients are only weakly affected by changes in the impurity density gradient to maintain quasi-neutrality. However as the impurity content is increased the effect on the main species gradients can be considerable, and it is difficult to anticipate how they will change to keep quasi-neutrality. We therefore carry out two different scans. In the first scan we maintain radial quasi-neutrality by modifying the main ion density gradient along with the impurity density gradient. In the second scan we keep the main ion density gradient fixed as the impurity density gradient is varied,

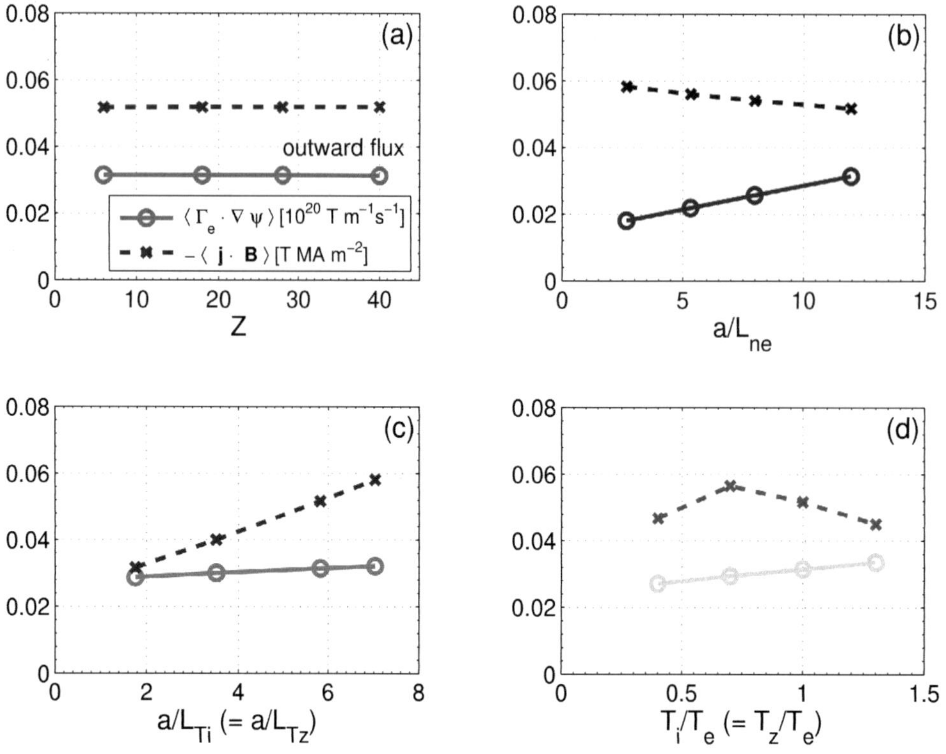

Figure 2. Main species radial flux (——○) and bootstrap current (- - - -×) as functions of impurity charge (a), electron density gradient (b), ion temperature gradient (c), and ion temperature (d) with electron temperature fixed. Note that since $\langle \Gamma_z \cdot \nabla\psi\rangle = 0$, due to ambipolarity $\langle \Gamma_e \cdot \nabla\psi\rangle = \langle \Gamma_i \cdot \nabla\psi\rangle$, and also that in (d) $\partial T_i/\partial r$ $(= \partial T_z/\partial r)$ is fixed which implies that a/L_{Ti} $(= a/L_{Tz})$ is varying.

violating radial quasi-neutrality. As shown in the figure, keeping the main species gradients fixed leads to an increase in impurity peaking, strength of the ambipolar electric field and main species outward radial flux with plasma effective charge, whereas the opposite trend is found when quasi-neutrality is considered. This suggests that increasing the impurity content should lead to an increase in these three quantities, but the effect is counteracted by the fact that the main species gradients are simultaneously decreased to maintain quasi-neutrality. Interestingly the size of the bootstrap current shows a similar decreasing trend with plasma effective charge in both cases, and the reduction is significant. For the case when the main species gradients are fixed there is even a sign change around $Z_{\text{eff}} = 2.15$. As the effective charge is likely to be larger than $Z_{\text{eff}} = 1.05$ in a real W7-X plasma, this could have implications on how the bootstrap current in stellarators should be calculated. Previous calculations of the bootstrap current in stellarators have almost always been made considering a pure plasma.

3. Conclusions and discussion

We used a recently developed continuum drift-kinetic solver, the SFINCS code, to study neoclassical impurity transport in stellarators. In particular, we studied a W7-X standard configuration magnetic equilibrium close to the plasma edge, considering a single impurity species

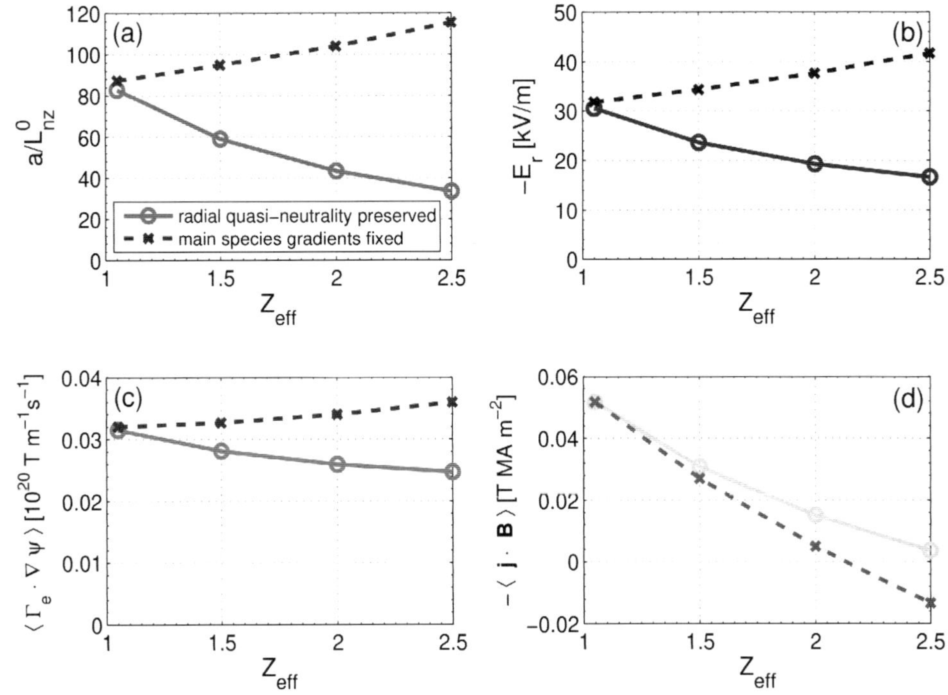

Figure 3. Impurity peaking factor (a), ambipolar radial electric field (b), main species radial flux (c), and bootstrap current (d) as functions of plasma effective charge. The analysis has been performed both maintaining radial quasi-neutrality by modifying the main ion density gradient along with the zero-flux impurity density gradient (——O), and also neglecting quasi-neutrality keeping the main species density gradients fixed (- - - -×).

in a hydrogen plasma. We find that the neoclassical impurity peaking factor can be very large at the studied radial location, particularly for high-Z impurities. The peaking is amplified by the large main species density and temperature gradients. The ambipolar electric field is in the ion root regime, and it is thus not surprising to find an impurity accumulation. Furthermore we find that the bootstrap current is significantly affected by the value of the plasma effective charge. This could indicate that it is important to consider the complete ion composition when calculating the bootstrap current, rather than simply employing a pure plasma.

We note that the present analysis suggests that impurity accumulation could be a significant problem in a W7-X plasma. However we make two final remarks on the study. Firstly, in [6] it was shown that it can be important to take into account the variation of the electrostatic potential on a flux-surface, $\Phi_1 = \Phi - \langle\Phi\rangle$, since impurities are particularly affected due to their higher charge. This is not done in our analysis and would require implementing extra terms in the kinetic equation solved by SFINCS. Secondly, since we calculate the neoclassical impurity peaking the contribution from turbulence to the impurity transport is completely neglected. Turbulence in stellarators is an area where much remains to be explored [2]. It is thus difficult to tell if the turbulent impurity transport would contribute significantly or possibly even dominate over the collisional impurity transport. In tokamak plasmas the impurity peaking can in many cases be determined almost solely by the turbulent transport, since the collisional transport is typically much lower than the turbulent transport, in contrast to stellarators. The impurity peaking factors found from a gyrokinetic approach in tokamaks are much smaller than the peaking factors presented in this work, typically $a/L_{nz}^{0} \sim \mathcal{O}(1)$. It is therefore not unlikely that turbulence could cause a reduction of the large impurity peaking we find here. The neoclassical pinch velocity of the baseline case in our study is estimated to be $V \sim 0.2\, v_{Ti}\, (\rho_i/a)^2$ (corresponding to 0.2 in gyro-Bohm units), which is of the same order as a typical turbulent impurity pinch in tokamaks.

Acknowledgments
The authors are grateful to J Geiger for providing the W7-X equilibrium data, and to I Pusztai, T Fülöp, C D Beidler and P Helander for input on the work. This project has received funding from the European Union's Horizon 2020 research and innovation programme under grant agreement number 633053. The views and opinions expressed herein do not necessarily reflect those of the European Commission.

References

[1] Kikuchi M, Lackner K and Tran M Q 2012 *Fusion physics* (Vienna: International Atomic Energy Agency)
[2] Helander P et al 2012 *Plasma Phys. Control. Fusion* **54** 124009
[3] Helander P 2014 *Rep. Prog. Phys.* **77** 087001
[4] Braun S and Helander P 2010 *Phys. Plasmas* **17** 072514
[5] Braun S 2010 Effect of Impurities on Kinetic Transport Processes in Fusion Plasmas, PhD thesis Max-Planck-Institut für Plasmaphysik `http://ub-ed.ub.uni-greifswald.de/opus/volltexte/2010/888/`
[6] García-Regaña J M, Kleiber R, Beidler C D, Turkin Y, Maaßberg H and Helander P 2013 *Plasma Phys. Control. Fusion* **55** 074008
[7] Landreman M, Smith H M, Mollén A and Helander P 2014 *Phys. Plasmas* **21** 042503
[8] Turkin Y, Beidler C D, Maaßberg H, Murakami S, Tribaldos V and Wakasa A 2011 *Phys. Plasmas* **18** 022505

Journal of Physics: Conference Series **561** (2014) 012013 doi:10.1088/1742-6596/561/1/012013

Turbulent transport close to marginal instability: role of the source driving the system out of equilibrium

C Norscini[1], P Ghendrih[1], T Cartier-Michaud[1], G Dif-Pradalier[1], D Milelli[1], Y Sarazin[1], J Abiteboul[2], D Estève[1], X Garbet[1], V Grandgirard[1], G Latu[1]

[1] CEA, IRFM, F-13108 Saint-Paul-lez-Durance, France

[2] Max-Planck-Institut für Plasmaphysik, Boltzmannstr. 2, D-85748 Garching, Germany

E-mail: claudia.norscini@cea.fr

Abstract. When modeling plasma turbulence, two different means of driving the system out of equilibrium are considered. On one hand, flux driven (FD) approaches are based on the idea that no scale separation can be assumed in a turbulent system. On the other hand, gradient driven (GD) approaches rely on the idea that the back-reaction of fluctuations on the mean profiles is not a critical ingredient for turbulence self-organization and saturation. We find that FD and GD systems strongly differ in regimes close to marginal stability. The characteristic non linear upshift is recovered in GD simulations but no comparable behavior is possible in FD case. Discrepancy between the various analysis in terms of diffusion and pinch velocities and between the models is also discussed.

1. Introduction

Modeling plasma turbulent transport remains a challenging task and many issues that drive fusion performance are unresolved. The complexity of the so-called anomalous transport, in the light of the recent results with tungsten walls [1], is the cause of major uncertainties and risk regarding the coming high performance experiences whether D-T experiments on JET or ITER $Q = 10$ goal [2]. State of the art simulations with gyro-kinetic codes require long computational time to yield realistic results to be compared in detail to a growing wealth of experimental measurements.

Among the issues that are presently addressed are those related to the way the system is driven out of equilibrium and the associated uncertainty regarding the predictive capability of these codes. Two means of addressing this issue are implemented in the codes.

On the one hand, some simulations rely on scale separation, i.e. the slow evolution of the mean profiles can be addressed on a different footing than the fluctuations. Such an assumption lead to developing a series of so-called *local* codes where the mean profiles are considered constant and set by the user. A recent focal point of this activity is evaluating the range of predicted fluxes and density fluctuation levels when the prescribed gradients are changed within the

experimental error bars. Together with an effort towards completeness of the turbulence models this opens the way to quite a broad range of tunable parameters. The effort of first principle modeling then leads to a large farming effort to explore the range of possible parameters and determine the best fit to experiments [3,4]. A procedure that seems more in line with empirical modeling.
On the other hand, a series of experimental results have shown the limit of the local assumption [5–8]. In the framework of understanding and explaining such *non-local* aspect of plasma transport, global Flux-Driven models (i.e. with no scale separation assumption) have been developed during the last twenty years [9–14]. *Non-local* aspects have been identified, such as avalanche-like events [12, 15, 16] or jet-like patterns, named as '$E \times B$ staircases' [17]. Large scale predator-prey cycles are also observed in flux-driven simulations [18]. As a consequence, the standard local diffusion process ansatz appeared to fall short in describing the large scale transport [12,17,19,20].
Analyzing the difference between *local* and *global*, as well as between Gradient and Flux Driven approaches can lead to a better understanding of turbulent transport and help in improving the predictive capability of the modeling effort. It is interesting to note that the impact of different forcing on transport is addressed by a wider research community than fusion plasma physics [21].

In this paper, we present a quantitative comparison of the two forcing mechanisms and their impact on the regimes explored by the system [21]. We focus our work on the transition regime from stability to turbulence. As a versatile tool we use the TOKAM2D code, a minimum turbulent model initially developed to address SOL turbulence. In Section 2, we describe the equations solved by the TOKAM2D code [12] in the original flux driven version(FD). We then define the fixed gradient (GD) version and the general method applied to compare the two cases, Section 3. The results of the comparison are addressed in Section 4. In Section 5 are found the conclusion and a discussion of the ongoing work.

2. The TOKAM2D model

The TOKAM2D system only addresses particle transport with given constant temperatures so that the pressure nT is in fact proportional to the density. The model is that of interchange governed turbulence [22]. It solves the density balance equation and the charge balance equation $\vec{\nabla}\cdot\vec{j} = 0$, $\mathbf{j}$ being the total current. The turbulent transport is described by the three fluid drifts, i.e the electric drift v_e^{elec} ,the diamagnetic drift v^{dia} and the polarization drift v^{pol} [12,22]. The model is considered in the flute approximation, with no variation along the magnetic field lines. This allows one to reduce the model to two dimensions, the radial direction x, $x = (r - a)/\rho_s$ where a is the plasma minor radius,r the radius of the considered magnetic surface and ρ_s the hybrid Larmor radius, and an angle that describes the poloidal angle on the magnetic surface y, $y = a\theta/\rho_s$ where θ is the actual poloidal angle. The only terms that contribute when averaging over the parallel coordinate are the divergence of the parallel particle flux flowing to the limiters $\nabla_{\|}j_{\|}$, governed by the sheath physics, and the divergence of the diamagnetic current $\nabla_{\perp}j^{dia}$ that takes into account the curvature of the magnetic field. Their derivation is treated in detail in [12].

Time normalization in the code is in $1/\omega_i$, hence the inverse of the ion Larmor frequency. The angle coordinate is periodic so that the average of any particle flux on a given magnetic surface yields the radial component of the flux. In this model, field lines connect region with different properties so that parallel loss terms govern transport that tends to equilibrate these regions. Both parallel loss terms to end plates in a quasi flute approximation or parallel transport taking into account the ballooned cross field transport lead to an outward particle flux and a current exchange. The latter is particularly important since it bounds the inverse cascade that would otherwise take place.

2.1. flux driven version

In the flux driven approach we assume that the density and the potential profile can fluctuate not only at small but also at large scale, i.e. the fluctuations size can be comparable to the size of the box. Such an assumption implies that the driving of the system is a particle source that is localized radially. In this way the initial profile is null and it will build up through the source a density gradient that will drive transport in bursts in the box. The equations solved by the FD version of TOKAM2D are

$$\partial_t n + [\phi, n] = D_\perp \Delta_\perp n - \sigma_n \Gamma + S \tag{1}$$

$$\partial_t W + [\phi, W] + g\frac{\partial_y n}{n} = \nu_\perp \Delta_\perp W + \sigma_\phi J \tag{2}$$

where:

$$W = \nabla_\perp^2 \phi \tag{3}$$

and where the parallel loss currents in the SOL case are defined as:

$$\Gamma = n \exp(\Lambda - \phi) \tag{4}$$

$$J = \Big(1 - \exp(\Lambda - \phi)\Big) \tag{5}$$

The particle balance equation Eq.(1) includes the leading order particle flux due to the electric drift velocity, which takes the form of the Poisson bracket between the normalized electric potential ϕ and the density, together with a source term S, the parallel particle sink Γ, defined for the SOL case in (4), and a small transverse diffusion $D_\perp$. It is to be noted that while the source term is localized radially, the loss terms is distributed in the simulation region. The particle flux that drives the turbulence therefore decays as the radial distance from the source increases. In these expressions, Λ is the plasma potential with respect to the reference potential of the wall, it includes the plasma floating potential and the biasing potential if applied. The control parameter of the parallel particle loss term σ_n is proportional to the normalized saturation current. The charge balance equation Eq.(2) takes the form of the evolution equation of the vorticity W defined in (3), with an electric drift convection, the Poisson bracket term between the electric potential and the vorticity, together with a diffusive damping of the convective motion with diffusion coefficient $\nu_\perp$. The curvature charge separation is simplified to a constant interchange term g. This g-term is the standard gravity term when addressing buoyancy effects in a neutral fluid. A similar term should be included in the density equation, however this term is a lower order term than the electric drift convection term. Consequently it is only retained in the charge conservation equation because the latter only includes lower order terms for the cross-field current, i.e. the curvature and polarization currents. The parallel loss current J, defined for the SOL case by (5), closes the current loop with a restoring term that tends to compensate charge separation and is all the more efficient when the cross-section is large.

3. How to compare different forcing terms?

The definition of gradient driven forcing is often implicit and thus confusing. Generally, the gradient driven approach is a simplified version to model turbulent transport based on the assumption that the fluctuation scale is much smaller than the mean profile scale. This opens the way to scale separation, we can study only the fluctuating terms and consider a constant mean profile. This approach is quite handy when comparing numerical and experimental results. In this case the density profile measured in tokamaks can be supplied as input for the simulation. Furthermore, this approach reduces drastically the computational time: the system starts already from a steady state condition, nullifying the time interval needed to reach confinement in the FD simulations.

3.1. The local like approach

In the first and most conventional GD version the profile is defined by an input coefficient $-Ln$ that defines the ratio of the mean density and the gradient of the latter and such value is considered constant all over the simulation domain. Splitting the density and potential into the fluctuating part $\tilde{n} = n - < n >_y$ and $\tilde{\phi} = \phi - < \phi >_y$, and the mean part such that $\bar{\phi} =< \phi >_y$ and $\bar{n} =< n >_y$, we can then rewrite the equations of the TOKAM2D for the FD version, hence :

$$\partial_t \bar{n} + \left\langle \left[\tilde{\phi}, \tilde{n}\right] \right\rangle = D_\perp \Delta_\perp \bar{n} - \sigma_n \bar{\Gamma} \quad (6)$$

$$\partial_t \bar{W} + \left\langle \left[\tilde{\phi}, \tilde{W}\right] \right\rangle = \nu_\perp \Delta_\perp \bar{W} + \sigma_\phi \bar{J} \quad (7)$$

The equations solved by the code for the fluctuating density and potential will be respectively equal to eq.1-eq.7 and eq.2-eq.7:

$$\partial_t \tilde{n} + \left[\tilde{\phi}, \tilde{n}\right] - \left\langle \left[\tilde{\phi}, \tilde{n}\right] \right\rangle + \left[\tilde{\phi}, \bar{n}\right] + \left[\bar{\phi}, \tilde{n}\right] = D_\perp \Delta_\perp \tilde{n} - \sigma_n \tilde{n} + \sigma_n \bar{\Gamma} \tilde{\phi} \quad (8)$$

$$\partial_t \tilde{W} + \left[\tilde{\phi}, \tilde{W}\right] - \left\langle \left[\tilde{\phi}, \tilde{W}\right] \right\rangle + \left[\tilde{\phi}, \bar{W}\right] + \left[\bar{\phi}, \tilde{W}\right] = -g \partial_y \frac{\tilde{n}}{\bar{n}} + \nu_\perp \Delta_\perp \tilde{W} + \sigma_\phi \tilde{\phi} \frac{\bar{\Gamma}}{\bar{n}}. \quad (9)$$

We then further assume $< \Phi >_y$ to be constant in the radial direction and equal to $\bar{\Phi} = \Lambda$, so that $\bar{\Gamma} = \bar{n}$, and $< n >_y$ to be constant in time with e-folding length $1/Ln$ in the radial direction. A more accurate presentation would introduce two evolution time scales and assume that the mean density evolves slowly compared to the fluctuations. Implicitly, one then assumes that the two transport terms $\left\langle \left[\tilde{\phi}, \tilde{n}\right] \right\rangle$ and $\left\langle \left[\tilde{\phi}, \tilde{W}\right] \right\rangle$ are small (they contribute on the long time scale evolution of the mean values). Consistently, one can consider to remove them in the GD evolution equations for the fluctuations, however, at the cost of losing a symmetry in the fluctuation evolution equations.

$$\partial_t \tilde{n} + \left[\tilde{\phi}, \tilde{n}\right] + \left[\tilde{\phi}, \bar{n}\right] = D_\perp \Delta_\perp \tilde{n} - \sigma_n \tilde{n} + \sigma_n \tilde{\phi} \quad (10)$$

$$\partial_t \tilde{W} + \left[\tilde{\phi}, \tilde{W}\right] = -g \partial_y \tilde{n} + \nu_\perp \Delta_\perp \tilde{W} + \sigma_\phi \tilde{\phi}. \quad (11)$$

In the latter set of equations, the field $\tilde{n}$ stands in fact for $\tilde{n}/\bar{n}$ hence the relative fluctuations.

3.2. A general comparison method

In order to define an appropriate method to compare the two approaches we decide to follow the same process applied for the comparison between experimental and numerical results. Namely, one takes the steady state density profile from the experiment as input for the simulation and then compare the computed and experimental flux. For our purpose, the flux driven simulation is compared to the fixed gradient one along the same line. We run long a FD simulation in order to define a steady state condition, then we calculate the mean density profile averaged in time and poloidal direction to smooth out the fluctuations. Such a profile $\bar{n}(r)$ or its gradient, normally defined as $1/Ln = -\partial_x \bar{n}/\bar{n}$, is then inserted as parameter for GD simulation, in which the source is set at zero.

The sketch shown in fig.(1) summarizes this comparison approach.

4. Different transport regimes at large scale

In order to decide which parameters to set for the comparison, we analyze the dispersion relation of the interchange instability. We want to be able to bring the system from a stable condition to a fully turbulent regime.

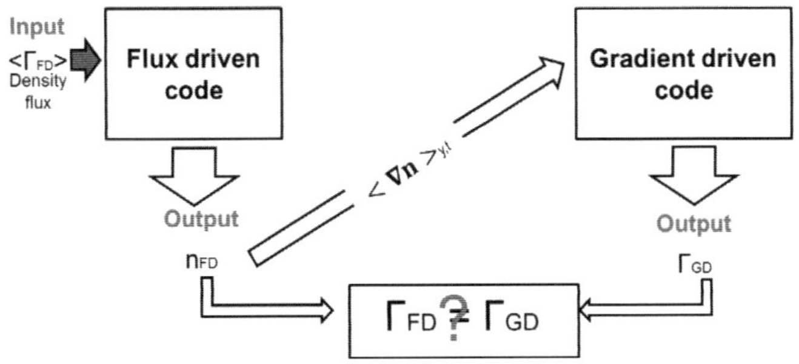

Figure 1. Sketch of the procedure used to compare the two different mechanisms that are used to drive the system out of equilibrium

Assuming that the fluctuations are small for both the density and potential field, we can then investigate the stability of the chosen reference state $\bar{n}$ and $\bar{\phi}$ when considering the linearized set of equations (10, 11).

$$\tilde{n}_k = \hat{n}_{kx,ky} \exp(i(k_x x + k_y y)) \exp(\Omega_k t) + cc \tag{12}$$
$$\tilde{\phi}_k = \hat{\phi}_{kx,ky} \exp(i(k_x x + k_y y)) \exp(\Omega_k t) + cc \tag{13}$$

where 'cc' means 'complex conjugates' and γ_k represents the real part of Ω_k. The perturbation will be excited if γ_k is positive, i.e. the system is unstable. Substituting eq.12 and eq.13 in eq.10 and eq.11 defined for the GD approach, we recover the following dispersion relation:

$$\Big(\Omega_k + Dk^2 + \sigma_n\Big)\hat{n} + \Big(i\frac{k_y}{L_n} - \sigma_n\Big)\hat{\phi} = 0 \tag{14}$$
$$\Big(-ig\frac{k_y}{k^2}\Big)\hat{n} + \Big(\Omega_k + \nu k^2 + \frac{\sigma_\phi}{k^2}\Big)\hat{\phi} = 0 \tag{15}$$

A non trivial solution is possible when Ω_k satisfies the second order equation:

$$\Omega_k^2 + B_k\Omega_k + C_k = 0 \tag{16}$$

where $B_k = (D+\nu)k^2+\sigma_n+\sigma_\phi/k^2$ and $C_k = (Dk^2+\sigma_n)(\nu k^2+\sigma_\phi/k^2)-(gk_y^2)/(L_n k^2)-i\sigma_n g\, k_y/k^2$. Finally if we consider no dissipation or loss term (D=0, ν=0, σ_n=0, σ_ϕ=0), hence $B_k = 0$ and $C_k = (gk_y^2)/(L_n k^2)$, one can state that the growth rate of the unstable modes is proportional to $\sqrt{g/Ln}$. Specifically, decreasing g or increasing Ln, we reduce the linear growth rate and consequently the characteristic time of the turbulence.

Table 1. parameter used in the following simulation both for flux and gradient driven case

D	ν	σ_n	σ_ϕ	N_x	N_y	L_x	L_y	t_{diag}
10^{-2}	10^{-2}	6.1 10^{-5}	6.1 10^{-5}	128	128	128	128	32

The parameters used for the following simulations are specified in table 1. For this set of parameters the maximum of the growth rate is at $ky_{max} = 8/128$ and $kx_{max} = 0$ and the critical value corresponds to $(g/Ln)_{crit}$ equal to 0.4 10^{-5}. If the ratio of the two quantities is smaller than $(g/Ln)_{crit}$, the growth rate is negative for all modes and the system is linearly stable. Conversely, if we set the system just above the critical value, only a few modes are linearly unstable, namely around ky_{max} and kx_{max}.

For the flux driven simulation, we cannot control the two terms, the gradient being in fact self consistently defined. For the GD case, both parameters can be set by the user and therefore we can study how and when the system departs from the linear one. We start analyzing the FD case in order to define which appropriate value or range of values of $1/Ln$ has to be used for the GD simulations.

A scan on the g-term is performed to analyze the response of the FD system close to criticality. The number of points of the simulation grid, N_x and N_y, have also been varied in order to test the sensitivity of the code to the grid-size. The increase of N_x and N_y only impacts the high frequency part of the spectrum. Since the transport analysis is based on time and space averages, the simulation output is readily observed not to exhibit relevant differences when increasing the grid-size.

4.1. FD CASE: self-organization close to marginal stability

First, we define an effective diffusion D_{eff} in order to quantify the turbulent transport [10,23]. The effective diffusivity is used to quantify the macroscopic transport.

$$\frac{<\Gamma_{turb}>_{t,y}}{\bar{n}} = -D_{eff} < \frac{\partial_x \bar{n}}{\bar{n}} >_{t,y} \tag{17}$$

A coarse graining approach with flux surface average and long time averaging is required to determine D_{eff} with a smooth behavior as shown in ref. [19]. Should one require that large scale transport follows a Fick's law, namely that D_{eff} is a scalar with a single value would require further coarse graining in the radial direction [19]. Another possible transport law can be considered by combining a diffusion D_V and a pinch velocity V.

$$\frac{<\Gamma_{turb}>_{t,y}}{\bar{n}} = -D_V < \frac{\partial_x \bar{n}}{\bar{n}} >_{t,y} + V \tag{18}$$

Compared to the expression of D_{eff} which is simply defined as the ratio $<\Gamma> Ln/<n>$, the latter transport law is based on a linear approximation of the data such that the slope is governed by D_V and the value at $1/Ln \to 0$ determines V. It is possible to generalize the linear fitting process by identifying within the data several regions where the linear fit holds, hence introducing a non-linear dependence of the transport law on the gradient. From fig.(2) one defines three different regions:

- a flat top region, region I, localized immediately after the source, in this region the flux is maximum and nearly constant,
- region II that connects the first and third region and where the particle flux decreases,
- region III, where the density gradient is positive and turbulence stabilized.

One can then identify the incremental transport coefficients, in region I from $x1 = 15$ to $x2 = 30$, width $d_I = 15$, $D_I \approx 0.076$ and is associated to an outward pinch velocity of $V_I \approx 0.008$, while in region II from $x2$ to $x3 = 80$, width $d_{II} = 50$, $D_{II} \approx 0.44$ and the outward pinch velocity is $V_{II} \approx 0.001$. Here the radial units are units of ρ_s, the pinch velocity in units of c_s, and correspond therefore to Mach numbers, and the diffusion coefficients are normalized to the Bohm diffusion coefficient, i.e. $\rho_s c_s$. For this data, the value of g-term is equal to 5.7 10^{-4}, a standard value for TOKAM2D simulations determined for Tore Supra parameters, and the value of $1/Ln$

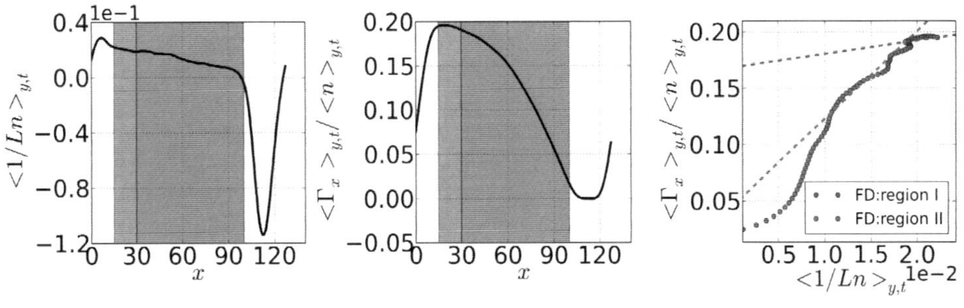

Figure 2. (a) Time and y-averaged density gradient $1/Ln$. (b) Time and y-averaged turbulent flux divided by the Time and y-averaged density $\Gamma/\bar{n}$,(c)$\Gamma/\bar{n}$ as a function of$1/Ln$. The turbulent flux Γ is normalized by $N_x/(2\pi)$.

varies from $-1.15\ 10^{-2}$ to $2.9\ 10^{-2}$ as indicated on fig.(2). The ratio $D/(dV)$ allows one to determine the dominant transport mechanism (since both act in the same direction). Hence, in region I, $D_I/(d_I V_I) \approx 0.625$ while in region II $D_{II}/(d_{II} V_{II}) \approx 8.8$; in first region the transport is governed by the convection while in the second one the diffusion dominates the transport. On fig.(3) a scan of the g-term allows one to see its impact on the radial transport coarse grained in time and on a flux surface. For all values of g, the plot of $< \Gamma > / < n >$ versus g/Ln is characterized by a sub-threshold transport governed by turbulent spreading into the stable region at weak density gradient, a sharp rise of transport in the vicinity of the threshold (indicative of profile stiffness at lowest values of g) followed by a less stiff regime. One can also observe a specific aspect of the effect of the g-term on the transport properties. In particular the stiffness of the turbulent response seems to decrease as g is increased (hence at smaller ρ_*). For g equal or smaller than $2\ 10^{-4}$, the flux and the density gradient only explore the vicinity of the threshold. Conversely, for the larger values of g, the flux dependence on $1/Ln$ recalls more conventional diffusion-convection regimes.

Let us analyze the transport regime for $g = 2\ 10^{-4}$, hence close to criticality where one can expect that self-organization dominates the transport response. Furthermore this regime is likely more relevant for ITER size devices. The characteristic time of turbulence is proportional to $(g/Ln - g/Ln_*)^{1/2}$, where g/Ln_* is the threshold value. Decreasing g, one thus expects a slower self-organization time of the turbulent patterns. This is readily seen on fig.(4) where is shown the contour plot of the evolution of the radial profile of flux surface averaged electric potential $\Phi_{zf}(r,t)$. One can observe that Φ_{zf} tends to develop long lasting structures, akin to the $E \times B$ staircases. The shearing effect associated to the zonal flow structure then acts as micro barriers, dumping the turbulence locally. Such structures seem to disappear for larger values of g. It appears that these micro transport barriers are mostly localized in the region I. One can notice that this region tends to shrink radially as g is increased.

4.2. Dimits shift region in gradient driven simulations

In the GD case we assume that the gradient is constant in the all the domain. Unlike the flux driven simulations that are characterized by a curve in the ($1/Ln$, $< \Gamma > / < n >$) plane, the gradient driven simulation yield a single point. In order to compare with the FD case we must therefore perform a scan of Ln for different values of g-term. This provides a means to compare the response of the system to the way it is driven out of equilibrium.

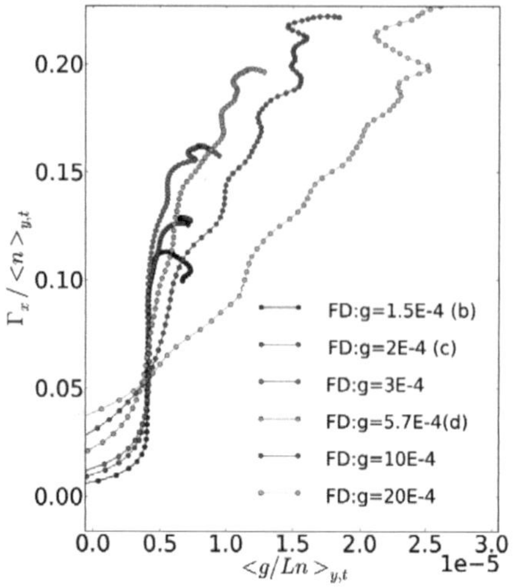

Figure 3. $<\Gamma>/<n>$ as a function of g/Ln, for g ranging from 1.5 10^{-4} to 10^{-4}. The turbulent flux Γ is divided by $N_x/(2\pi)$.

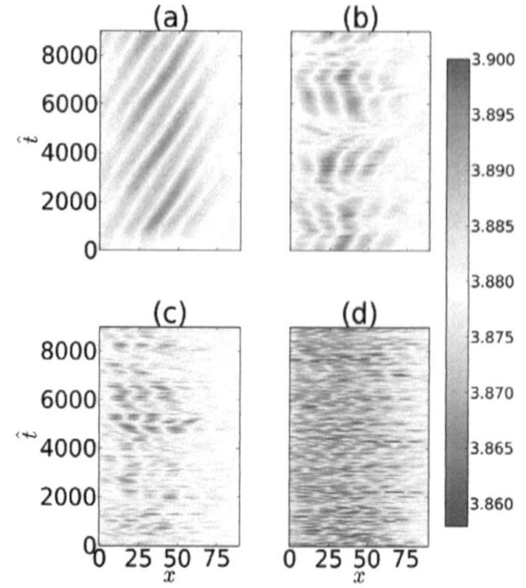

Figure 4. Contour plot of the y-averaged electric potential for different values of g: (a) 1.2 10^{-4}, (b) 1.5 10^{-4}, (c) 2 10^{-4}, and (d) 5.7 10^{-4}

Let us first compare the case with $g = 2.2\ 10^{-4}$ and we scan $1/Ln$ from $L^* = 0.02$ (corresponding to the critical value) to a maximum set at 0.05. In fig.(5), the two curves of $<\Gamma>/<n>$ in terms of the driving term g/Ln are displayed. One finds an unexpected feature of the local (fixed gradient) approach. Just above the threshold, in the region referred to as the Dimits region [24], the system is linearly unstable but the flux is not governed by the conventional Fick's law since it first decays as the drive is increased. The system only recovers the standard behavior with increasing $1/Ln$ when the latter exceeds 2.5 10^{-4}. This new onset value plays the role of a secondary non-linear threshold.

The decrease of the flux can be explained by coupling the interchange instability and the Kelvin Helmholtz (KH) instability. The interchange instability linearly grows with g/Ln and is maximum for small values of kx. It thus grows into a nonlinear state of streamer like modes. These become KH unstable which leads to the build-up of Zonal flows. Such a mechanism leads to a predator-prey behavior. Acting as weak transport barriers, these flows reduce the perturbation amplitude and stabilize the non-linear mode. Further increasing $1/Ln$ also increases the number of linearly unstable modes until the interchange instability becomes dominant and the system is fully turbulent.

Such upshift is not present in the flux driven simulation. We thus find a different behavior of the turbulent flux predicted by the two ways the system is driven out of equilibrium in the vicinity of the threshold, the so-called Dimits shift region. The systems approach to the stable region is markedly different. For the FD case the immediate positive response of the flux for $1/Ln - 1/Ln_* > 0$ shows that the Dimits shift cannot be recovered. We can state that the observed non linear upshift is characteristic of local simulations. In the FD case we cannot constrain the gradient profile and the system strongly differs an ideal predator-prey behavior although some features of the latter reduced model still hold.

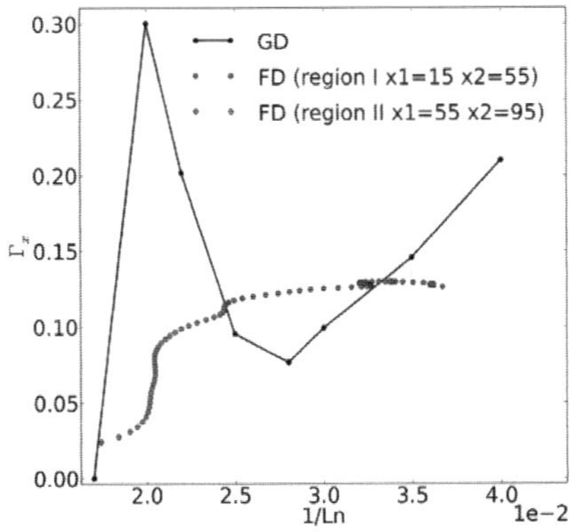

Figure 5. Γ as a function of $1/Ln$ for $g = 2\ 10^{-4}$ in FD and GD approaches. The values of Γ are divided by $N_x/(\bar{n}2\pi)$ in the FD approach while these values are divided by $N_x/(2\pi)$ in the GD case.

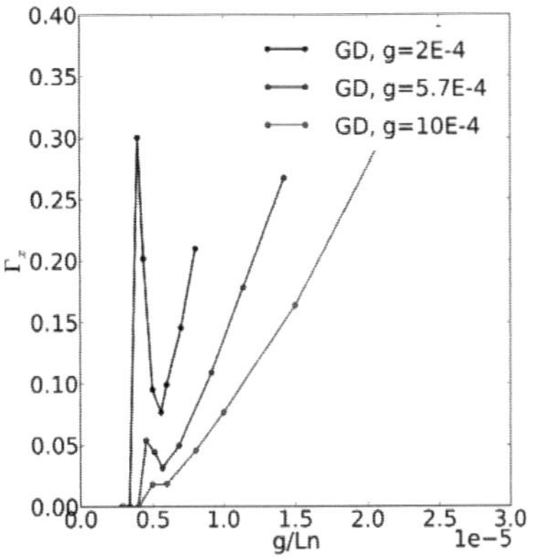

Figure 6. $\Gamma/\bar{n}$ as a of function of g/Ln, for $g = [2\ 10^{-4}, 10^{-3}]$ in the GD case. The values of Γ are divided by $N_x/(2\pi)$.

On the one hand, one can notice on fig.(5) that in region I of flux driven simulation the effective diffusivity defined as $- < \Gamma_I >_{y,t,dx} / < Ln_I >_{y,t,x}$, as previously the subscript I stands for region I, is comparable to the effective diffusivity of the local case for Ln equal to $< Ln_I >_{y,t,d_I}$. On the other hand, the flux calculated in GD simulations is far more sensitive to small variations of $1/Ln$ than in FD ones. From fig.(6) we can see that the effective diffusivity grows with the curvature term, in agreement with the FD case, but, in the local case, the vicinity of the threshold is still subject to a Dimits shift like effect, although its impact is less stringent on large scale transport as g is increased. When we increase the g-term we do not recover a strong variation of the flux as in the case presented in fig.(5), but we can see still identify a region with a sudden variation on the incremental diffusivity.

5. Discussion and conclusion

We have analyzed in this paper the impact of various forcing on turbulent transport. Particularly we have focused our attention on understanding and quantifying how two different means of forcing the system out of equilibrium modify transport properties in the vicinity of marginal criticality. A characteristic aspect of the Gradient Driven(GD) case is the presence of a non linear upshift of the turbulence threshold. A comparable aspect cannot be found for the Flux Driven (FD) case. Such results can be explained by the fact that in the Gradient Driven case it is possible to constrain the transport in regimes that are not explored self consistently by the system in the flux driven case. In strongly turbulent, hence well above the Dimits shift [24] for GD and at large interchange drive, it is possible to find comparable values of the effective diffusivity for the two different cases. This is all the more striking that the patterns that develop close to the source in the FD case are reminiscent of staircases reported in FD gyro-kinetic simulations [17]. The latter are long living structures that behave like transport barriers. The observation of similar effective diffusivity in a region with self-regulation of transport as in the local GD case is thus rather surprising. However, when considering linear approximations of the particle flux in terms of the density gradient length -transport modeled in terms of a diffusion coefficient complete by a pinch term- a clear departure is found as reported previously [19]. Incremental transport analysis further underlines the latter feature.

These results are being further investigated, in particular regarding the interplay between the various regions and patterns that develop in the radial direction. Key issues are: a ρ_* analysis in terms of a scan of the box size, a more detailed analysis of the source region and its neighborhood to understand its specific effect and the analysis of the particular role of the stable region via spreading or coupling of the larger modes to stable regions.

All these issues are relevant when analyzing the gyro-kinetic simulations which is the ultimate goal of this work. In particular, quantifying the difference between flux and gradient driven gyrokinetic simulations [25, 26] is needed. Various source term models have been implemented in TOKAM2D that allow one to drive the FD systems with prescribed profiles that depart in particular from a constant gradient length. As with Krook source terms, the effect of a local restoring force as well as larger scale patterns of the restoring forces must be investigated and their effect quantified. The case of prescribed thermal bath boundary conditions completed by a Krook like restoring force which is the standard drive in gyrokinetic codes, is an example of such a way of driving the system out of equilibrium based on two characteristic scales in time and space. Such a source term is now implemented in the GYSELA code allowing to step the present analysis to the case of gyrokinetic turbulence.

Acknowledgments
The authors want to thank Ch. Passeron, D. Moulton and P. Hill and acknowledge fruitful interactions at the Festival de Théorie, Aix en Provence, 2013.

References

[1] M. Beurskens, L. Frassinetti, C. Challis, C. Giroud, S. Saarelma, B. Alper, C. Angioni, P. Bilkova, C. Bourdelle, S. Brezinsek, P. Buratti, G. Calabro, T. Eich, J. Flanagan, E. Giovannozzi, M. Groth, J. Hobirk, E. Joffrin, M. Leyland, P. Lomas, E. de la Luna, M. Kempenaars, G. Maddison, C. Maggi, P. Mantica, M. Maslov, G. Matthews, M.-L. Mayoral, R. Neu, I. Nunes, T. Osborne, F. Rimini, R. Scannell, E. Solano, P. Snyder, I. Voitsekhovitch, P. de Vries, and J.-E. Contributors, "Global and pedestal confinement in jet with a be/w metallic wall," *Nuclear Fusion*, vol. 54, no. 4, p. 043001, 2014.

[2] M. Shimada, D. Campbell, V. Mukhovatov, M. Fujiwara, N. Kirneva, K. Lackner, M. Nagami, V. Pustovitov, N. Uckan, J. Wesley, N. Asakura, A. Costley, A. Donn, E. Doyle, A. Fasoli, C. Gormezano, Y. Gribov, O. Gruber, T. Hender, W. Houlberg, S. Ide, Y. Kamada, A. Leonard, B. Lipschultz, A. Loarte, K. Miyamoto, V. Mukhovatov, T. Osborne, A. Polevoi, and A. Sips, "Chapter 1: Overview and summary," *Nuclear Fusion*, vol. 47, no. 6, p. S1, 2007.

[3] N. T. Howard, A. E. White, M. Greenwald, M. L. Reinke, J. Walk, C. Holland, J. Candy, and T. Grler, "Investigation of the transport shortfall in alcator c-mod l-mode plasmas," *Physics of Plasmas (1994-present)*, vol. 20, no. 3, pp. –, 2013.

[4] D. Told, F. Jenko, T. Grler, F. J. Casson, E. Fable, and A. U. Team, "Characterizing turbulent transport in asdex upgrade l-mode plasmas via nonlinear gyrokinetic simulations," *Physics of Plasmas (1994-present)*, vol. 20, no. 12, pp. –, 2013.

[5] TFR Group, "Pellet injection experiments on the tfr tokamak," *Nuclear Fusion*, vol. 27, no. 12, p. 1975, 1987.

[6] K. W. Gentle, W. L. Rowan, R. V. Bravenec, G. Cima, T. P. Crowley, H. Gasquet, G. A. Hallock, J. Heard, A. Ouroua, P. E. Phillips, D. W. Ross, P. M. Schoch, and C. Watts, "Strong nonlocal effects in a tokamak perturbative transport experiment," *Phys. Rev. Lett.*, vol. 74, pp. 3620–3623, May 1995.

[7] X. L. Zou, A. Géraud, P. Gomez, M. Mattioli, J. L. Sgui, F. Clairet, C. D. Michelis, P. Devynck, T. D. de Wit, M. Erba, C. Fenzi, X. Garbet, C. Gil, P. Hennequin, F. Imbeaux, E. Joffrin, G. Leclert, A. L. Pecquet, Y. Peysson, R. Sabot, and T. F. Seak, "Edge cooling experiments and non-local transport phenomena in tore supra," *Plasma Physics and Controlled Fusion*, vol. 42, no. 10, p. 1067, 2000.

[8] S. Inagaki, T. Tokuzawa, N. Tamura, S.-I. Itoh, T. Kobayashi, K. Ida, T. Shimozuma, S. Kubo, K. Tanaka, T. Ido, A. Shimizu, H. Tsuchiya, N. Kasuya, Y. Nagayama, K. Kawahata, S. Sudo, H. Yamada, A. Fujisawa,

K. Itoh, and the LHD Experiment Group, "How is turbulence intensity determined by macroscopic variables in a toroidal plasma?," *Nuclear Fusion*, vol. 53, no. 11, p. 113006, 2013.

[9] P. H. Diamond and T. S. Hahm, "On the dynamics of turbulent transport near marginal stability," *Physics of Plasmas (1994-present)*, vol. 2, no. 10, pp. 3640–3649, 1995.

[10] B. A. Carreras, C. Hidalgo, E. Sánchez, M. A. Pedrosa, R. Balbín, I. García-Cortés, B. van Milligen, D. E. Newman, and V. E. Lynch, "Fluctuation-induced flux at the plasma edge in toroidal devices," *Physics of Plasmas (1994-present)*, vol. 3, no. 7, pp. 2664–2672, 1996.

[11] X. Garbet and R. E. Waltz, "Heat flux driven ion turbulence," *Physics of Plasmas (1994-present)*, vol. 5, no. 8, pp. 2836–2845, 1998.

[12] Y. Sarazin and P. Ghendrih, "Intermittent particle transport in two-dimensional edge turbulence," *Physics of Plasmas (1994-present)*, vol. 5, no. 12, pp. 4214–4228, 1998.

[13] P. Ghendrih, Y. Sarazin, G. Attuel, S. Benkadda, P. Beyer, G. Falchetto, C. Figarella, X. Garbet, V. Grandgirard, and M. Ottaviani, "Theoretical analysis of the influence of external biasing on long range turbulent transport in the scrape-off layer," *Nuclear Fusion*, vol. 43, no. 10, p. 1013, 2003.

[14] S. Y, V. Grandgirard, E. Fleurence, X. Garbet, P. Ghendrih, P. Bertrand, and G. Depret, "Kinetic features of interchange turbulence," *Plasma Physics and Controlled Fusion*, vol. 47, no. 10, p. 1817, 2005.

[15] Y. Sarazin, V. Grandgirard, J. Abiteboul, S. Allfrey, X. Garbet, P. Ghendrih, G. Latu, A. Strugarek, and G. Dif-Pradalier, "Large scale dynamics in flux driven gyrokinetic turbulence," *Nuclear Fusion*, vol. 50, no. 5, p. 054004, 2010.

[16] S. Jolliet and Y. Idomura, "Plasma size scaling of avalanche-like heat transport in tokamaks," *Nuclear Fusion*, vol. 52, no. 2, p. 023026, 2012.

[17] G. Dif-Pradalier, P. H. Diamond, V. Grandgirard, Y. Sarazin, J. Abiteboul, X. Garbet, P. Ghendrih, A. Strugarek, S. Ku, and C. S. Chang, "On the validity of the local diffusive paradigm in turbulent plasma transport," *Phys. Rev. E*, vol. 82, p. 025401, Aug 2010.

[18] G. Darmet, P. Ghendrih, Y. Sarazin, X. Garbet, and V. Grandgirard, "Intermittency in flux driven kinetic simulations of trapped ion turbulence," *Communications in Nonlinear Science and Numerical Simulation*, vol. 13, no. 1, pp. 53 – 58, 2008. Vlasovia 2006: The Second International Workshop on the Theory and Applications of the Vlasov Equation.

[19] P. Ghendrih, C. Norscini, F. Hasenbeck, G. Dif-Pradalier, J. Abiteboul, T. Cartier-Michaud, X. Garbet, V. Grandgirard, Y. Marandet, Y. Sarazin, P. Tamain, and D. Zarzoso, "Thermodynamical and microscopic properties of turbulent transport in the edge plasma," *Journal of Physics: Conference Series*, vol. 401, no. 1, p. 012007, 2012.

[20] V. Naulin, "Turbulent transport and the plasma edge," *Journal of Nuclear Materials*, vol. 363-365, pp. 24 – 31, 2007. Plasma-Surface Interactions-17.

[21] B. Saint-Michel, B. Dubrulle, L. Marié, F. Ravelet, and F. Daviaud, "Evidence for forcing-dependent steady states in a turbulent swirling flow," *Phys. Rev. Lett.*, vol. 111, p. 234502, Dec 2013.

[22] X. Garbet, L. Laurent, J.-P. Roubin, and A. Samain, "A model for the turbulence in the scrape-off layer of tokamaks," *Nuclear Fusion*, vol. 31, no. 5, p. 967, 1991.

[23] C. C. Petty and T. C. Luce, "Scaling of heat transport with collisionality," *Physics of Plasmas (1994-present)*, vol. 6, no. 3, pp. 909–921, 1999.

[24] A. M. Dimits, G. Bateman, M. A. Beer, B. I. Cohen, W. Dorland, G. W. Hammett, C. Kim, J. E. Kinsey, M. Kotschenreuther, A. H. Kritz, L. L. Lao, J. Mandrekas, W. M. Nevins, S. E. Parker, A. J. Redd, D. E. Shumaker, R. Sydora, and J. Weiland, "Comparisons and physics basis of tokamak transport models and turbulence simulations," *Physics of Plasmas (1994-present)*, vol. 7, no. 3, pp. 969–983, 2000.

[25] T. Görler, X. Lapillonne, S. Brunner, T. Dannert, F. Jenko, S. K. Aghdam, P. Marcus, B. F. McMillan, F. Merz, O. Sauter, D. Told, and L. Villard, "Flux- and gradient-driven global gyrokinetic simulation of tokamak turbulencea)," *Physics of Plasmas (1994-present)*, vol. 18, no. 5, pp. –, 2011.

[26] M. Nakata and Y. Idomura, "Plasma size and collisionality scaling of ion-temperature-gradient-driven turbulence," *Nuclear Fusion*, vol. 53, no. 11, p. 113039, 2013.

Grid-free tree-code simulations of the plasma-material interaction region

C Salmagne[1,2], D Reiter[2] and P Gibbon[3,]

[1] German Research School for Simulation Sciences GmbH, Jülich, Germany
[2] Institute of Energy and Climate Research, IEK-4, Forschungszentrum Jülich, Jülich, Germany
[3] Institute for Advanced Simulation, Jülich Supertcomputing Centre, Forschungszentrum Jülich, Jülich, Germany

E-mail: c.salmagne@fz-juelich.de

Abstract. A fully kinetic grid-free model based on a Barnes-Hut tree code is used to self-consistently simulate a collisionless plasma bounded by two floating walls. The workhorse for simulating such plasma wall transition layers is currently the PIC method. However, the present grid-free formulation provides a powerful independent tool to test it [1] and to possibly extend particle simulations towards collisional regimes in a more internally consistent way. Here, we use the grid-free massively parallel Barnes-Hut tree-code PEPC - a well established tool for simulations of Laser-plasmas and astrophysical applications - to develop a 3D ab initio plasma target interaction model. With our approach an electrostatic sheath naturally builds up within the first couple of Debye lengths close to the wall rather than being imposed as a prescribed boundary condition.

We verified the code using analytic results [2] as well as 1D PIC simulations [3]. The model was then used to investigate the influence of inclined magnetic fields on the plasma material interface. We used the code to study the correlation between the magnetic field angle and the angular distribution of incident particles.

1. Introduction

A magnetically confined plasma in a fusion device interacts with the surrounding material at the first wall as well as in the divertor region. The degradation of the plasma facing components is significantly influenced by the properties of the plasma particles close to the walls. Erosion rates, for example, depend non-linearly on the energy of bombarding particles so that a detailed knowledge of particle properties is essential.

If a plasma is in contact with a surface, the plasma particles impinge upon this surface and charge it up electrically. Due to the higher mobility, electrons hit the surface first and charge it negatively. Thus, the potential at the wall Φ_w is negative relative to the plasma potential in the center $\Phi_0 = \Phi(x = 0)$. The potential drop and the associated electric field slow down electrons and decrease the electron flux out of the plasma on the one hand and accelerate ions on the other. If the surface is floating, the electron and ion flux is balanced, resulting in zero net current in equilibrium. Due to Debye shielding, the largest part of the potential drop occurs over the first couple of Debye lengths λ_D close to the wall. This area is called *electrostatic sheath*, *collector sheath* or sometimes just *sheath*. The large electrical fields in the sheath have a strong influence on the properties of particles hitting the adjacent wall.

The codes mainly used in plasma-edge-modeling – large scale fluid models – lack the kinetic physics and the resolution necessary to handle the sheath area. Therefore, macroscopic boundary conditions at the walls, or more specifically at the sheath entrance, have to be prescribed for these codes. However, for many realistic configurations, commonly used simple 1D Bohm sheath conditions are not valid. By contrast, a kinetic model of the plasma wall interaction region can describe the sheath region in microscopic detail and with the necessary spatial and temporal resolution. A fully kinetic simulation can possibly provide correct boundary conditions for fluid or gyro-kinetic codes operating on larger spatio-temporal scales.

In kinetic particle simulations, an ensemble of simulation particles is investigated. The particle trajectories are advanced according to the self consistent electric (or more generally electromagnetic) field. In order to compute the electric field for one particle, the contribution to this field from all other particles is needed. Thus, the computation of the self-consistent field requires $\mathcal{O}(n^2)$ operations. There are several grid-free as well as grid-based methods used for particle simulations whose common purpose is the reduction of the complexity of the force computation to $\mathcal{O}(n\log(n))$ or even further. The Barnes-Hut tree code [4] is one well-established grid-free scheme, the Fast Multipole Method (FMM) [5] is another. On the grid-based side, Particle-In-Cell (PIC) is the most widely used.

The simulation model for the plasma-wall transition area presented here is based on the grid-free Barnes-Hut tree-code PEPC (*P*retty *E*fficient *P*arallel *C*oulomb solver)[6]. PEPC has been successfully used for simulations of a variety of systems such as laser-plasma-interactions or gravitational problems [7, 8, 9] and can efficiently make use of modern many processor super computers [7, 10].

The paper is organized as follows. In section 2, the code and the simulated system are described. In section 3 results of simulations with and without an oblique magnetic field are presented. For latter configuration, analytic [2] and numerical [3] results are known and we use those to verify our model. Results for systems with a constant external magnetic field with varied field inclination are also shown. Finally, conclusions are presented in section 4.

2. Model of the plasma-wall transition region

The sketch of the simulated plasma-wall transition region and a typical plasma potential without an external magnetic field is given in Fig. 1. The system is filled with a plasma over the range $-L \le x \le L$. A volumetric source emits particles in the source region ($-L_s \le x \le -L_c$ and $L_c \le x \le L_s$). In between the two parts of the source region ($-L_c < x < L_c$), there is a collisional, source-free plasma where particles may undergo collisions. The region $x > L_s$ and $x < -L_s$ is also source-free. According to Emmert et al [2] the plasma potential is constant for $L_s < x < L_{se}$ and $-L_{se} < x < -L_s$. At $x \ge L_{se}$ and $x \le -L_{se}$, we find the electrostatic sheath. It has a width of several Debye lengths. Within the sheath, the strongest electric fields are present. The main part of the total potential drop Φ_w is localized here.

Most one-dimensional analytic models for the collisionless plasma-wall transition region are based on a kinetic description of the ions and assume that electrons follow a Boltzmann relation (see e.g. [2, 11, 12]). For the desired steady-state solutions the kinetic equation for ions reduces to

$$\frac{\partial}{\partial x}[f(x,E)v(x,E)] = S(x,E), \tag{1}$$

with x the position, $E = 1/2mv^2 + q\Phi(x)$ the total Energy, v the ion velocity, $f(x,E)$ the ion distribution function and $S(x,E)$ the ion source function. The total ion Energy E is taken to be a constant of motion. The density is obtained by integrating $S(x,E)/v(x,E)$ over the (x,E) phase space. Inserting the particle densities in the Poisson-equation gives an integro-differential equation for the potential that, in general, has to be solved numerically.

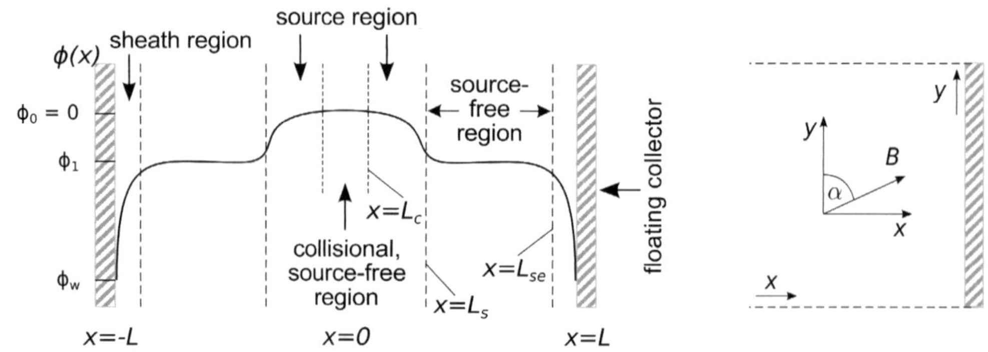

Figure 1. Schematic drawing of the simulation system along the x-axis and definition of the magnetic field angle α in the $xy-$plane

The used source velocity distribution along the magnetic field is the same as used by Emmert et al. [2] and can be written as

$$S(x, v_{\|}) = S_0 h(x) \left(\frac{m|v_{\|}|}{kT_0} \right) \exp\left(\frac{-mv_{\|}^2}{2kT_0} \right). \tag{2}$$

S_0 is the source strength, $h(x)$ is a given spatial source profile, which is a constant inside the source region in our case and T_0 is the ion source temperature. This type of source function is applicable, if the plasma has a non-drifting Maxwellian velocity distribution far from the wall[12].

The collisional, source-free region in the system center plays a major role in the present particle simulations. In a collisionless system there is no mechanism to refill the high energetic tail of the electron distribution function that is depopulated due to the loss of these electrons at the walls. A mechanism to achieve the necessary repopulation has to be added to the simulation model. In our model, the collisional region is condensed to a plane at $x = 0$. Electrons that cross this plane are rethermalized. That is, their velocity in x-direction is resampled according to (2).

To simulate the described system, a specially adapted version of the PEPC code is used. Starting with an empty system, a constant number of electrons and ions per time-step is inserted. Particles hitting the walls are removed from the system and their charge is added to the total charge of the wall. When a steady-state is reached, the particle influx from the source is balanced by an equal particle outflux towards the walls.

For all simulations, the ion to electron mass ratio is set to $\mu = m_i/m_e = 100$ and temperature ratio is $\tau = T_{i0}/T_{e0} = 1$. The artificially low mass ratio reduces the ratio of the ion Lamor radius

$$r_L = \frac{m_i v_\perp}{q_i B} \tag{3}$$

to the Debye length λ_D as well as the ion traversal time

$$t_{trav,i} = \frac{L}{\sqrt{kT_i/m_i}}. \tag{4}$$

$t_{trav,i}$ is proportional to the time that is needed until the system reaches steady state. Thus, a lower ion to electron mass ratio reduces the simulation time. It also influences the particle

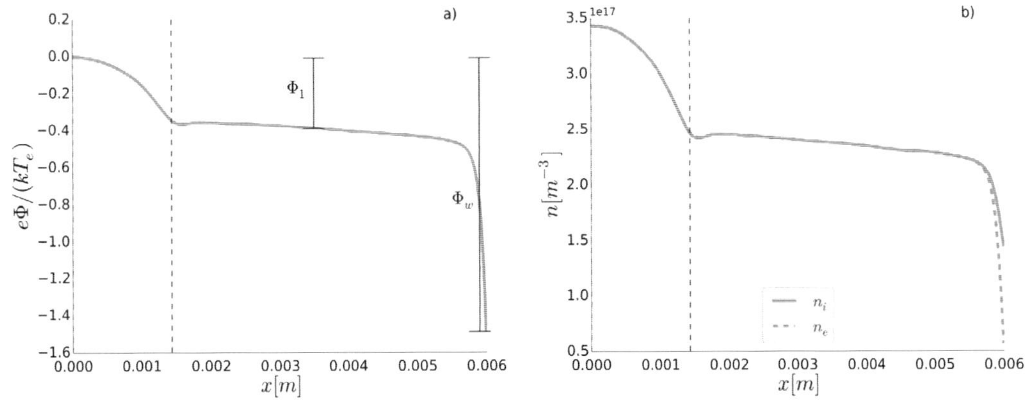

Figure 2. Potential (a) and density (b) profiles along the x-axis for $x \geq 0$. The end of the source region is indicated by the dashed black lines.

fluxes as well as the potential drop and has to be considered when transferring the simulation results to experimental plasmas. The super-particle factor f_{sup}, the factor that describes the ratio of simulation particle mass (charge) and the particle mass (charge) is set to $f_{sup} = 400$. The electron and ion source temperature is set to $kT_{e0} = kT_{i0} = 10\,\mathrm{eV}$, the magnetic field to $B = 0.33\,\mathrm{T}$. The system size is $2L_x \times 2L_y \times 2L_z = 12\,\mathrm{mm} \times 1\,\mathrm{mm} \times 1\,\mathrm{mm}$. We use periodic boundary conditions in y and z direction. With the resulting densities and the electron source temperature we get $\lambda_D = 4.7 \times 10^{-5}\,\mathrm{m}$ for the Debye length λ_D and thus $L_x \approx 125\lambda_D$ for all simulations. The equilibrium particle numbers are approximately $N_I \gtrsim N_e \approx 7 \times 10^6$. For advancing the particle trajectories we use a standard Boris solver [13] and a time-step of $\omega_p \Delta t \approx 0.1$. We use a source velocity distribution of type (2) for electrons as well as for ions which corresponds to scheme '1a' in [3].

3. Modeling results

Code verification To verify the developed code, we can relate to analytic [2] as well as numerical results [3]. Both are one dimensional and do not take a magnetic field into account. However, if the magnetic field is perpendicular to the wall, it does not alter the plasma properties because parallel transport is not influenced by the magnetic field.

The potential and density profile of a simulated plasma are shown in figure 2. Both profiles show a drop within the source region. The ions are accelerated due to the potential drop and satisfy the Bohm criterion at the end of the source region. Within the source-free section the profiles are almost flat. There is a small remaining potential drop. The chosen super-particle factor $f_{sup} > 1$ leads to an exaggerated collisionality within the simulated plasma and the collisions cause the slight slope of the potential profile in the source-free region. We need to further investigate the exact relations between this effect and the mentioned parameter. However, for the chosen configuration, the effect is very small. To take the potential drop in this region into account, we define the source potential drop between the center of the system ($x = 0$) and the center of the source-free region ($x = 0.0035\,\mathrm{m}$) instead of the source exit. For the source potential drop we find $\Phi_1 = (0.39 \pm 0.01)\,\mathrm{e}/(\mathrm{kT_e})$. The larger part of the total potential Φ_w drop is located in the sheath close to the wall. While the main plasma is quasi-neutral in the source region as well as in the source free region, ion and electron densities differ in the sheath region. On the scale of several Debye lengths a positive space charge shields the plasma from the negative wall. The total potential drop in our simulation, $\Phi_w = (1.59 \pm 0.01)\,\mathrm{e}/(\mathrm{kT_e})$, is

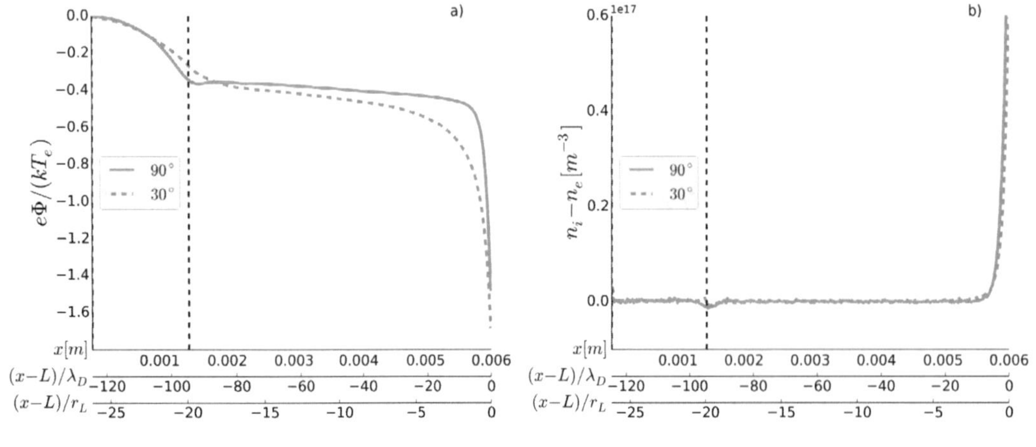

Figure 3. Potential profile (a) along the x-axis for $x \geq 0$ for a plasma with a perpendicular magnetic field and with 30° inclination. The graph on the right (b) shows the space charge of the plasma for both cases.

in agreement with $\Phi_w = (1.575 \pm 0.011)\,\mathrm{e}/(\mathrm{kT_e})$, the value in table I from [3]. The grid-free model as well as the PIC model both yield a potential drop which is higher than the value from analytic theory [2] which is $\Phi_w = 1.51\,\mathrm{e}/(\mathrm{kT_e})$. In the analytic solution leading to this value, the assumption that the electrons follow a Boltzmann relation is made. This requires high electron-electron collisionality and implies the assumption $m_I \gg m_e$ which is not the case for both simulations, where $m_I = 100m_e$. In addition, rethermalization only takes place at $x = 0$ in the simulation, so that the electron velocity distribution is not Maxwellian for all x. Thus, a small difference was to be expected.

Oblique magnetic fields Next, we study a case with a magnetic field with varying inclination angle α. We change the inclination angle from $\alpha = 90°$ to $\alpha = 30°$ in 15° steps. The basic structure of the potential of a plasma-wall transition area with an oblique magnetic field was first shown by Chodura in 1982 [14]. Adjacent to the electrostatic sheath region a potential drop on the length scale of the ion Lamor radius r_L develops. This area is the so called magnetic pre-sheath. While the total potential drop only changes weakly with the angle α, the ratio of the sheath and pre-sheath potential drop changes more significantly.

Figure 3 a) shows the potential profiles of a case with a perpendicular magnetic field and a case with an inclination of 30°. The potential drop in the source region is slightly larger for the case with inclined magnetic field. Within the source free region the potential of the case with inclined magnetic field starts to decrease a few r_L from the wall, while the potential of the 90° case drops a few λ_D from the wall. Figure 3 b) shows the difference of the ion and electron density for both cases. For both cases, quasi-neutrality holds throughout the whole plasma except for the electrostatic sheath close to the wall. Thus, the potential of the plasma with inclined magnetic field starts to decrease within the quasi-neutral plasma. The total potential drop in the 30° case is a little larger, but similar. This agrees with the results in [14]. Chodura simulated the plasma from a plasma interface at one end of the simulation domain to the wall at the other end. To ensure a smooth transition at the plasma interface, he uses the boundary condition

$$v_{ix0} \geq C_s \sin\alpha, \qquad C_s = \sqrt{\frac{\gamma_i T_i + \gamma_e T_e}{m_i}}, \tag{5}$$

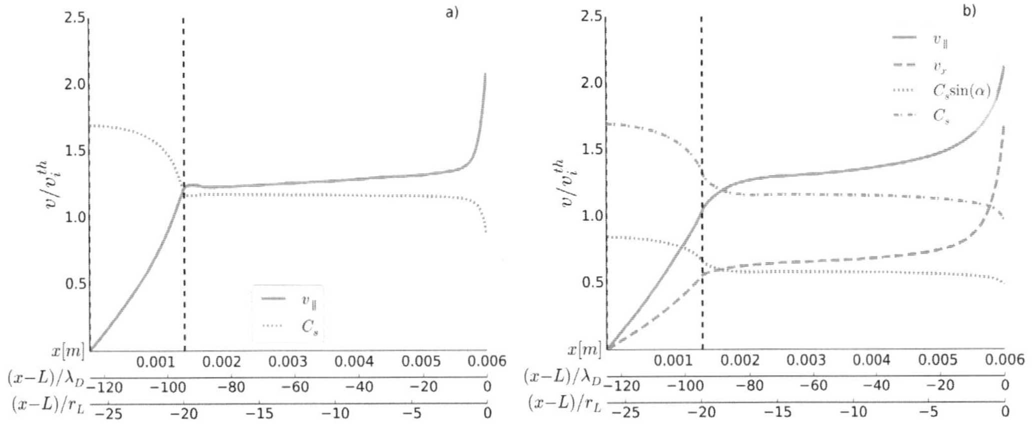

Figure 4. Ion velocity profiles along the x-axis for $x \geq 0$ for a plasma with perpendicular magnetic field (a) and 30° inclination angle (b). For the computation of the ion sound speed C_s the parameters $\gamma_i = 2$, $\gamma_e = 1$ are used.

where v_{ix0} is the ion flow velocity of the instreaming ions in x direction and C_s is the ion sound speed. This condition is equivalent to the condition

$$v_{i\|0} \geq C_s. \tag{6}$$

In the present work, we simulate a plasma that has a stagnation point in the center, and the self-consistently computed electric field within the source region accelerates the plasma so that it meets Chodura's boundary condition right outside of the source region as a result, rather than as a boundary condition. This can be seen in figure 4 for the 90° case as well as for the 30° case. Figure 4 b) also shows, that in addition to the aforementioned condition, the inequality $v_{ix0} \geq C_s$, the Bohm criterion, is fulfilled at the entrance of the electrostatic sheath. Figure 5 a) shows the plasma potential at different positions in the plasma. The position x_1 is localized in the center of the source free region. As expected, the potential Φ_1 does only depend weakly on α. The potential at the wall Φ_w changes slightly more with the magnetic field angle α, but the variation is still small, as stated before. At the intermediate point x_3 though the changes of the potential Φ_3 are stronger. x_3 is located directly at the entrance of the sheath region. The potential there is significantly influenced by α because of the magnetic pre-sheath adjacent to the electrostatic sheath.

The average energy of particles hitting the wall shown in figure 5 b) depends on the total potential drop Φ_w. Thus, the variation of the energy is also small. The ion energy E_i grows with decreasing α and increasing Φ_w. The electron energy changes inversely and its variation is even smaller.

The variation of the average angle of incidence $\beta_{i/e}$ is also shown in figure 5 b). While the changes are small for $\alpha = 90°, 75°$ and $60°$ they grow with shallower angles, especially for ions. For large magnetic field angels α the ion velocity in x-direction is larger than that in the yz-plane due to the acceleration by the electric field in the electrostatic sheath. This leads to larger angles of incidence ($\beta = 90°$ is a perpendicular hit with $v_x \gg v_y, v_z$). With shallower angles, the velocity in the y-direction grows. Thus, the angle of incidence gets smaller. The angle of the electrons does not vary that strongly. It is also closer to $\beta = 45°$ which indicates a more isotropic velocity distribution.

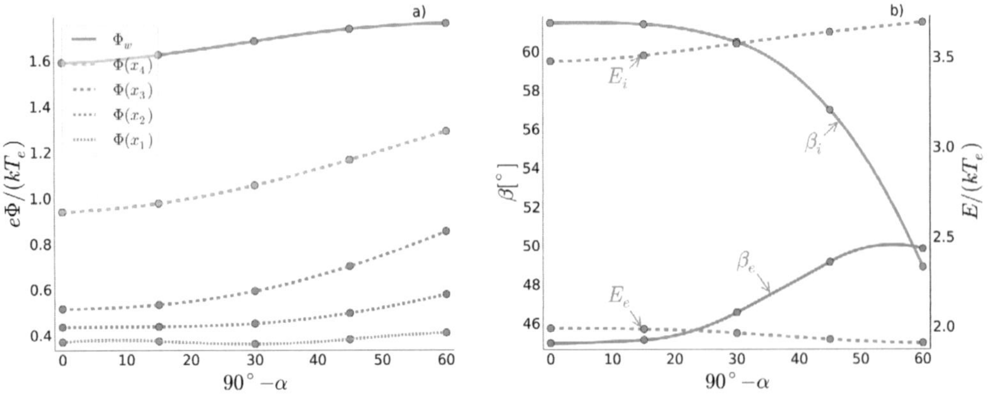

Figure 5. a) Potential relative to Φ_0 at different distances $\Delta x = x - L$ from the surface. These are $\Delta x_1 = -62.8\lambda_D = -13.1r_L$, $\Delta x_2 = -16.6\lambda_D = -3.5r_L$, $\Delta x_3 = -5.25\lambda_D = -1.1r_L$, $\Delta x_4 = -1.47\lambda_D = -0.3r_L$. Φ_w is measured directly at the wall. b) Average particle energy and angle of incidence of particles hitting the surface

4. Conclusions

We successfully established a grid-free simulation model based on the parallel Barnes-Hut tree-code PEPC. Using this tool, we simulated an almost collisionless plasma-wall transition region with and without an oblique magnetic field. The results of the cases with a perpendicular magnetic field are in agreement with an analytic solution as well as with PIC simulations, validating the applicability of the grid-free method for simulations of the plasma-wall interaction zone.

The results for plasmas with oblique magnetic field also confirm PIC results. We furthermore computed the 'boundary conditions' at the pre-sheath entrance self-consistently and these turn out to be consistent with assumptions made in the original work on the magnetic pre-sheath.

The remaining slope of the potential profiles in the source free region indicates that collisions alter the simulated systems, which would not be expected due to the used plasma parameters. Further research on this effect and its connection to numerical parameters and particle resolution has to be conducted.

Acknowledgments

The authors gratefully acknowledge Benedikt Steinbusch for assistance with PEPC performance issues, as well as the helpful support by Jülich Supercomputing Centre and the JSC staff. This work was supported via the JARA-HPC project JIEK41 on the Jülich supercomputers JUQUEEN and JUROPA as well as via the IFERC project TreePS.

References

[1] Matyash K, Schneider R, Sydora R and Taccogna F 2008 *Contributions to Plasma Physics* **48** 116–120
[2] Emmert G A, Wieland R M, Mense A T and Davidson J N 1980 *Physics of Fluids* **23** 803
[3] Procassini R J, Birdsall C K and Morse E C 1990 *Physics of Fluids B: Plasma Physics* **2** 3191
[4] Barnes J E and Hut P 1986 *Nature* **324** 446–449
[5] Greengard L and Rokhlin V 1987 *Journal of Computational Physics* **73** 325–348
[6] PEPC website URL `http://www.fz-juelich.de/ias/jsc/pepc` accessed August 2014
[7] Winkel M, Speck R, Hübner H, Arnold L, Krause R and Gibbon P 2012 *Computer Physics Communications* **183** 880–889
[8] Gibbon P 2005 *Physical Review E* **72** 026411

[9] Gibbon P, Speck R, Karmakar A, Arnold L, Frings W, Berberich B, Reiter D and Maek M 2010 *IEEE Transactions on Plasma Science* **38** 2367–2376
[10] Speck R, Arnold L and Gibbon P 2011 *Journal of Computational Science* **2** 138–143
[11] Bissell R C and Johnson P C 1987 *Physics of Fluids* **30** 779
[12] Scheuer J T and Emmert G a 1988 *Physics of Fluids* **31** 3645
[13] Boris J P 1970 The Acceleration Calculation from a Scalar Potential, Princeton University, Plasma Physics Laboratory Report MATT-769
[14] Chodura R 1982 *Physics of Fluids* **25** 1628

Journal of Physics: Conference Series **561** (2014) 012015 doi:10.1088/1742-6596/561/1/012015

X-point modelling in linear configurations using BOUT++

B W Shanahan and B D Dudson

York Plasma Institute, Department of Physics, University of York, Heslington, York YO10 5DD

E-mail: bws502@york.ac.uk

Abstract. Magnetic X-point configurations in tokamak geometries are critical in determining edge and scrape off layer (SOL) dynamics, and hence particle and heat flux onto plasma facing components. Alternative configurations have been proposed which aim to reduce fluxes to material surfaces, but their performance depends on cross-field transport in the region of the null point which is currently poorly understood. There is therefore a need for theoretical and experimental studies of turbulence in X-point magnetic configurations. In conventional 3D turbulence simulations of tokamaks, a field-aligned coordinate system is used, which introduces numerical instabilities at the null point due to zero volume elements. As a result, X-point dynamics are often interpolated based on nearby flux surfaces, which could exclude relevant physics. Here we simulate X-point configurations in linear geometries using a non-field-aligned coordinate system, and present results of 3D drift-wave turbulence and flow simulations in X-point configurations using an isothermal model which evolves density, vorticity, parallel velocity and parallel current density. Simulations have been performed to explore the feasibility of experimentally studying X-point configurations in linear plasma devices which indicate that even a modest coil set carrying 300A should produce a measurable effect on the driftwave turbulence and plasma profiles near the null region. The energy dynamics of the system are also explored and indicate that an X-point causes ohmic dissipation in higher mode number turbulence.

1. Introduction

There is growing concern that conventional divertor configurations in future devices could experience unacceptably high heat fluxes, especially during transient events [1, 2]. As a result, there have been studies into alternative concepts which could reduce the heat loads on divertor targets [3, 4]. The effectiveness of these configurations, however, are often sensitive to cross field transport in the null region, which is currently poorly understood [5].

Many of the current 3D plasma turbulence simulations do not directly model magnetic X-points, as they employ a field-aligned coordinate system [6, 7, 8]. The purely toroidal field at X-points creates numerical instability due to zero volume elements and therefore cannot be explicitly included in a field-aligned coordinate system. A common method is to interpolate plasma quantities from either side of the X-point region. This approach, however, could potentially exclude important physics associated with this region, such as mixing in the null region [9]. As a region of complicated plasma dynamics which can greatly affect the edge and SOL flows, the X-point region is an area which merits more detailed numerical study.

Alternatives to field-aligned coordinates, such as the Flux Coordinate Independent approach [10, 11] also ameliorates this issue, though its efficiency and accuracy are currently under investigation. While this is a functionality currently being implemented into `BOUT++` , this paper will address a simpler method of employing a cartesian coordinate system.

This paper is arranged as follows. The geometry and fluid model used in the following turbulence simulations are given in Section 2. Section 3 compares the turbulence characteristics and energy dynamics of a simple linear device to those of a device with X-point azimuthal fields, and Section 4 asserts our conclusions and future work.

2. Simulation geometry and model

2.1. Geometry and coordinate system

As an initial study into X-point simulation in BOUT++, we have considered a machine as shown in Figure 1. The machine parameters are chosen to correspond to a hypothetical university-scale experiment capable of producing X-points for tokamak-relevant scenarios. This machine will allow for direct investigation of fundamental plasma physics associated with tokamak divertor regions and allow simple diagnostic access to a null region, an area which is currently poorly understood. The machine parameters were chosen based on those achieved in previous machines [12, 13]. This device is proposed to produce low temperature, high density plasmas relevant to detachment and divertor-relevant scenarios. The electron temperature will be simulated as $T_{e0} \sim 2eV$, which is typical for linear devices as parallel streaming limits the maximum electron temperatures which can be achieved. The machine will achieve densities of up to $10^{19}m^{-3}$, as in previous machines [12, 14], for detachment-relevant scenarios. The initial plasma minor radius was considered to be 5cm. The axial magnetic field is $B_0 \sim 0.2 - 0.3T$. An additional azimuthal field is created by longitudinal wires which can create X-point fields of approximately $B_{ext} \sim 0.02T$ at the edge of the simulation domain, which corresponds to a current of about $300A$ in the internal X-point wires, which are considered to be 18-turn coils. More exotic configurations such as the snowflake can easily be reproduced by the inclusion of more longtiudinal wires. It should be noted that the interchange drive will not be present in this configuration, as there is no curvature. While this means that this machine will have different turbulent characteristics to a tokamak, it is still capable of producing tokamak-relevant studies applicable to divertor areas.

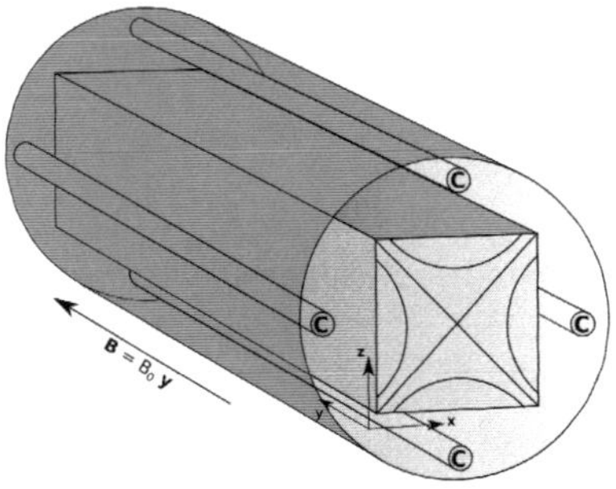

Figure 1. Schematic of proposed device showing vessel and X-point coils. Cartesian coordinate system measuring 30cm x 30cm x 3m shown inlaid within the vessel. The four X-point coils are labelled with a 'c'.

In most 3 dimensional turbulence simulations, a field-aligned coordinate system is used in the interest of numerical efficiency. This method cannot be used to explicitly model X-points, where the field is completely toroidal, as this introduces zero-volume elements. To ameliorate these issues, we have imposed a cartesian coordinate system, as shown in the centre of the machine pictured in Figure 1. The implementation of this coordinate system ensures that there are no zero volume elements, as each dimension is perpendicular to the other two, and therefore X-points (and any other singularities) can be effectively incorporated.

Figure 1 also illustrates the shape of the external field used to simulate X-point scenarios. This field was considered to be of the form shown in equation 1 [5], where $\theta = tan^{-1}\left(\frac{z}{x}\right)$ is the azimuthal angle. The magnitude and specific geometry of this field can be arbitrarily chosen by altering A_0, the exponent of r and corresponding coefficient of θ.

$$A_{ext} = A_0 r^2 cos(2\theta) \tag{1}$$

The expression in equation 1 then creates a pertubed field, as $\mathbf{B} = \nabla \times A_{ext}\mathbf{y}$. The implementation of this perturbed magnetic field is described in section 2.3.

2.2. Isothermal Model

An isothermal cold-ion fluid model initially constructed for plasma blob studies [15, 16] is extended here to perform initial studies of X-point dynamics. The model is electrostatic, inviscid, and employs the Boussinesque approximation [17, 18]. It should also be noted that the results presented here are performed in a linear geometry ($\xi \equiv \nabla \times \frac{\mathbf{b}}{B} \to 0$), and therefore any terms involving curvature are neglected in the following analysis. Furthermore, the isothermal electron temperature T_{e0} is 2eV. While this model includes several simplifications, it still captures relevant physics such as Kelvin-Helmholtz and driftwave turbulence, an important class of instabilities in tokamak edge plasmas [19], as it is a ubiquitous instability. The equations which are solved are given as follows in SI units:

Density continuity equation:

$$\frac{dN}{dt} = \nabla_{\|}\frac{J_{\|}}{e} - N_0\nabla_{\|}u_{\|} \tag{2}$$

Vorticity:

$$\rho_s^2 N_0 \frac{d\Omega}{dt} = \nabla_{\|}\frac{J_{\|}}{e} \tag{3}$$

Parallel velocity:

$$\frac{du_{\|}}{dt} = -\frac{c_s^2}{N_0}\nabla_{\|}n \tag{4}$$

Ohm's law:

$$J_{\|} = \frac{\sigma_{\|}T_{e0}}{eN_0}(\nabla_{\|}N - N_0\nabla_{\|}\phi) \tag{5}$$

Where $\Omega \equiv \nabla_{\perp}^2\phi$ is vorticity, total derivatives are split via $\frac{d}{dt} = \frac{\partial}{\partial t} + \mathbf{u_E}\cdot\nabla + \mathbf{u}_{\|}\cdot\nabla$, and parallel derivatives are evaluated using $\nabla_{\|} = \mathbf{b}\cdot\nabla$. In the above equations, $\rho_s = \frac{c_s}{\Omega_i}$ is the Bohm gyroradius. These equations are normalized such that density (N) is normalized to $N_0 = 5\times10^{18}m^{-3}$, speeds are normalized to the sound speed, and $\phi = \frac{e\Phi}{T_{e0}}$ is the normalized electrostatic plasma potential. The boundary conditions for the simulations presented here will be discussed in section 3.

This model differs from that used in reference [16] in that it incorporates parallel ion free streaming, $u_{\|}$, as parallel flows are vital when determining the effects of X-points. Additionally, energy conservation required the restriction that N is considered constant (N_0) in terms where it is not differentiated. See section 3.3 for a full discussion on the energy dynamics within the system.

Journal of Physics: Conference Series **561** (2014) 012015 doi:10.1088/1742-6596/561/1/012015

2.3. Numerical Methods

This model was solved for the system described in section 2.1 using a resolution of 1.15mm ($1.25\rho_s$) in the plane perpendicular to **B** (x, z), and 8cm ($90.4\rho_s$) in direction parallel to **B** (y). The longitudinal direction was chosen to have a relatively low resolution, as fine structure dynamics were expected to be more dominant in the perpendicular planes. Previous studies of linear devices [20] have also indicated that low axial wave-numbers are dominant in linear configurations. Additionally, Alfvén lengthscales ($\frac{V_A}{\Omega_{ci}}$) are about 20cm, and therefore these dynamics are adequately resolved, as Alfvén wave transit times are shorter than perpendicular drift dynamics. The parallel dynamics introduced by the imposed X-point field were implemented by altering the $\mathbf{b} \cdot \nabla$ operator such that $\nabla_{\|}(f) \rightarrow \nabla_{\|}(f) - \left[\frac{A_{\|}}{B}, f\right]$, where the brackets indicate Poisson brackets, which were implemented using the Arakawa method [21]. Time integration was implemented using the implicit time integration solver CVODE, within the SUite of Nonlinear abd DIfferential/ALgebraic equation Solvers (SUNDIALS) [22]. Finally, the Laplacian solver, which calculates potential (ϕ) from vorticity (Ω), in `BOUT++` was altered to invert using discrete sine transforms in the z direction (perpendicular to **B** in the azimuthal plane), which eliminates the periodicity inherent in typical Laplacian inversion utilizing Fourier transforms.

3. Simulation analysis

3.1. Implementation and behavior

The model represented by equations 2 - 5 was implemented into `BOUT++` and simulations were performed with and without an externally applied X-point field. The plasma source was modeled as a constant flow at the sound speed into the domain from one of the longitundal boundaries ($y = 0$). The opposing logitundinal boundary was considered to have zero-gradient boundary conditions in cases both with and without an imposed X-point. In cases with an X-point, sheath boundary conditions pertinent to presheath entrance as found in reference [23] were implemented. These boundary conditions are modified Bohm boundary conditions which account for the oblique magnetic fields at the Chodura sheath. As these boundary conditions rely on an oblique magnetic field, they could not be imposed in a basic linear case without an externally applied field. As such, these cases were constrained to zero-gradient boundary conditions on all fields in the directions perpendicular to the magnetic field.

Figure 2 shows the time evolution of the total energy of the fluctuations, which will be discussed later in Section 3.3. The simulations begin with an initial perturbation, and fluctuations grow until reaching a saturated, turbulent stage. This stage is shown in as the flat section in Figure 2. All analysis presented in this work has been conducted within this regime.

Figure 3 illustrates the mean radial density profiles both with (solid) and without (dotted) an X-point at various locations along the length of the device. The radial profiles in each case is taken along a diagonal ($x = z$) line across the azimuthal plane which lies on the X-point. Figure 3 indicates that the introduction of a magnetic X-point introduces an asymetric off-axis peak of density and a narrowing of the total profile.

The narrowing of the profiles induced by the X-point as shown in Figure 3 can be attributed to the externally applied azimuthal magnetic field, which is shown in Figure 4. The profile is distorted due to the external magnetic field which maps a circular input profile to an elongated profile at the opposite (sheath) end. This can be described by simply following field lines. As one moves along the axis of the machine, the perturbed field shifts the total magnetic field such that it is closer to the null point in two directions (top right and bottom left in Figure 4). This therefore creates a narrowing of the profiles in that direction. Likewise the opposite occurs in the other two directions, creating a broadened radial density distribution. Finally, Figure 3 also indicates that the radial density profiles are smoother without an X-point field applied. This

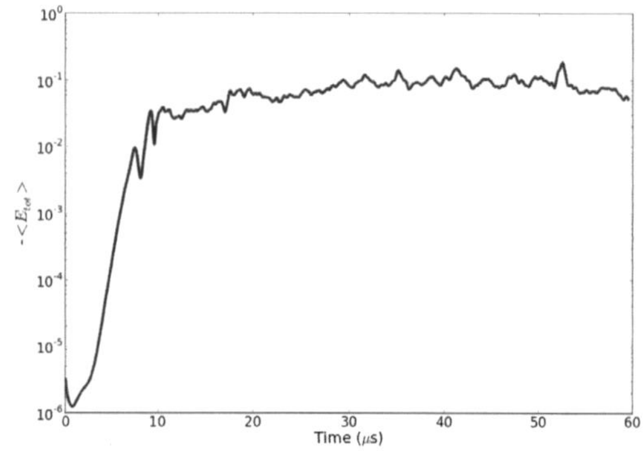

Figure 2. Time evolution of the volume averaged total dissipation $(-\langle E_{tot}\rangle)$ in a case without an X-point. For reference, the ion gyroperiod is 59ns

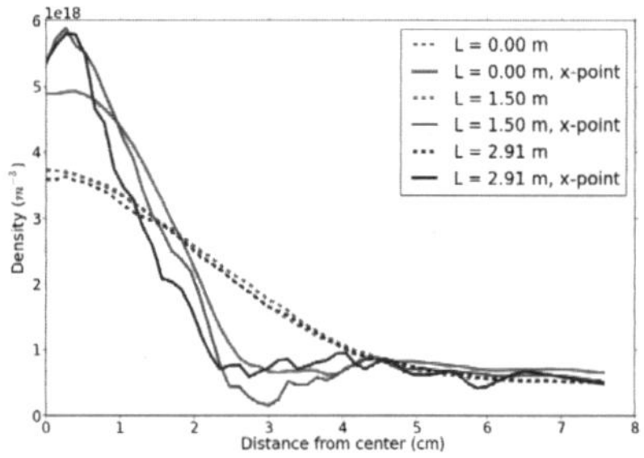

Figure 3. Time averaged radial density profiles along a diagonal in the azimuthal plane with (solid) and without (dotted) an X-point applied.

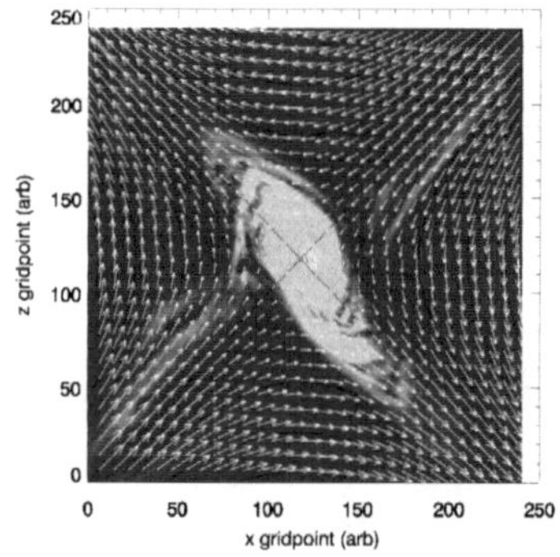

Figure 4. Density contour at y=1.5m with overlaid magnetic field.

could perhaps be explained by considering the cross field transport introduced by the X-point which inhibits the azimuthal flow of plasma, forcing coherent structures to dissolve.

3.2. Transport and cross correlation

To study the parallel dynamics of the system, cross correlation analysis was employed. The inbuilt cross correlation function in IDL was used to compute the cross correlation (P_{fg}) of two quantities f and g using the method shown in equation 6. Here k is the index for each population and M is the number of elements in said population.

$$P_{fg} = \sum_{k=0}^{M-1}(f_k - \overline{f})(g_k - \overline{g}) \left\{ \left[\sum_{k=0}^{M-1}(f_k - \overline{f})^2 \right] \left[\sum_{k=0}^{M-1}(g_k - \overline{g})^2 \right] \right\}^{-\frac{1}{2}} \tag{6}$$

Using this method, one can determine the correlation between two quantities at different points within the domain. This serves as a convenient method to visualize flow dynamics, as shown in Figure 5, which shows the two dimensional density cross correlations in an azimuthal plane at the center of the machine ($y = 1.5m$) when an X-point field is applied. This figure illustrates the effects of an induced X-point on parallel dynamics, as flows are apparent along field lines. The reference point in Figure 5 is shown in black near the null region.

Figure 6 plots the cross field flux (Γ) when an X-point is applied, given by $\Gamma = \langle N v_\perp \rangle$ where $v_\perp = \frac{|\mathbf{b}\times\nabla\widetilde{\phi}|}{B}$ and $\widetilde{\phi} = \phi - \langle\phi\rangle$. The cross field flux indicates the regions of high cross field transport, which is an important factor in the effectiveness of novel divertor configurations such as the snowflake [5].

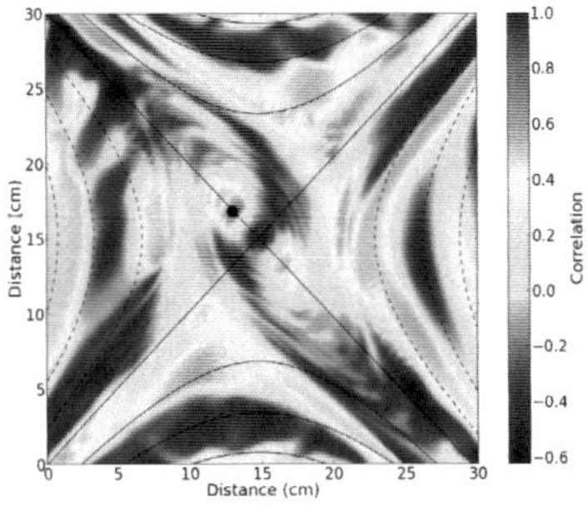

Figure 5. Two dimensional cross correlation at y=1.5m, reference point shown as black dot.

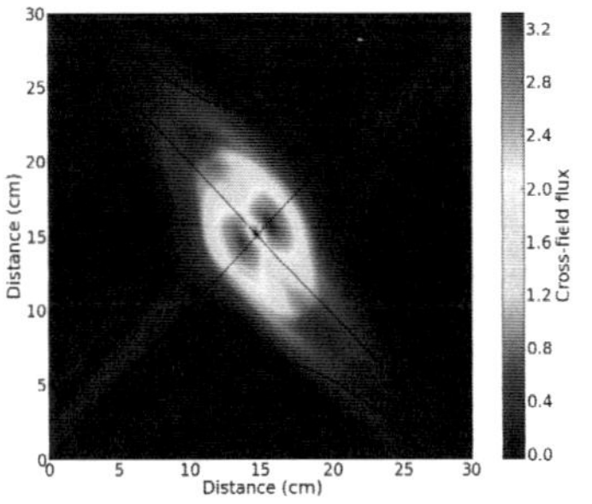

Figure 6. Cross field flux at y=1.5m, indicating that the center of the machine is the dominant area for cross-field transport.

Figures 5 and 6 indicate two characteristics of note. Firstly, Figure 5 indicates that away from the null point, flows remain along field lines, being highly correlated to positions parallel to field lines, but adjacent areas perpendicular to field lines are less correlated. Secondly, Figure 6 indicates that the null point is the main area of cross-field transport allowing plasma to flow from one region to another.

Furthermore, the effect of the imposed X-point on the system's turbulence can be examined using basic synthetic diagnostics. Specifically, one can look at the turbulence at a constant radius from the center of the plasma column, in this case considered to be $r = 5cm$, and measure the phase shift between potential and density fluctuations. Figures 7 and 8 indicate this relationship measured on turbulence with logitundinal fluctuations (driftwave-like), which was chosen due to its ubiquitous nature. [19, 24]

It is clear in Figures 7 and 8 that the induced X-point introduces a broadening in the turbulent spectrum. This is to be expected, as the perpendicular, perturbed X-point field lines inhibit the transport of turbulent structures, and therefore decorrelate adjacent fluctuations.

3.3. Energy dynamics

An analysis of the energy dynamics within the system was performed similar to that found in reference [25]. In this work, the system is spectrally-decomposed in the azimuthal and axial directions, and the energy of each mode is analyzed to determine the energy transfer channels and dissipation. The energy of each Fourier mode where $\vec{k} = (m, n)$, with m being the poloidal mode number and n the axial mode number, is:

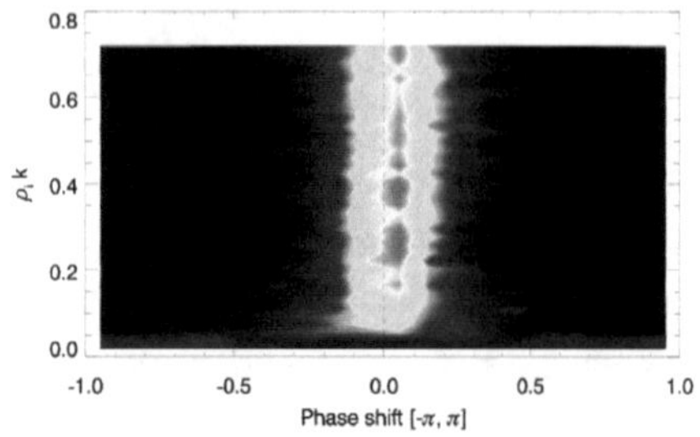

Figure 7. Phase shift comparison of density and potential perturbations in driftwave-like modes in a basic linear geometry.

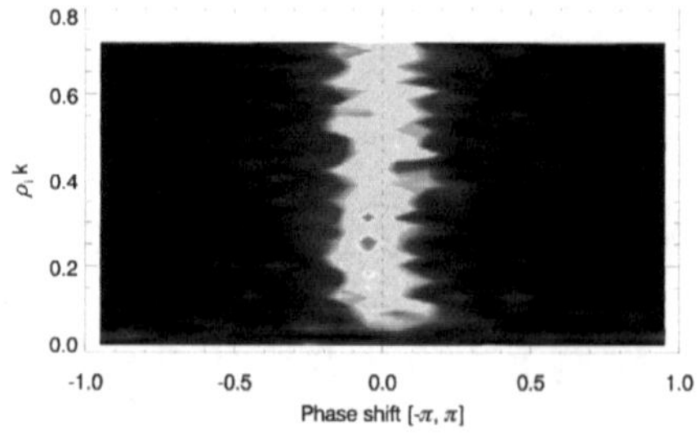

Figure 8. Phase shift comparison of density and potential perturbations in driftwave-like modes with an applied X-point.

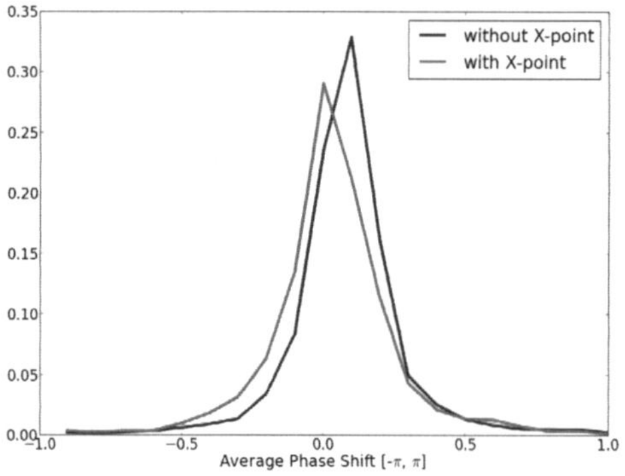

Figure 9. Phase shift profiles averaged over $\rho_i k$.

$$E_{tot}(\vec{k}) = \frac{1}{2}\left\langle \underbrace{\left|N_{\vec{k}}\right|^2}_{(a)} + \underbrace{\frac{1}{2}\left|u_{\parallel,\vec{k}}\right|^2}_{(b)} + \underbrace{\rho_s^2 \frac{1}{2}\left[\left|\frac{\partial \phi_{\vec{k}}}{\partial r}\right|^2 + \frac{m^2}{r^2}\left|\phi_{\vec{k}}\right|^2\right]}_{(c)} \right\rangle \tag{7}$$

Where N_k, $u_{\parallel,\vec{k}}$ and $\phi_{\vec{k}}$ are the Fourier-transposed density, parallel velocity, and potential, respectively. The analysis presented here occurs in the turblent, steady state phase described in Figure 2. The individual terms in equation 7 indicate the (a) internal energy, (b) parallel kinetic energy, and (c) perpendicular kinetic energy of the system in Fourier space. The evolution of each Fourier mode can be described as [25]:

$$\frac{\partial E_f(\vec{k})}{\partial t} = Q_f(\vec{k}) + C_f(\vec{k}) + D_f(\vec{k}) \tag{8}$$

Here, f indicates each field $(N, u_\parallel, \phi, j_\parallel)$, and $Q_f(\vec{k})$, $C_f(\vec{k})$, and $D_f(\vec{k})$ stand for the nonconservative energy forces (i.e. external sources and sinks), linear energy transfer channels, and dissipation terms respectively. There are only conservative energy forces in our system, so $Q_f(\vec{k}) = 0$. The exact expressions for $C_f(\vec{k})$ and $D_f(\vec{k})$ are:

$$C_N(\vec{k}) = \frac{1}{eN_0^2} Re \left\langle ik_\parallel j_{\parallel,\vec{k}} N_{\vec{k}}^* - \frac{1}{N_0} ik_\parallel u_{\parallel,\vec{k}} N_{\vec{k}}^* \right\rangle \tag{9}$$

$$C_\phi(\vec{k}) = \frac{1}{eN_0^2} Re \left\langle ik_\parallel j_{\parallel,\vec{k}} \phi_{\vec{k}}^* \right\rangle \tag{10}$$

$$C_{u_\parallel}(\vec{k}) = \frac{1}{N_0} Re \left\langle ik_\parallel N_{\vec{k}} u_{\parallel,\vec{k}}^* \right\rangle \tag{11}$$

$$C_{j_\parallel}(\vec{k}) = \frac{1}{eN_0^2} Re \left\langle ik_\parallel N_{\vec{k}} j_{\parallel,\vec{k}}^* - iN_0 k_\parallel \phi_{\vec{k}} j_{\parallel,\vec{k}}^* \right\rangle \tag{12}$$

$$D_{j_\parallel}(\vec{k}) = Re \left\langle \frac{-j_\parallel^2}{\sigma_\parallel T_{e0}} \right\rangle \tag{13}$$

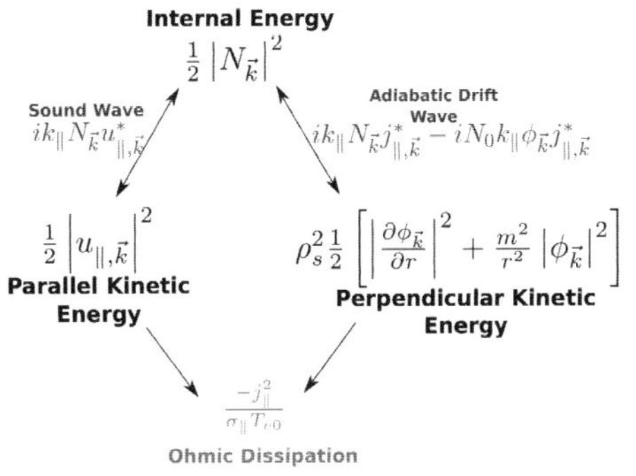

Figure 10. Energy dynamics diagram indicating the transfer of energy via sound waves and adiabatic drift waves.

Furthermore, our system conserves energy except for ohmic dissipation, $\sum_f C_f(\vec{k}) = 0$, as shown in Figure 10. As a result, equation 8 elucidates the wavenumbers at which energy is dissipated from the system, as the only terms remaining are those found in $D_f(\vec{k})$. Figures 11 and 12 plot the nonconservative rate of change of the total energy, $\frac{\partial E_f(\vec{k})}{\partial t}$, as a function of wavenumber in cases with and without an X-point.

These results are similar to those found in [25] in that the energy dynamics (in this case dissipation) is localized to small longitudinal mode numbers. It should be noted that the nonconservative rate of change of our total energy is negative, as the only contribution is from the ohmic dissipation, $D_{j_\parallel}$. Furthermore, and somewhat unexpectedly, an X-point reduces the dissipation in higher n-numbers, indicating that the azimuthally-applied X-point perturbs the strictly parallel mode number dissipation. This could perhaps be explained by considering the distortion of the profiles as shown in Figure 4.

Figures 13 and 14 show the nonconservative rate of change of the total energy at selected longitudinal mode numbers as a function of azimuthal mode number. It is clear from these plots that the n=0 mode is dominant across the azimuthal mode number spectrum both with and without an X-point. While the introduction of an X-point induces a shift of dissipation to lower n-numbers, Figures 13 and 14 indicate that dissipation increases for each of those lower-n modes.

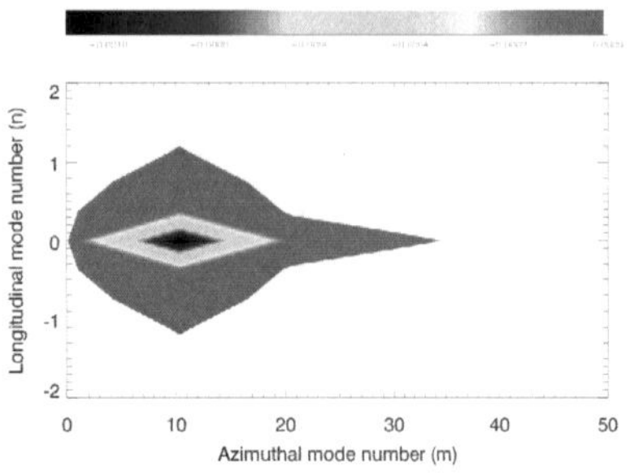

Figure 11. Total spectral nonconservative energy loss without an X-point

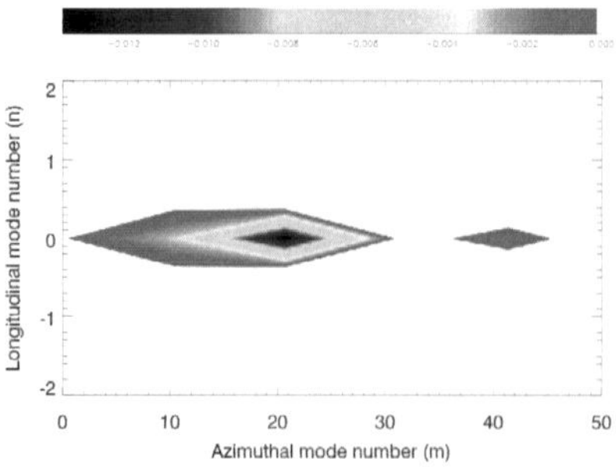

Figure 12. Total spectral nonconservative energy loss with an X-point

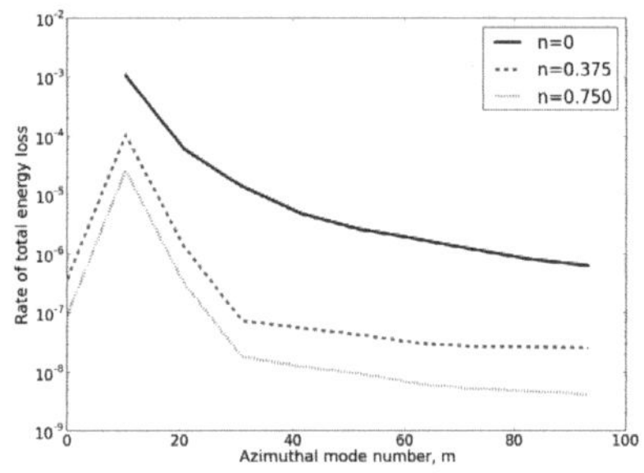

Figure 13. Specific longitudinal mode number traces of the total spectral nonconservative energy loss without an X-point

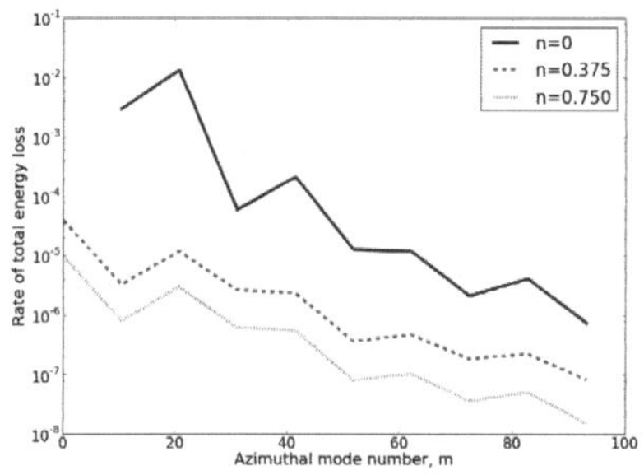

Figure 14. Specific longitudinal mode number traces of the total spectral nonconservative energy loss with an X-point

4. Conclusions and future work

BOUT++ can capably handle direct X-point simulations by imposing a simple cartesian coordinate system and altering parallel operators to include perturbed external fields. The introduction of a modest magnetic X-point perturbs the turbulence characteristics of the plasma column and alters the energy dynamics within the system. The simulations presented here indicate that a simple langmuir probe in such a device would be able to distinguish an altered driftwave phase shift, and measure a perturbed density profile exhibiting transport along the X-point field lines.

The Flux Coordinate Independent (FCI) method currently being implemented into BOUT++ could potentially offer a more efficient method to capably handle magnetic X-point regions. In addition, the implementation of a more complex model which includes temperature evolution will allow studies of heat and energy convection which could elucidate physics relevant to novel divertor configurations. With this model and the implementation of the FCI method, full tokamak simulations including explicit magnetic X-point simulation could be accomplished using BOUT++ .

References

[1] G Federici, A Loarte, and G Strohmayer. Assessment of erosion of the iter divertor targets during type i elms. *Plasma physics and controlled fusion*, 45(9):1523, 2003.

[2] A Zhitlukhin, N Klimov, I Landman, J Linke, A Loarte, M Merola, V Podkovyrov, G Federici, B Bazylev, S Pestchanyi, et al. Effects of elms on iter divertor armour materials. *Journal of nuclear materials*, 363:301–307, 2007.

[3] S I Krasheninnikov, A Y Pigarov, and D J Sigmar. Plasma recombination and divertor detachment. *Physics Letters A*, 214(5):285–291, 1996.

[4] P M Valanju, M Kotschenreuther, S M Mahajan, and J Canik. Super-x divertors and high power density fusion devices. *Physics of Plasmas*, 16(5):056110–056110, 2009.

[5] D D Ryutov, R H Cohen, T D Rognlien, and M V Umansky. A snowflake divertor: a possible solution to the power exhaust problem for tokamaks. *Plasma Physics and Controlled Fusion*, 54(12):124050, 2012.

[6] B D Dudson, M V Umansky, X Q Xu, P B Snyder, and H R Wilson. BOUT++: A framework for parallel plasma fluid simulations. *Computer Physics Communications*, 180:1467–1480, 2009.

[7] P Ricci, F D Halpern, S Jolliet, J Loizu, A Mosetto, A Fasoli, I Furno, and C Theiler. Simulation of plasma turbulence in scrape-off layer conditions: the gbs code, simulation results and code validation. *Plasma Physics and Controlled Fusion*, 54(12):124047, 2012.

[8] B D Scott. Free-energy conservation in local gyrofluid models. *Physics of Plasmas (1994-present)*, 12(10):–, 2005.

[9] W A Farmer and D D Ryutov. Axisymmetric curvature-driven instability in a model divertor geometry. *Physics of Plasmas (1994-present)*, 20(9):–, 2013.

[10] F Hariri and M Ottaviani. A flux-coordinate independent field-aligned approach to plasma turbulence simulations. *Computer Physics Communications*, 184(11):2419 – 2429, 2013.

[11] F Hariri, P Hill, M Ottaviani, and Y Sarazin. The flux-coordinate independent approach applied to x-point geometries. *Physics of Plasmas (1994-present)*, 21(8):–, 2014.

[12] B Mihalji, P K Browning, and K J Gibson. Spatially resolved spectroscopy of detached recombining plasmas in the university of manchester linear system divertor simulator. *Physics of Plasmas (1994-present)*, 14(1):–, 2007.

[13] J C Sandeman, P Uddholm, J A Elliott, and M G Rusbridge. Experiments on drift wave launching: I. dispersion curves and damping rates. *Plasma Physics and Controlled Fusion*, 39(1):159, 1997.

[14] W Gekelman, H Pfister, Z Lucky, D Bamber, Jand Leneman, and J Maggs. Design, construction, and properties of the large plasma research devicethe lapd at ucla. *Review of Scientific Instruments*, 62(12):2875–2883, 1991.

[15] J R Angus, M V Umansky, and S I Krasheninnikov. Effect of drift waves on plasma blob dynamics. *Physical Review Letters*, 108(21):215002, 2012.

[16] N R Walkden, B D Dudson, and G Fishpool. Characterization of 3d filament dynamics in a mast sol flux tube geometry. *Plasma Physics and Controlled Fusion*, 55(10):105005, 2013.

[17] J R Angus and M V Umansky. Modeling of large amplitude plasma blobs in three-dimensions. *Physics of Plasmas (1994-present)*, 21(1):–, 2014.

[18] G Q Yu, S I Krasheninnikov, and P N Guzdar. Two-dimensional modelling of blob dynamics in tokamak edge plasmas. *Physics of Plasmas (1994-present)*, 13(4):–, 2006.

[19] B D Scott. The nonlinear drift wave instability and its role in tokamak edge turbulence. *New Journal of Physics*, 4(1):52, 2002.

[20] M. V. Umansky, P. Popovich, T. A. Carter, B. Friedman, and W. M. Nevins. Numerical simulation and analysis of plasma turbulence the large plasma devicea). *Physics of Plasmas (1994-present)*, 18(5):–, 2011.

[21] A Arakawa and V R Lamb. Computational design of the basic dynamical processes of the {UCLA} general circulation model. In J Chang, editor, *General Circulation Models of the Atmosphere*, volume 17 of *Methods in Computational Physics: Advances in Research and Applications*, pages 173 – 265. Elsevier, 1977.

[22] Advanced computational software collection, us doe.

[23] J Loizu, P Ricci, F D Halpern, and S Jolliet. Boundary conditions for plasma fluid models at the magnetic presheath entrance. *Physics of Plasmas (1994-present)*, 19(12):–, 2012.

[24] B Scott. *Plasma Physics and Controlled Fusion*, 45:A385–398, 2003.

[25] B Friedman, T A Carter, M V Umansky, D Schaffner, and B Dudson. Energy dynamics in a simulation of lapd turbulence. *Physics of Plasmas (1994-present)*, 19(10):–, 2012.

Linear stability studies including resistive wall effects with the CASTOR/STARWALL code

E. Strumberger, S. Günter, P. Merkel, and C. Tichmann
Max Planck Institute for Plasma Physics, Boltzmannstr. 2, 85748 Garching, Germany
E-mail: ers@ipp.mpg.de

Abstract. CASTOR/STARWALL is a code package for linear stability studies consisting of the CASTROR_3DW code, the STARWALL code, and a hybrid version of both codes named CASTOR3D that is currently under development. The latter solves an extended eigenvalue problem consisting of the plasma part defined by the weak form of the perturbed single-fluid MagnetoHydroDynamic (MHD) equations, and a vacuum part which is derived from an energy functional. This new code can describe wall resistivity and plasma inertia simultaneously. Furthermore, besides the so far used straight field line coordinates, additionally, general flux coordinates are implemented which are more appropriate for the description of instabilities located close to the separatrix (e.g. Edge Localized Modes (ELMs)). The formulation of the eigenvalue problem in the new coordinates is not limited to axisymmetric equilibria, so that, finally, stability studies of resistive and rotating, 3D equilibria should be possible. In this paper we will sketch the theory, report on the progress of the code development, and present first results.

1. Introduction

The interaction of the plasma with currents induced in external conducting structures plays an important role for many tokamak instabilities, such as Vertical Displacement Events (VDEs), Resistive Wall Modes (RWMs), etc. Linear stability studies yield the growth rates and mode structures of instabilities in the initial linear phase. They are numerically less expensive, more robust and flexible than non-linear calculations. This allows quick parameter studies and is very useful for benchmarking the linear phase of non-linear simulations, e.g. made with the non-linear JOREK-STARWALL code [1, 2].

There exist several linear stability codes, e.g. CASTOR [3, 4, 5], CASTOR_3DW [6], MARS-F [7], CAS3D [8], STARWALL [9], and CarMa [10], which use various methodes to take external conducting structures into account. The response of ideal conducting structures to unit perturbations of the plasma can be computed independently of the eigenvalue problem of the plasma interior. The resulting response matrix is the only information needed to compute the boundary terms of the eigenvalue problem. This method is, for example, used in the CASTOR and CASTOR_3DW codes. While the CASTOR code is limited to axisymmetric, closed walls, the CASTOR_3DW code has been extended to 3D wall geometry for which the STARWALL code provides the vacuum response matrix. However, the response of resistive conducting structures depends on the growth rate. For slowly growing RWMs the plasma inertia can be neglected. In that case the plasma response to unit perturbations is computed, and the resulting plasma response matrix is used for the solution of the vacuum problem. This method is, for example, used in the CarMa code [10], and in the STARWALL code [9]. The latter makes use of the potential energy of the plasma, which is computed with the CAS3D stability code for given unit perturbations. In contrast to the CarMa code, which is able to treat volumetric 3D conducting structures, but is restricted to axisymmetric equilibria, the STARWALL code can also be applied to 3D equilibria. However, it uses the 'thin wall approximation' and is restricted to ideal plasmas without flow. In order to study stability properties of realistic configurations, the separation of the problems as described above is usually not possible. That is, resistive walls with complex 3D geometry and plasma inertia (e.g. in case of a toroidally rotating plasma) have to be considered simultaneously. In the CarMa code this problem is solved by the so-called 'backward' coupling strategy [11]. There, the dynamic response of the conducting structures is integrated in the MHD equations.

The CASTOR3D, which is currently under development, follows another strategy. It is a hybrid of the CASTOR_3DW code and the STARWALL code. This code solves an extended eigenvalue problem consisting of the plasma part defined by the weak form of the perturbed single-fluid MHD equations (CASTOR_3DW code), and a vacuum part which is derived from an energy functional (STARWALL code). Furthermore, besides the so far used straight field line coordinates, additionally, general flux coordinates are implemented, which are more appropriate for the description of instabilities located close to the separatrix (e.g. ELMs). The matrix elements derived for the new coordinates are not limited to axisymmetric equilibria, so that, finally, stability studies of resistive and rotating, 3D equilibria should be possible.

The paper is organized as follows. The theory is briefly described in section 2. The present status of the code and first results are subject of section 3, and, finally, an outlook is given in section 4.

2. Theory

2.1. Coordinate systems

Because of the reasons mentioned in the introduction, general flux coordinates describing also 3D equilibria are currently implemented in the CASTOR3D code. These coordinates, named (s, v, u), describe a right-handed coordinate system with s being the radial coordinate, and v, u being toroidal and poloidal angle-like coordinates, respectively. While the original CASTOR code [3, 4] and its subsequent versions use straight field line coordinates, being a special choice of flux coordinates, the CASTOR3D code supports any flux coordinate system. Furthermore, either the square root of the normalized poloidal flux, ρ_{pol}, or the square root of the normalized toroidal flux, ρ_{tor}, can be chosen as radial coordinate s. Using an equilibrium input provided by the NEMEC code [12], $s = \rho_{tor}$, and v, u being the angle-like coordinates of the NEMEC code, are the logical choice for these coordinates.

2.2. The basic equations

The linearized, ideal, single fluid MHD equations for the perturbations read

$$\lambda\rho = -\vec{v}\cdot\nabla\rho_0 - \rho_0\nabla\cdot\vec{v},$$

$$\lambda\rho_0\vec{v} = -\nabla(\rho_0 T + \rho T_0)/m + (\nabla\times\vec{B}_0\times\vec{B} + \nabla\times\vec{B}\times\vec{B}_0)/\mu_0,$$

$$\lambda T = -\vec{v}\cdot\nabla T_0 - (\Gamma - 1)T_0\nabla\cdot\vec{v}, \tag{1}$$

$$\lambda\vec{B} = \nabla\times(\vec{v}\times\vec{B}_0),$$

with the time dependence of the perturbed quantities taken to be $\sim e^{\lambda t}$, and $\lambda = \gamma + i\omega$ being the complex eigenvalue. The quantities ρ_0, T_0, and $\vec{B}_0$ denote density, temperature, and magnetic field of the unperturbed equilibrium. Here the fluid velocity $\vec{v}_0$ is assumed to be zero. The compressibility of the plasma is described by the coefficient Γ, with $\Gamma = 5/3$ set during the following computations. The quantities m and μ_0 denote ion mass and vacuum permeability, respectively. For the corresponding perturbed quantities $\rho, T, \vec{v}$, and $\vec{B} = \nabla \times \vec{A}$ an analogue ansatz [3, 4] is made as in the original CASTOR code. That is,

$$\rho = \frac{1}{s}\hat{\rho}, \qquad T = \frac{1}{s}\hat{T}, \qquad \vec{A} = -i\hat{A}_s \nabla s + \hat{A}_v \nabla v + \hat{A}_u \nabla u,$$

$$\vec{v} = v^s \vec{r}_{,s} + v^v \vec{r}_{,v} + v^u \vec{r}_{,u} = \frac{R^2}{\sqrt{g}} \hat{v}_s \vec{r}_{,s} + \frac{R^2}{i\iota f} \hat{v}_v \vec{B}_0 + \frac{R^2}{i\sqrt{g}} \hat{v}_u \vec{r}_{,u} \,. \tag{2}$$

Here $\sqrt{g}$ is the Jacobian, $\hat{f} = \partial\Phi/\partial s$ the derivative of the toroidal flux Φ, and ι the rotational transform. The variables $\hat{\rho}, \hat{v}_s, \hat{v}_u, \hat{v}_v, \hat{T}, \hat{A}_s, \hat{A}_u$, and $\hat{A}_v$ are collected in the symbolic vector $\vec{w} = (\hat{\rho}, \hat{v}_s, \hat{v}_u, \hat{v}_v, \hat{T}, \hat{A}_s, \hat{A}_u, \hat{A}_v)$. Each of its components is expressed by a complex Fourier series in poloidal and toroidal direction, and a finite element interpolation, $h^{\bar{j}}_{\bar{k},\bar{p}}(s)$, in radial direction:

$$w_{\bar{k}}(s,u,v) = \sum_{\bar{m},\bar{n},\bar{j},\bar{p}} \alpha^{\bar{k},\bar{p}}_{\bar{m},\bar{n},\bar{j}} h^{\bar{j}}_{\bar{k},\bar{p}}(s) e^{2\pi i(\bar{m}u+\bar{n}v)} + \alpha^{*\bar{k},\bar{p}}_{\bar{m},\bar{n},\bar{j}} h^{\bar{j}}_{\bar{k},\bar{p}}(s) e^{-2\pi i(\bar{m}u+\bar{n}v)}. \tag{3}$$

In (3) $\bar{k}$ labels the eight variables, $\bar{m}, \bar{n}$ the poloidal and toroidal Fourier harmonics, and $\bar{j}$ the radial grid points. The interpolating functions are either quadratic ($\bar{k}$=1,3,4,5,6) or cubic ($\bar{k}$=2,7,8) Hermite polynomials. $\bar{p} = 1, 2$ labels the two different quadratic, $h^{\bar{j}}_{\bar{p}}$, and cubic Hermite polynomials, $H^{\bar{j}}_{\bar{p}}$, at each grid point. In 3D geometry the conjugate complex term in (3) becomes necessary, because the two orthogonal solutions have different eigenvalues in non-axisymmetric configurations.

In case of the equation for the radial velocity component v^s the formulation of the weak form leads to surface integrals of the form

$$W_s = -\int_0^1 dv \int_0^1 du H^j_h e^{\pm 2\pi i(mu+nv)} R^2 \vec{B}_0 \cdot \vec{B}|_{s=1}, \tag{4}$$

where $\vec{B}$ has to fulfill the boundary condition, $\vec{n}_p \vec{B} = \vec{n}_p \vec{b}$, with $\vec{n}_p = \nabla s / |\vec{r}_{,v} \times \vec{r}_{,u}|$ being the normal vector on the plasma boundary in outward direction, and $\vec{b}$ being the perturbation of the vacuum field.

In order to determine the vacuum field perturbation in the presence of an ideal or resistive wall a variational method is used with the Lagrangian [9]

$$\mathcal{L} = \frac{\mu_0}{8\pi} \int_{S_p} df_p \int_{S_p} df_{p'} \frac{\vec{K}_p \cdot \vec{K}_{p'}}{|\vec{r}_p - \vec{r}_{p'}|} + \frac{\mu_0}{4\pi} \int_{S_p} df_p \int_{S_w} df_w \frac{\vec{K}_p \cdot \vec{K}_w}{|\vec{r}_p - \vec{r}_w|} \tag{5}$$

$$+ \frac{\mu_0}{8\pi} \int_{S_w} df_w \int_{S_w} df_{w'} \frac{\vec{K}_w \cdot \vec{K}_{w'}}{|\vec{r}_w - \vec{r}_{w'}|} + \frac{1}{2\lambda\sigma_w d_w} \int_{S_w} df_w \vec{K}_w \cdot \vec{K}_w + \int_{S_p} df_p (\vec{n}_p \cdot \vec{\xi}) \vec{n}_p \cdot (\vec{K}_p \times \vec{B}_0).$$

The divergence-free surface currents, $\vec{K}_{p,w} = \vec{n}_{p,w} \times \nabla \Phi^{p,w}$, generate the magentic field perturbation in the vacuum domain, where Φ^p and Φ^w are the current potentials. The indices

p and w label plasma boundary and wall, respectively. The wall is characterized by its specific resistivity σ_w, and its width d_w. The first three terms of the Lagrangian (5) are the energies produced by surface currents in the plasma boundary and the wall. The fourth term is the energy dissipated in the resistive wall, while the boundary condition at the plasma boundary, $\vec{n}_p \times \vec{A} = -(\vec{n}_p \cdot \vec{\xi})\vec{B}_0$, leads to the last term. Furthermore, the displacement vector $\vec{\xi}$ is related with the velocity perturbation $\vec{v}$ and the eigenvalue λ via

$$\vec{\xi} = \frac{\vec{v}}{\lambda}. \tag{6}$$

In case of the plasma boundary or a closed wall with smooth geometry a Fourier ansatz

$$\Phi(u,v) = \sum_{m,n} \hat{\phi}^s_{mn} \sin(2\pi(mu+nv)) + \hat{\phi}^c_{mn} \cos(2\pi(mu+nv)) \tag{7}$$

is made for the corresponding current potentials, while a wall with complex, multiply-connected structure is discretized into triangles. Then $\vec{K}$ is assumed to be constant on a triangle and is defined by the values of the current potential at the vertices of the triangle. That is,

$$\vec{K}_\Delta = \frac{\hat{\phi}_1 \vec{r}_{23} + \hat{\phi}_2 \vec{r}_{31} + \hat{\phi}_3 \vec{r}_{12}}{|\vec{r}_{21} \times \vec{r}_{32}|}. \tag{8}$$

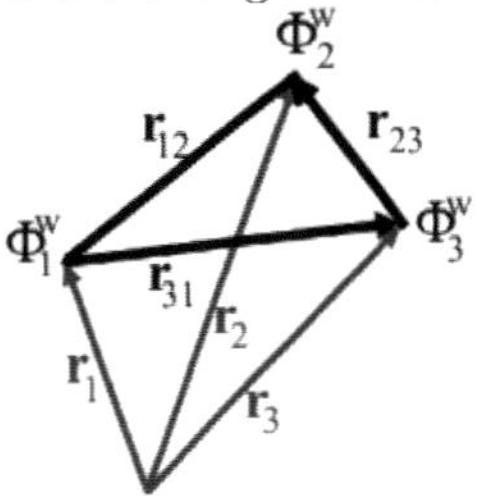

Fig. 1: Triangle wall element.

The variation of the Lagrangian (5) with respect to the variables $\hat{\phi}^{sc}_{mn}$ and/or $\hat{\phi}_i$, and the use of (6) yields a coupled set of linear equations

$$\lambda(\mathbf{M}_{pp}\vec{x}_p + \mathbf{M}_{pw}\vec{x}_w) = -\mathbf{R}_{ps}\vec{x}_s, \qquad \lambda(\mathbf{M}_{wp}\vec{x}_p + \mathbf{M}_{ww}\vec{x}_w) = -\mathbf{R}_{ww}\vec{x}_w. \tag{9}$$

with the matrices $\mathbf{M}_{pp}, \mathbf{M}_{pw}$, and $\mathbf{M}_{ww}$ resulting from its first three terms, $\mathbf{R}_{ww}$ from its fourth, and $\mathbf{R}_{ps}$ from its last term. The symbolic vector, $\vec{x}_p = (\{\hat{\phi}^{p,s}_{mn}\}, \{\hat{\phi}^{p,c}_{mn}\})$, represents the Fourier coefficents of Φ^p, while $\vec{x}_w = (\{\hat{\phi}^{w,s}_{mn}\}, \{\hat{\phi}^{w,c}_{mn}\})$ or $\vec{x}_w = (\{\hat{\phi}^w_i\})$ ($i = 1, .., N_\Delta$, with N_Δ being the number of vertices), represent Φ^w. Finally, $\vec{x}_s = (\{\alpha^{\bar{k},\bar{p}}_{\bar{m},\bar{n},\bar{j}}\}, \{\alpha^{*\bar{k},\bar{p}}_{\bar{m},\bar{n},\bar{j}}\})$ is related to the components of the symbolic vector $\vec{w}$ at the plasma boundary.

2.3. The extended eigenvalue problem

The combination of the weak form of the perturbed MHD equations and equations (9), the latter resulting from the variation of the Lagrangian (5), yields the extended eigenvalue problem

$$\lambda \begin{pmatrix} \mathbf{B}_{ll} & \mathbf{B}_{ls} & \mathbf{0}_{lp} & \mathbf{0}_{lw} \\ \mathbf{B}_{sl} & \mathbf{B}_{ss} & \mathbf{0}_{sp} & \mathbf{0}_{lw} \\ \mathbf{0}_{pl} & \mathbf{0}_{ps} & \mathbf{M}_{pp} & \mathbf{M}_{pw} \\ \mathbf{0}_{wl} & \mathbf{0}_{ws} & \mathbf{M}_{wp} & \mathbf{M}_{ww} \end{pmatrix} \begin{pmatrix} \vec{x}_l \\ \vec{x}_s \\ \vec{x}_p \\ \vec{x}_w \end{pmatrix} = \begin{pmatrix} \mathbf{A}_{ll} & \mathbf{A}_{ls} & \mathbf{0}_{lp} & \mathbf{0}_{lw} \\ \mathbf{A}_{sl} & \mathbf{A}_{ss} & \mathbf{R}_{sp} & \mathbf{R}_{sw} \\ \mathbf{0}_{pl} & -\mathbf{R}_{ps} & \mathbf{0}_{pp} & \mathbf{0}_{pw} \\ \mathbf{0}_{wl} & \mathbf{0}_{ws} & \mathbf{0}_{wp} & -\mathbf{R}_{ww} \end{pmatrix} \begin{pmatrix} \vec{x}_l \\ \vec{x}_s \\ \vec{x}_p \\ \vec{x}_w \end{pmatrix}. \tag{10}$$

Here the sub-matrices $\mathbf{A}_{ll}, \mathbf{A}_{ls}, \mathbf{A}_{sl}, \mathbf{A}_{ss}$ and $\mathbf{B}_{ll}, \mathbf{B}_{ls}, \mathbf{B}_{sl}, \mathbf{B}_{ss}$, and the sub-vectors $\vec{x}_l$ and $\vec{x}_s$ are related to the weak form of the MHD equations describing the interior (index l) and/or the plasma boundary (index s), respectively. The magnetic field perturbation $\vec{B}$ in the surface integral (4) is obtained from the surface currents in the plasma boundary and the wall

$$\vec{B} = \vec{B}_p + \vec{B}_w = \frac{\mu_0}{4\pi}\int_{S_p} df'\vec{K}_{p'} \times \frac{(\vec{r}_p - \vec{r}_{p'})}{|\vec{r}_p - \vec{r}_{p'}|^3} + \frac{\mu_0}{4\pi}\int_{S_p} df'\vec{K}_{w'} \times \frac{(\vec{r}_w - \vec{r}_{w'})}{|\vec{r}_w - \vec{r}_{w'}|^3}. \quad (11)$$

Inserting (11) and the corresponding expressions for the surface currents $\vec{K}_{p'}$ and $\vec{K}_{w'}$ into (4) yields the matrices $\mathbf{R}_{sp}$ and $\mathbf{R}_{sw}$ in (10). The remaining elements of (10) have already been explained in the previous section.

3. Results

In the following we present first results obtained for an ideal external kink mode in the presence of an ideal, and a resistive wall by solving the extendend eigenvalue problem. For these computations the straight field line coordinates are used. Furthermore, the growth rates of internal kink modes are calculated using the new coordinates.

3.1. Solution of the extended eigenvalue problem

The results of the CASTOR3D code are compared with analytical results calculated for a large aspect-ratio, low-β, circular cross-section tokamak plasma surrounded by a thin, uniform, resistive shell. In Ref. [13] plasma and shell are approximated by a cylinder of radius a, and a thin, concentric cylindrical shell of radius $r_w > a$, respectively. Then, the wall time τ_w is defined by $\tau_w = \mu_0\sigma_w d_w r_w$ with σ_w and d_w being the wall conductivity and thickness, respectively. In this analytical model the dispersion relation for a m, n resistive wall mode is $\gamma\tau_w = \Delta_w^m$, with

$$\Delta_w^m = \frac{2m}{(r_c/r_w)^{2m} - 1}. \quad (12)$$

Here r_c is the critical radius which corresponds to the ideal stability boundary. That is, an ideal wall located at $r_w < r_c$ stabilizes the mode. Then, the normalized growth rate $\bar{\gamma}$ results to

$$\bar{\gamma} = \mu_0\sigma_w d_w\gamma = \frac{2m}{r_w((r_c/r_w)^{2m} - 1)}. \quad (13)$$

For the numerical calculations we assume an aspect ratio of $A = 10$, and a q-profile analogue to the one defined in Ref. [13] with q_0=1.3 and q_a=2.9 being the values of the safety factor at magnetic axis and plasma boundary, respectively. This equilibrium is unstable with respect to a m=3,n=1 ideal external kink mode, and the critical radius numerically results to $r_c = 1.28073$ m.

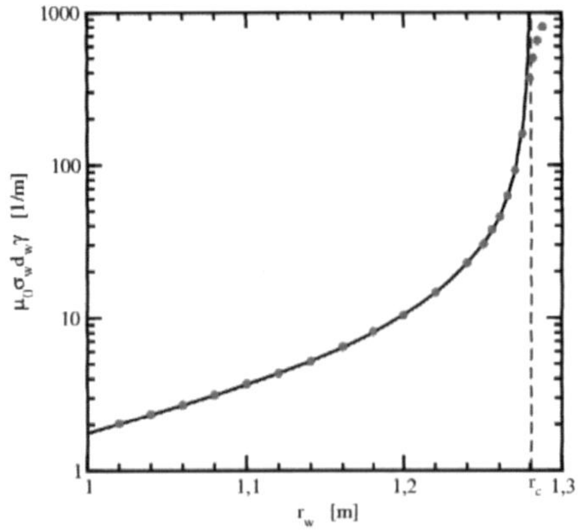

Fig. 2: Comparison of the analytically computed normalized growth rate (black line) with the CASTOR3D results (red dots) as function of the wall position. The plasma boundary is located at $r_w = 1$ m. The dashed blue vertical line indicates the critical radius r_c.

Journal of Physics: Conference Series **561** (2014) 012016 doi:10.1088/1742-6596/561/1/012016

The analytical and the numerical results agree very well as long as the resistive wall is located well inside r_c. Very close, but still within this radius the analytical growth rate rapidly increases, while the increase of the numerical growth rate is limited by the plasma inertia. The solution of the extended eigenvalue problem is valid for every distance of the wall, as shown in Fig 3. There, the growth rates as function of the wall position are plotted for an ideal, and a resistive wall.

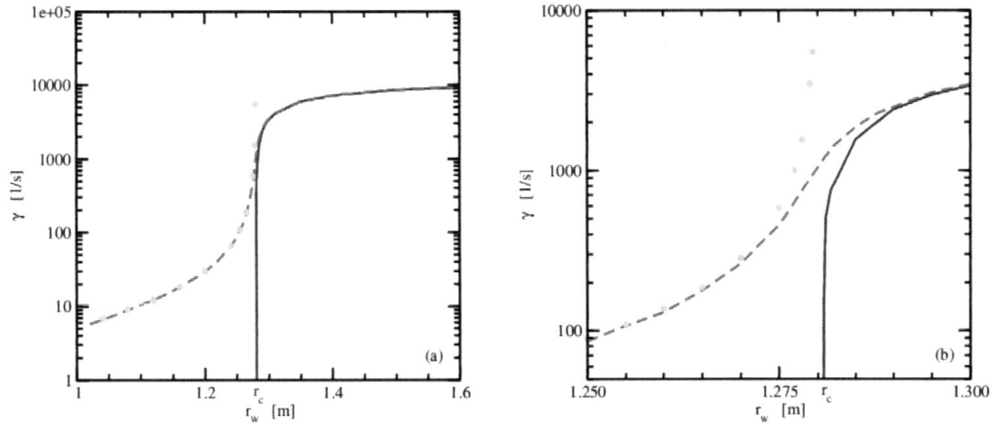

Fig. 3: (a) Growth rates of an external kink mode as function of the position of an ideal wall (blue solid line), and a resistive wall (red dashed line), both computed with the CASTOR3D code. The green stars show the results obtained with the STARWALL code in case of the resistive wall. (b) Enlargement of the region around the critical radius r_c.

In the case of the ideal wall the growth rate converges towards the no wall limit for large wall distances, and it shows an almost singular decrease, when the wall approaches the ideal stability boundary. However, in the case of a resistive wall, the growth rate increases continuously from very small values typically for a RWM to the high values of the no wall limit. The results of the STARWALL code agree with the solution of the extended eigenvalue problem for small growth rates, but due to the neglection of the kinetic energy the code fails when the wall position approaches or exceeds the critical radius.

3.2. Use of general flux coordinates

The CASTOR_3DW code has been extended to the computation of growth rates in the presence of 3D wall configurations. Although the toroidal mode coupling has been implemented, its use is still restricted to axisymmetric equilibria. Furthermore, the used straight field line coordinates are not a suitable choice if the modes are located close to the separatrix. In Fig. 4 flux surfaces up to the plasma boundary are plotted for a simple test equilibrium with circular cross-section (Fig. 4a), and an ASDEX Upgrade (AUG)-like equilibrium (Fig. 4b). For the latter a flux surface close but inside the separatrix has been chosen as plasma boundary. For both equilibria poloidal coordinate lines are plotted for the straight field line coordinates, and for the 'NEMEC' coordinates. While the latter show only a slight bending close to the separatrix, the poloidal coordinate lines of the straight field line coordinates become almost parallel to the radial ones. For these reasons, the matrix elements in the weak form of the ideal MHD equations have been derived in general 3D form and implemented in the CASTOR3D code.

The two test equilibria considered are unstable with respect to a $n = 1$ (circular equilibrium) and a n=3 (AUG-like equilibrium) internal kink mode, respectively. The growth rates of these modes are determined for both, straight field line and 'NEMEC' coordinates, as shown in Figs 5 and 6. There, eigenfunctions and growth rates of the modes are compared. Because of the

Journal of Physics: Conference Series **561** (2014) 012016 doi:10.1088/1742-6596/561/1/012016

different poloidal coordinates the Fourier spectra of the eigenfunctions are different, but the leading harmonics are the same for either coordinate system. Furthermore, the eigenvalues agree very well within 0.24% for the $n = 1$, and 1% for the $n = 3$ mode.

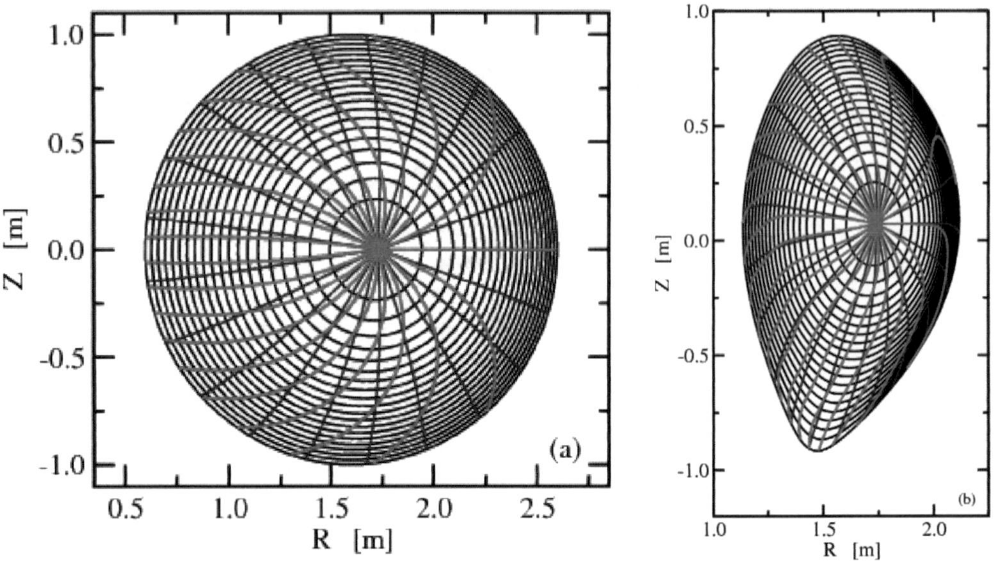

Fig. 4: Flux surfaces (black), poloidal coordinate lines for 'NEMEC' coordinates (blue), and straight field line coordinates (red) for a plasma with circular cross-section (a), and an AUG-like equilibrium (b).

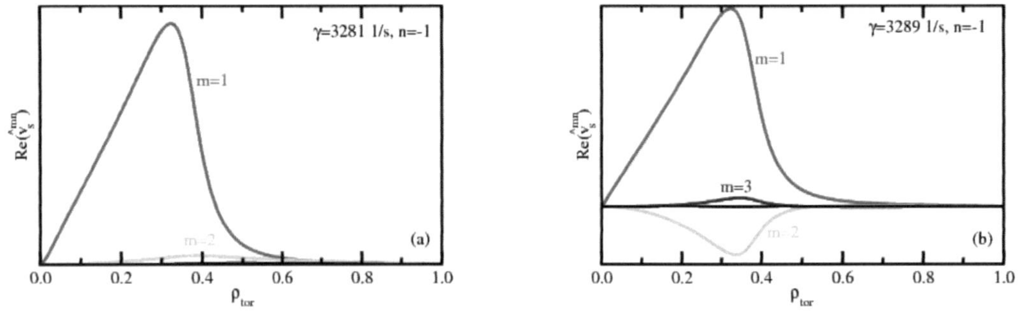

Fig. 5: Circular equilibrium: Real parts of the harmonics, $\hat{v}_s^{mn}$, of the eigenfunctions in dependence of ρ_{tor} for straight field line coordinates (a), and 'NEMEC' coordinates (b).

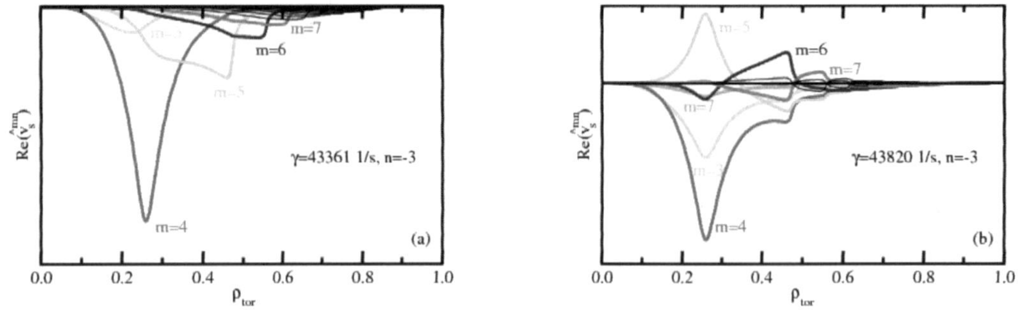

Fig. 6: AUG-like equilibrium: Real parts of the harmonics, $\hat{v}_s^{mn}$, of the eigenfunctions in dependence of ρ_{tor} for straight field line coordinates (a), and 'NEMEC' coordinates (b).

Since, in case of these internal modes, the CASTOR3D code works very well, it will be tested for external and edge localized modes in a further step.

4. Outlook

The CASTOR3D code is a hybrid of the CASTOR_3DW and the STARWALL codes. The new implementations, namely the solution of the extended eigenvalue problem and the use of general flux coordinates, yield very promising first results. However, there is still a plenty of analytical, programming and computational work to do. That is, further matrix elements describing plasma resistivity, plasma flow, etc., have to be derived in 3D, general form. The STARWALL part has to be adapted to the new coordinates, and further conducting structures, e.g. coils, have to be implemented in the CASTOR3D code. Although the 3D form of the matrix elements was used for the computations presented in the previous section, these computations were performed for axisymmetric equilibria. That is, terms that only play a role in 3D geometry still have to be tested. In order to perform stability studies for realistic axisymmetric and 3D equilibria, and complex, multiply-connected conductor structures, an efficient organization and parallelization of the CASTOR3D code is necessary. Since a non-hermitian eigenvalue problem has to be solved, a parallel eigenvalue solver of the Scalable Library for Eigenvalue Problem Computations (SLEPc) [14] is currently implemented. All new implementations will be carefully tested and benchmarked with other codes, e.g. the CAS3D code (ideal, 3D stability studies) and the non-linear JOREK/STARWALL code. The goal is a fully 3D, linear stability code that includes physcial effects, such as plasma resistivity, plasma rotation, viscosity, etc., and which simultaneously takes 3D, resistive wall structures into account.

Acknowledgments

This project has received funding from the European Union's Horizon 2020 research and innovation programme under grant agreement number 633053. The views and opinions expressed herein do not necessarily reflect those of the European Commission.

References

[1] Hölzl M, Merkel P, Huysmans G T A, Nardon E, McAdams R and Chapmann I 2012 *JPCS* 012010
[2] Hölzl M, Merkel P, Huysmans G T A, Atanasiu C, Lackner K, Nardon E, Strumberger E, McAdams R, Liu F, Chapmann I and Fil A 2014 *JPCS* (*Preprint* `http://arxiv.org/abs/1408.6379`)
[3] Huysmans G T A, Goedbloed J P and W K 1993 *Phys. Fluids B* **5** 1545
[4] Kerner W, Goedbloed J P, Huysmans G T A, Poedts S and Schwarz E 1998 *J. Comput. Phys.* **142** 271
[5] Strumberger E, Merkel P, Sempf M and Günter S 2008 *Phys. Plasmas* **15** 056110 URL `http://dx.doi.org/10.1063/1.2884579`
[6] Strumberger E, Merkel P, Tichmann C and Günter S 2011 *38th EPS Conf. on Plasma Phys.* ECA Vol. 35G P5.082 (Strasbourg, France)
[7] Liu Y, Bondeson A, Gregoratto D, Fransson C M, Gribov Y and Paccanella R 2004 *Nuclear Fusion* **44** 77
[8] Nührenberg C 1996 *Phys. Plasmas* **3** 2401
[9] Merkel P and Sempf M *Fusion Energy 2006* (Proc. 21st Int. Conf., Chengdu, 2006) (Vienna: IAEA), CD-ROM file TH/P3-8 URL `http://www-naweb.iaea.org/napc/physics/FEC/FEC2006/html/index.htm`
[10] Portone A, Villone F, Liu Y, Albanese R and Rubinacci G 2008 *Plasma Phys. Control. Fusion* **50** 085004 URL `http://dx.doi.org/10.1088/0741-3335/50/8/085004`
[11] Liu Y and Villone F 2009 *Plasma Physics and Controlled Fusion* **51** 115008 URL `http://dx.doi.org/10.1088/0741-3335/51/11/115008`
[12] Hirshman S P, van Rij W I and Merkel P 1986 *Comput. Phys. Commun.* **43** 143
[13] Fitzpatrick R and Yu E P 1998 *Phys. Plasmas* **5** 2340
[14] Romin J E, Campos C, Romero E and Tomás A 2014 Slepc users manual Technical report dsic-ii/24/02 Universidad Politecnica de Valencia URL `http://www.grycap.upv.es/selpc`

Role of Trapped Electrons on Global Gyrokinetic Linear Stability of Collisionless Microtearing Modes

Aditya K Swamy, R Ganesh
Institute for Plasma Research, Bhat, Gandhinagar, India
E-mail: ganesh@ipr.res.in

J. Chowdhury
Dept of Physics, University of Colorado, Boulder, CO, 80309, USA

S. Brunner, J. Vaclavik, L. Villard
CRPP, EPFL, 1015 Lausanne, Switzerland

Abstract. Unstable collisional MicroTearing Mode (MTMs) have been found in experiments of high-β Spherical Tokamaks and are believed to be driven by drift resonance of trapped electrons. It has been recently shown that at large aspect ratio, the magnetic drift resonance of highly passing electrons is a minimal mechanism to drive the collisionless MTM unstable. In this work, a preliminary study of inclusion of trapped electrons in large aspect ratio tokamaks indicate that for the reference parameters investigated, the collisionless MTM retain their essential mode structures, while growth rates are only moderately affected.

1. Introduction

Microtearing modes (MTMs) are low frequency, high-n electromagnetic microinstabilities in Tokamak plasmas. The mode draws free energy from the electron temperature gradient and is excited above a threshold gradient. Thus, if found unstable, it is envisaged to open up an important channel of electron transport. These modes exhibit tearing parity, i.e. odd parity in electrostatic potential and even parity in parallel magnetic vector potential and rotate in the electron diamagnetic direction. Early analytical work predicted these modes to be unstable only in collisional plasmas [1, 2, 3]. Recent gyrokinetic simulations have found unstable collisional MTMs in various magnetic confinement configurations such as spherical tokamaks with very high β values in collisional or semi-collisional regime [4, 5, 6, 7, 8], standard tokamaks, such as ASDEX-Upgrade [9, 10], and weakly collisional Reversed Field Pinch plasmas [11, 12]. These linear gyrokinetic simulations, using a local flux-tube implementation, have thrown light on several characteristics of the mode, typically for high β and relatively moderate electron temperature gradient length scales. The connection to experimentally observed heat fluxes in devices through linear and nonlinear simulations is only beginning to be explored [8, 9, 5, 13].

Recent investigations have found that MTM is unstable in the absence of collisions as well [12, 14, 15], triggering a need to understand collisionless drive mechanisms. In the work of Ref. [14] pertaining to a spherical tokamak with high-β plasma, the magnetic drift resonance of trapped electrons is found to drive the collisionless MTM instability. At larger aspect ratios, the fraction of trapped particles is relatively small, and it has been found that collisionless MTM is driven unstable by the magnetic drift resonance of highly passing electrons [15]. Ions dynamics, on the other hand, do not affect the mode significantly. For typical β in tokamaks, it has been shown that the mode requires a high temperature gradient threshold to be unstable, and that the threshold is downshifted at higher β. For the ρ^* (ρ/a, normalized ion radius) value considered, the global mode structure is found to be sensitive to profile variations in Ref. [15]. The trapped fraction was ignored with a view to gain insight into the nature of MTMs. As the trapped population ($\sim \sqrt{2\epsilon}$) constitutes a significant fraction, it is pertinent to study the role of trapped electron dynamics in the MTM instability mechanism.

Such a study is carried out using the linear global gyrokinetic electromagnetic microstability code EM-GLOGYSTO, for a circular flux surface, axisymmetric equilibrium in toroidal coordinates (r, θ, ϕ). The model assumptions and formulation are detailed in Refs. [15, 16, 17, 18, 19, 20] and references thereof. In this work, many of the recent results described in Ref. [15] will be elaborated, where it is shown that passing electron alone suffice to drive MTM unstable in large aspect ratio tokamaks. In the later part, the effect of trapped electrons is addressed. As before, Shafranov shift is neglected for simplicity, while the $B_{\|}$ fluctuations, which become important at very high β are also neglected in the present work. Section 2 contains the various profiles and input parameters used for the study, and the discussion mentioned above. Section 3 contains the important conclusions and future directions.

2. Results & Discussion

2.1. Input Parameters

The equilibrium profiles and plasma parameters used in this work are as in Table 1, Ref. [15]. For details, the reader is referred to Section IIA therein. The corresponding profiles are plotted in Fig. 1. The radial position where η_j peaks is represented as $s = s_0$ and $\beta(s_0)$ as β_0 . Frequencies and growth rates are normalized to $\omega_0 = c_s/a$. For parameters throughout this work, we have $T_{0i} = T_{0e}$, thus $c_s = v_{th}$, $\omega_0 \simeq 1.7 \times 10^6\ s^{-1}$. Studies in the range $0.0 < \beta_0 < 0.1$ and $1 < a/L_T < 12.5$ relevant to the operational regimes in tokamaks [22, 23, 24, 25] show that, qualitatively, the nature of the mode in the poloidal cross-section, including the tearing parity of the eigenmode structures, radial extent etc. remain similar across the parameter space, within both cases viz. (i) excluding the trapped electrons, (ii) including the trapped electrons.

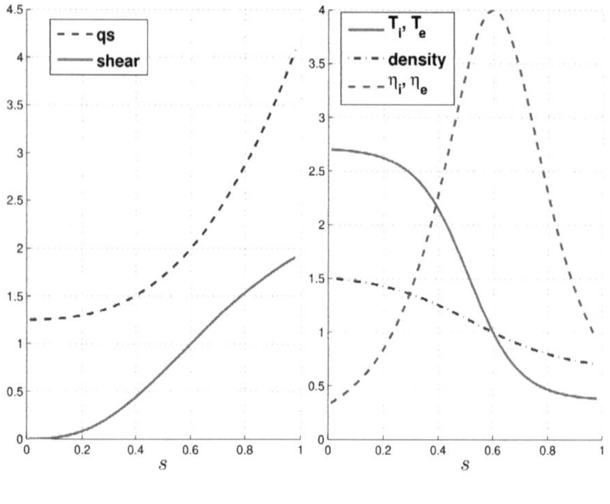

Figure 1. Equilibrium profiles

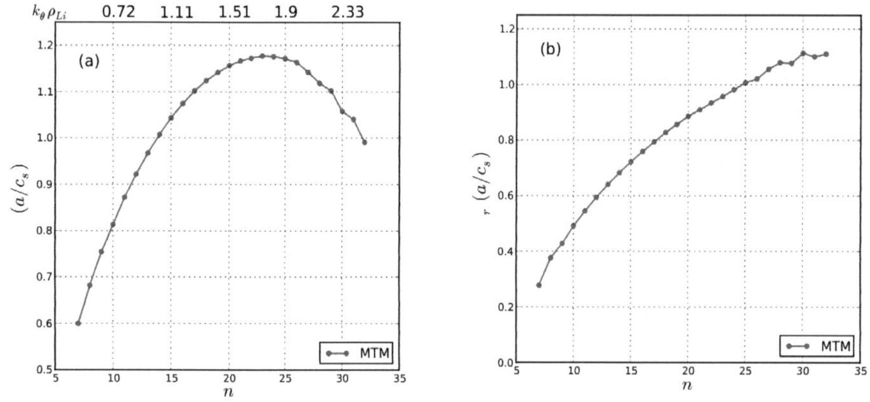

Figure 2. (a) Growth rates and (b) frequencies for MTMs, versus toroidal mode number n, excluding trapped electrons. With same data at β_0=5%, $\eta_{e,i}$=4, as in Fig.2, Ref. [15]. Mode-averaged $< k_\theta \rho_{Li} >$ values are indicated for the corresponding mode numbers on the upper x-axis (for definition, see again Sec II.B of Ref. [15]).

2.2. Mode Structures excluding trapped electrons

The MTM instability typically has been thought to require a collisional drive and thus studies have focused more in high-β spherical tokamaks that allow for relatively high density and have strong toroidal effects. The scattering of trapped electrons in the trapped-passing boundary in velocity space increases the effective drive mechanism according to the model of Catto-Rosenbluth, thus accentuating the instability. On the other hand, in lower-β large aspect ratio plasmas, the trapped electron fraction is relatively smaller. Thus passing electron dynamics, in particular, the resonance of the

Journal of Physics: Conference Series **561** (2014) 012017 doi:10.1088/1742-6596/561/1/012017

higher harmonics of transit frequency, magnetic drift resonance of passing electrons are more at play in destabilizing the collisionless MTM. The global mode characteristics are described in Ref. [15]. The mode shows a clear tearing parity, and is electromagnetic since it grows more unstable with β. The mode is seen to be most unstable at short, sub-ion scale wavelengths. Toroidal mode number $n = 23$ is seen to be the most unstable from Fig. 2

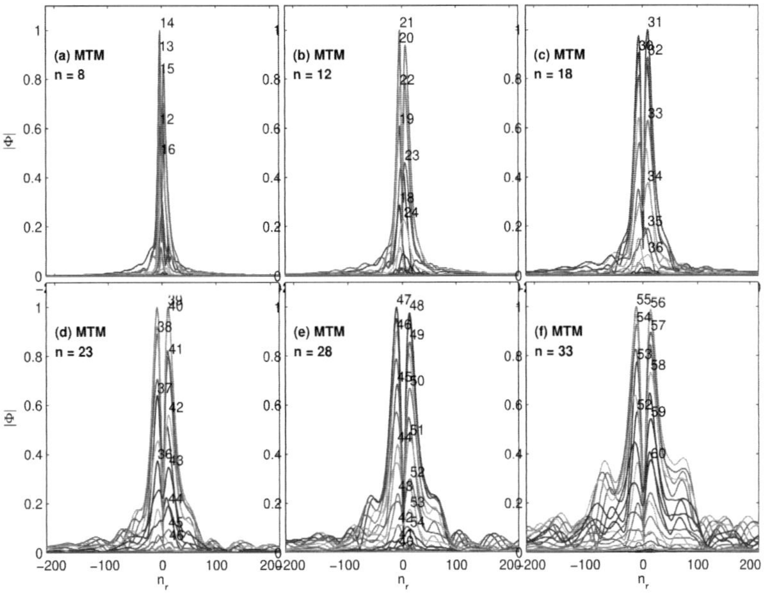

Figure 3. Fourier space plots for MTM $\tilde{\varphi}$ fluctuations $n = 8, 12, .., 33$, excluding trapped electrons.

In Figs. 3, 4 the Fourier space plots for $n = 8, 12...33$ are shown (For the real space plots, see Figs.7-11, Ref. [15]). Figs. 5, 6 show the contours of $\tilde{\varphi}$ and $\tilde{A}_{\parallel}$, respectively. It is found that, typically, about 20 poloidal modes m couple to form the converged global mode. The symmetry of the poloidal components of $\tilde{\varphi}$ in reciprocal space with a zero amplitude for $n_r = 0$ translates to the odd symmetry, whereas the maximum for $\tilde{A}_{\parallel}$ to even symmetry, in radial direction. It is interesting to note that the amplitude for $\tilde{A}_{\parallel}$ has a sharper fall with n_r, indicating that the number of Fourier modes are more than sufficient and the eigenmode is well resolved. The $\tilde{\varphi}$ spectrum is broad, whereas that of $\tilde{A}_{\parallel}$ is relatively narrow. In real space, $\tilde{\varphi}$ is localized and $\tilde{A}_{\parallel}$ more extended, (See. Ref. [15]) consistent with similar results reported in Ref. [10]. The fluctuations arise from very short scale dynamics of electrons, such as in Trapped Electron Mode [21] and requires high resolution and computational resources for convergence, particularly for higher n. MTM mode characteristics are known to be complex, because of the dependence on different parameters. In Fig. 7, the $n = 10, 12$ modes are seen to remain more unstable at lower β. The critical β is seen to be different for different n, with $n = 12$ having threshold near 2.5% and $n = 10$ having a threshold less than 2.5%

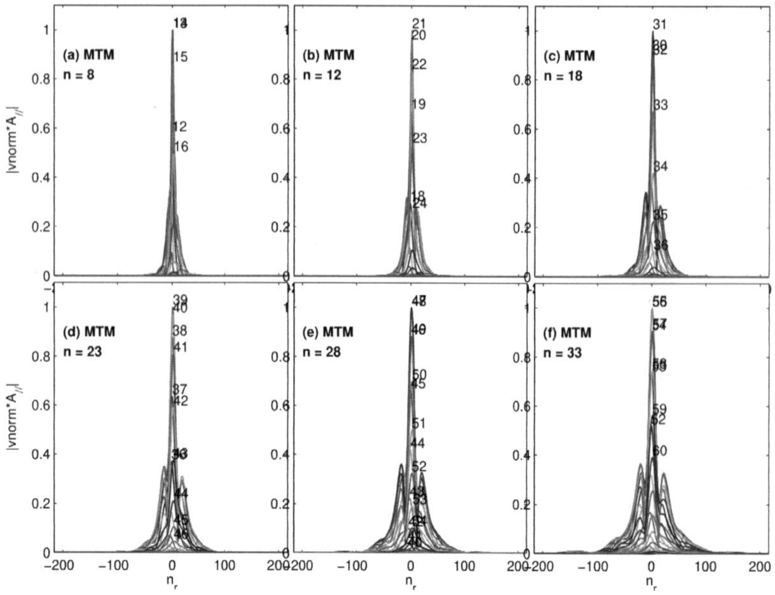

Figure 4. Fourier space plots for MTM $\tilde{A}_{\|}$ fluctuations $n = 8, 12, .., 33$, excluding trapped electrons.

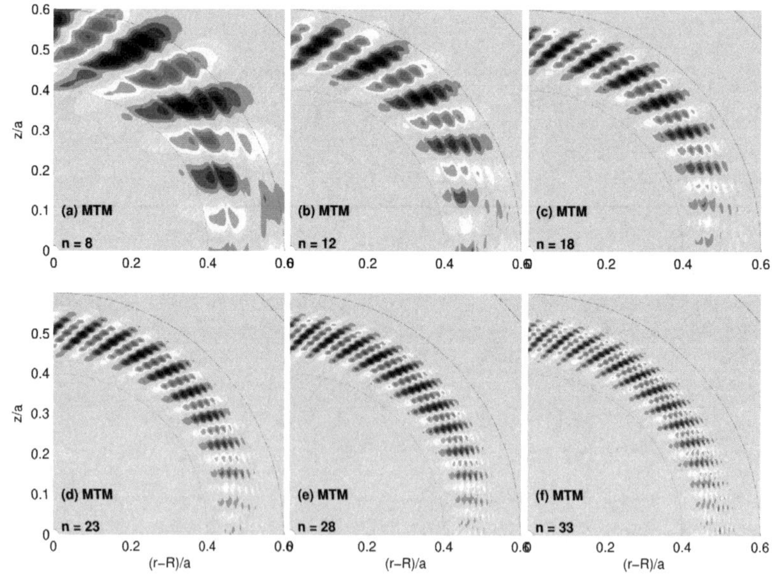

Figure 5. Contour plots for MTM $\tilde{\varphi}$ fluctuations $n = 8, 12, .., 33$, excluding trapped electrons. For clarity, only the first quadrant is shown. (See. Ref. [15] for full poloidal section.) The $\tilde{\varphi}$ exhibits odd-parity in poloidal angle about $\theta = 0$ ($y = 0$ axes)

2.3. Mode Structures with trapped electrons

Inclusion of the non-adiabaic contribution of trapped electron is addressed here. In Figs. 8, 9 the radial Fourier spectrum of $\tilde{\varphi}$ and $\tilde{A}_{\|}$ for each of the poloidal Fourier

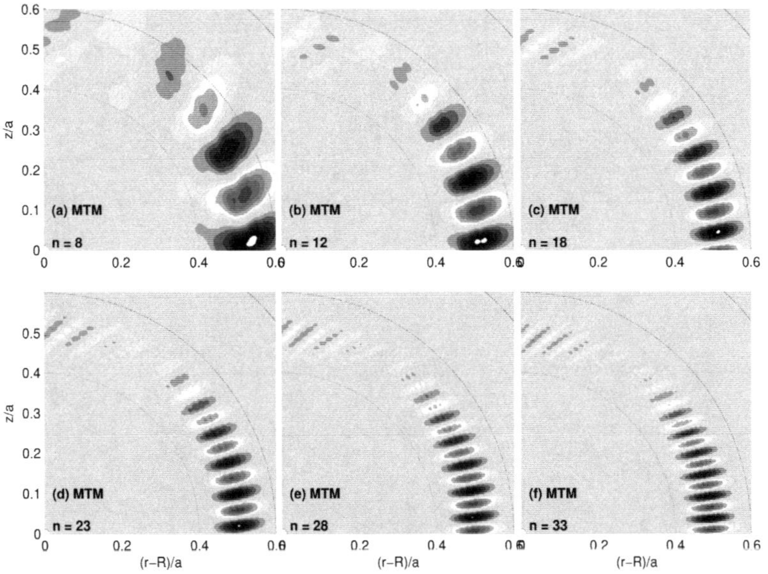

Figure 6. Contour plots for MTM $\tilde{A}_\parallel$ fluctuations $n = 8, 12, .., 33$, excluding trapped electrons. For clarity, only the first quadrant is shown. (See. Ref. [15] for full poloidal section.) The $\tilde{A}_\parallel$ exhibits even-parity in poloidal angle about $\theta = 0$ ($y = 0$ axes)

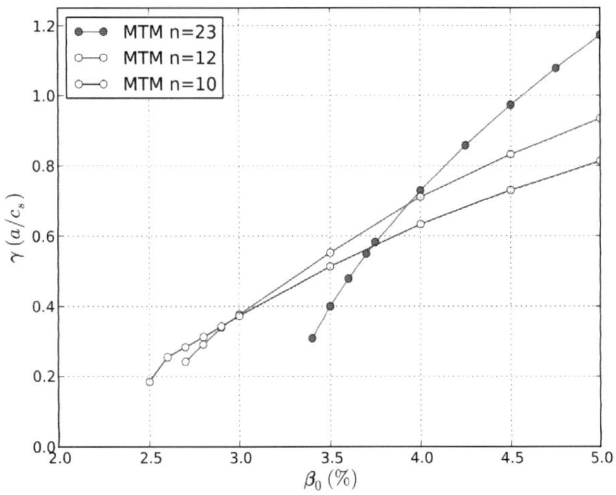

Figure 7. Dependence of MTM growth rate on β_0 for $n = 10, 12$ and 23, excluding trapped electrons. $n = 10$ mode has a β_{crit} less than 2.5%.

components m of the eigenfunctions are plotted. Figs. 10, 11 show the corresponding contour plots. The non-adiabatic response of the trapped elctrons increases the amplitude of higher radial harmonics in $\tilde{\varphi}$ resulting in a broader spectrum in the radial direction and sharper structures on the mode rational surfaces, as seen from the contours in Figs. 10 and more clearly noticed in the plot for $n = 8$.

On the other hand, the eigenfunction $\tilde{A}_{\|}$ is unchanged with the inclusion of trapped electrons. This is expected since $\tilde{j}_{\|}$ is not changed by the trapped electron population. The trapped electrons do not seem to affect the overall mode structures of $\tilde{\varphi}$ and $\tilde{A}_{\|}$ and preserve the parities. For the fastest growing mode $n = 23$, the eigenfrequency (ω_r, γ) (in units of c_s/a) without trapped electrons is $(0.96, 1.17)$, whereas the inclusion of trapped electrons leads to $(1.03, 1.23)$. Similarly, for $n = 12$, the eigenfrequency is $(0.59, 0.92)$ without and $(0.58, 0.83)$ with trapped electrons. The radial and poloidal extent, poloidal variation of the amplitude envelopes are nearly the same, with and without the trapped electrons.

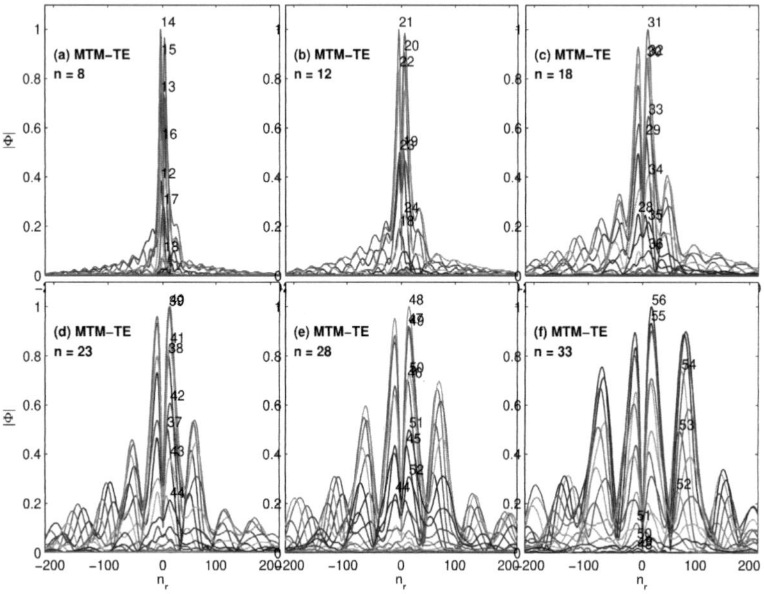

Figure 8. Fourier space plots for MTM $\tilde{\varphi}$ fluctuations $n = 8, 12, .., 33$. Comparison with Fig. 3 shows that inclusion of nonadiabatic trapped electrons bring small scale structures in $\tilde{\varphi}$.

3. Conclusions

For large aspect ratio, collisionless tokamak plasmas, we have presented a high resolution 2D global gyrokinetic stability study of MTM considering passing ions and electrons and the effect of inclusion of the trapped electrons on the mode structures. The model is electromagnetic in nature but neglects parallel magnetic field perturbations, which are important only at very high β. Completely collisionless Microtearing Modes were recently found unstable in large aspect ratio hot tokamaks in the finite temperature gradient region with a finite-beta plasma. The magnetic drift resonance of the passing electron population provides the main destabilizing mechanism, even though the trapped

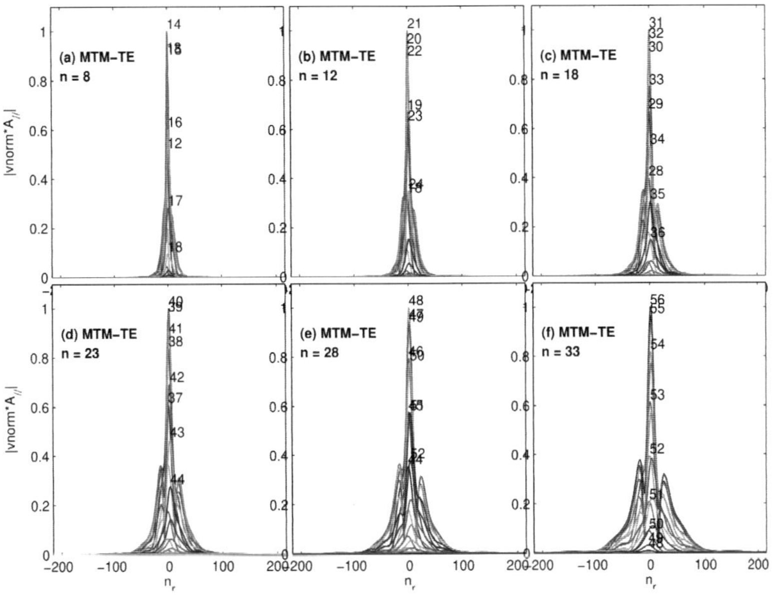

Figure 9. Fourier space plots for MTM $\tilde{A}_{\parallel}$ fluctuations $n = 8, 12, .., 33$, with trapped electrons

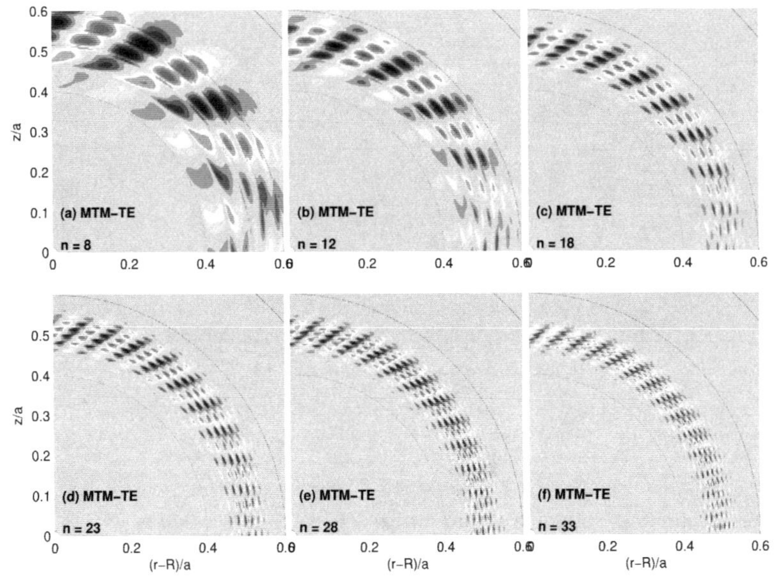

Figure 10. Contour plots for MTM $\tilde{\varphi}$ fluctuations $n = 8, 12, .., 33$. For clarity, only the first quadrant is shown. Sharp radial structures are seen in $\tilde{\varphi}$ due to trapped electrons. The odd symmetry with respect to θ is preserved. As n increases, the modes are seen more localised in r

electron population constitutes a significant proportion. For MTM with trapped electrons neglected, global calculations showed that $n = 23$ is the most dominant mode,

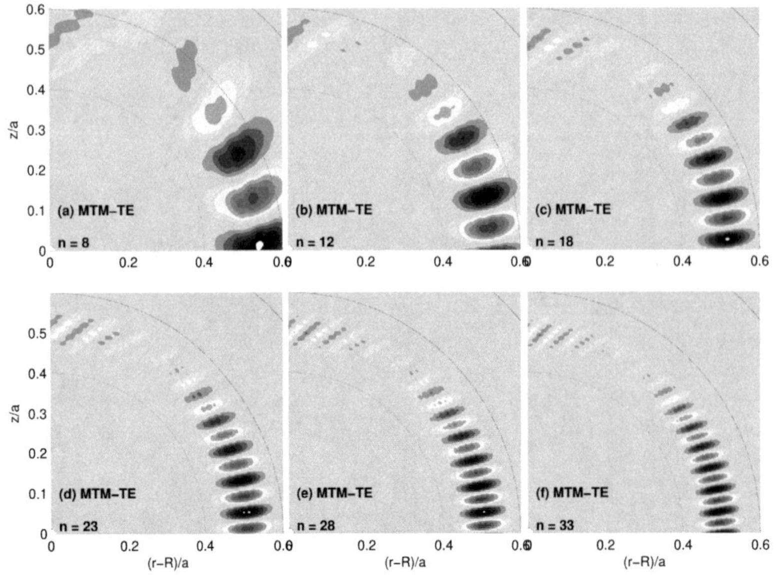

Figure 11. Contour plots for MTM $\tilde{A}_{\parallel}$ fluctuations $n = 8, 12, .., 33$. For clarity, only the first quadrant is shown. Inclusion of trapped electrons does not change $\tilde{A}_{\parallel}$ as shown in Fig. 6. The symmetry is preserved.

at $\beta_0 = 5\%$ and $a/L_{Te} \sim 5.0$, and has a critical β value of about 3.3%. Importantly however, the dependence of the growth dates on β is distinct for different toroidal modes n.

The tearing parity of $\tilde{\varphi}$ and $\tilde{A}_{\parallel}$ are preserved with the inclusion of trapped electrons. The nonadiabatic response of the trapped electrons is observed to excite higher radial Fourier harmonics of $\tilde{\varphi}$ fluctuations, leading to sharp structures in the radial direction. The structure of $\tilde{A}_{\parallel}$ is not strongly affected by the inclusion of trapped electrons, as can be expected. The contribution from trapped electron dynamics moderately affects the linear growth rates of the instability, changing by about 10%. A more detailed study of the influence of trapped electron dynamics and other plasma parameters on MTM will be communicated elsewhere.

[1] R. D. Hazeltine, D. Dobrott, and T. S. Wang, Phys. Fluids 18, 1778 (1975).
[2] J. F. Drake and Y. C. Lee, Phys. Fluids 20, 1341 (1977).
[3] P. J. Catto and M. N. Rosenbluth, Phys. Fluids 24, 243 (1981).
[4] D. J. Applegate, C. M. Roach, J W Connor, S C Cowley, W Dorland, R J Hastie, and N Joiner, Plasma Phys. Controlled Fusion 49(8), 1113 (2007).
[5] W. Guttenfelder, J. Candy, S. M. Kaye, W. M. Nevins, E. Wang, R. E. Bell, G. W. Hammett, B. P. LeBlanc, D. R. Mikkelsen, and H. Yuh, Phys. Rev. Lett. 106, 155004 (2011)
[6] W. Guttenfelder, J. Candy, S. M. Kaye, W. M. Nevins, R. E. Bell, G. W. Hammett, B. P. LeBlanc, and H. Yuh, Phys. Plasmas 19, 022506 (2012)
[7] W. Guttenfelder, J. Candy, S. M. Kaye, W. M. Nevins, R. E. Bell, G. W. Hammett, B. P. LeBlanc, and H. Yuh, Phys. Plasmas 19, 056119 (2012)
[8] K. L. Wong, S. Kaye, D. R. Mikkelsen, J. A. Krommes, K. Hill, R. Bell, and B. LeBlanc, Phys.

Rev. Lett. 99, 135003 (2007)
[9] H. Doerk, F. Jenko, M. J. Pueschel, and D. R. Hatch, Phys. Rev. Lett. 106, 155003 (2011)
[10] H. Doerk, F. Jenko, T. Görler, D. Told, M. J. Pueschel, and D. R. Hatch, Phys. Plasmas 19, 055907 (2012)
[11] D. Carmody, M. J. Pueschel, and P.W. Terry, Phys. Plasmas 20, 052110 (2013)
[12] I. Predebon and F. Sattin, Phys. Plasmas 20, 040701 (2013)
[13] S. Moradi, I. Pusztai, W. Guttenfelder, T. Fülöp and A. Mollén, Nucl. Fusion 53 063025 (2013)
[14] D Dickinson, C M Roach, S Saarelma, R Scannell, A Kirk and H R Wilson, Plasma Phys. Control. Fusion 55 (2013) 074006
[15] Aditya K. Swamy, R. Ganesh, J. Chowdhury, S. Brunner, J. Vaclavik, and L. Villard, Phys. Plasmas 21, 082513 (2014)
[16] S. Brunner, M. Fivaz, T. M. Tran and J. Vaclavik, Phys. Plasmas. **5**(11), 3929 (1998)
[17] S. Brunner, Ph.D thesis, EPFL, Switzerland, Thesis no : 1701 (1997)
[18] G. L. Falchetto, J. Vaclavik and L. Villard, Phys. Plasmas **10**, 1424, (2003)
[19] R. Ganesh, P. Angelino, J. Vaclavik and L. Villard, Phys. Plasmas, **11**, 3106 (2004)
[20] P. Angelino, Ph.D thesis, EPFL, Switzerland, Thesis no : 3559 (2006)
[21] J. Chowdhury, R. Ganesh S. Brunner, J. Vaclavik, L. Villard, and P. Angelino, Phys. Plasmas, **16**, 052507 (2009)
[22] E. A. Belli and J. Candy, Phys. Plasmas 17, 112314 (2010)
[23] L. Vermare, C Angioni, A Bottino, A G Peeters and ASDEX Upgrade Team, J. Phys. Conf Series 123, 012040(2008)
[24] E. Wang, X. Xu, J. Candy, R.J. Groebner, P.B. Snyder, Y. Chen, S.E. Parker, W. Wan, Gaimin Lu and J.Q. Dong, Nucl. Fusion 52 (2012) 103015
[25] R.J. Groebner, C.S. Chang, J.W. Hughes, R. Maingi, P.B. Snyder, X.Q. Xu, J.A. Boedo, D.P. Boyle, J.D. Callen, J.M. Canik, I. Cziegler, E.M. Davis, A. Diallo, P.H. Diamond, J.D. Elder, D.P. Eldon, D.R. Ernst, D.P. Fulton, M. Landreman, A.W. Leonard, J.D. Lore, T.H. Osborne, A.Y. Pankin, S.E. Parker, T.L. Rhodes, S.P. Smith, A.C. Sontag, W.M. Stacey, J. Walk, W. Wan, E.H.-J. Wang, J.G.Watkins, A.E. White, D.G. Whyte, Z. Yan, E.A. Belli, B.D. Bray, J. Candy, R.M. Churchill, T.M. Deterly, E.J. Doyle, M.E. Fenstermacher, N.M. Ferraro, A.E. Hubbard, I. Joseph, J.E. Kinsey, B. LaBombard, C.J. Lasnier, Z. Lin, B.L. Lipschultz, C. Liu, Y. Ma, G.R. McKee, D.M. Ponce, J.C. Rost, L. Schmitz, G.M. Staebler, L.E. Sugiyama, J.L. Terry, M.V. Umansky, R.E.Waltz, S.M. Wolfe, L. Zeng and S.J. Zweben, Nucl. Fusion 53 (2013) 093024

Journal of Physics: Conference Series **561** (2014) 012018 doi:10.1088/1742-6596/561/1/012018

Hamiltonian derivation of a gyrofluid model for collisionless magnetic reconnection

E Tassi

Aix-Marseille Université, Université de Toulon, CNRS Centre de Physique Théorique UMR 7332, 13288 Marseille, France

E-mail: tassi@cpt.univ-mrs.fr

Abstract. We consider a simple electromagnetic gyrokinetic model for collisionless plasmas and show that it possesses a Hamiltonian structure. Subsequently, from this model we derive a two-moment gyrofluid model by means of a procedure which guarantees that the resulting gyrofluid model is also Hamiltonian. The first step in the derivation consists of imposing a generic fluid closure in the Poisson bracket of the gyrokinetic model, after expressing such bracket in terms of the gyrofluid moments. The constraint of the Jacobi identity, which every Poisson bracket has to satisfy, selects then what closures can lead to a Hamiltonian gyrofluid system. For the case at hand, it turns out that the only closures (not involving integro/differential operators or an explicit dependence on the spatial coordinates) that lead to a valid Poisson bracket are those for which the second order parallel moment, independently for each species, is proportional to the zero order moment. In particular, if one chooses an isothermal closure based on the equilibrium temperatures and derives accordingly the Hamiltonian of the system from the Hamiltonian of the parent gyrokinetic model, one recovers a known Hamiltonian gyrofluid model for collisionless reconnection. The proposed procedure, in addition to yield a gyrofluid model which automatically conserves the total energy, provides also, through the resulting Poisson bracket, a way to derive further conservation laws of the gyrofluid model, associated with the so called Casimir invariants. We show that a relation exists between Casimir invariants of the gyrofluid model and those of the gyrokinetic parent model. The application of such Hamiltonian derivation procedure to this two-moment gyrofluid model is a first step toward its application to more realistic, higher-order fluid or gyrofluid models for tokamaks. It also extends to the electromagnetic gyrokinetic case, recent applications of the same procedure to Vlasov and drift-kinetic systems.

1. Introduction

Fluid models represent a widespread and effective tool for investigating important phenomena in fusion plasmas such as instabilities, turbulence and reconnection events. Indeed, fluid models offer a considerable advantage, in terms of required computational resources, with respect to kinetic models. On the other hand, compared to these, they obviously suffer from limitations in the range of scales and frequencies of the phenomena that they can describe. The derivation of sophisticated fluid models aiming at partially remedying such limitations is an active line of research since a long time. An essential part of the problem is related to the closure adopted to truncate the fluid hierarchy of equations obtained by taking moments of a parent kinetic model. A considerable effort, directed also toward applications to space plasma turbulence, has been carried out to derive fluid and gyrofluid models (see, e.g. Refs. [1–8]) by taking moments of kinetic or gyrokinetic models and imposing closures designed to satisfy desired properties, such as for instance consistency with kinetic linear theory or energy conservation. Relatively little attention has been paid, however, to ensure that the adopted closures satisfy a further criterion, which is the preservation of a Hamiltonian structure in the derivation of the non-dissipative part of the model. In other words, the parent kinetic model, in its non-dissipative limit, is supposed to possess a Hamiltonian structure, and, unless forcing or dissipative terms are voluntarily added when applying the fluid closure, the resulting fluid model is also supposed to possess a Hamiltonian structure. Apart from being an important feature from a fundamental point of view, the preservation of a Hamiltonian structure provides also some practical advantages. On one hand, it provides an unambiguously defined conserved total energy (the Hamiltonian functional). Moreover, given that fluid models for plasmas are often expressed in Eulerian form, their Hamiltonian structure is typically noncanonical (see, e.g. Ref. [9]), which implies the existence of additional conserved quantities, denoted as Casimirs. Their knowledge can provide useful information on the nonlinear dynamics of the fluid model. In the presence of dissipation, the decay rate and the cascade of such invariants can provide a further way to characterize plasma turbulence. Moreover, they can be used as additional invariants to test the conservation properties of numerical codes. Namely because with Hamiltonian plasma models one has to deal with noncanonical Poisson brackets, a delicate point in preserving the Hamiltonian structure throughout the derivation, is that of not violating the Jacobi identity. The latter is one of the properties defining a Poisson bracket. Although it is always satisfied for canonical Poisson brackets, showing its validity for noncanonical Poisson brackets is often far from obvious.

Recently, derivation procedures that preserve the Hamiltonian structure of the parent model have been applied to kinetic and drift-kinetic models [10–12]. A key element in these procedures consists of inserting a generic fluid closure relation in the bracket obtained from the parent kinetic model, and then imposing the constraint that the Jacobi identity be satisfied. Such constraint selects fluid closures that automatically lead to Hamiltonian fluid models. The purpose of the present paper is two-fold. In the first place, we show in Sec. 2 the existence of a Hamiltonian structure for a simple gyrokinetic electromagnetic model in slab geometry. In the second place, in Sec. 3 we apply the procedure of Ref. [11] to this model in order to derive a two-moment Hamiltonian gyrofluid model for magnetic reconnection driven by electron inertia. This type of reconnection is believed to be relevant for sawtooth oscillations in tokamaks (see, e.g. Ref. [13]). Because of the relative simplicity of the models under consideration, and of the limited number of fluid moments involved, the present results cannot be directly relevant for a realistic modelling of tokamak plasmas. They provide, however, an explicit example of a structure-preserving derivation which is a first step toward applications to more realistic higher-order fluid models. They also provide new knowledge about the Hamiltonian structures of the more basic gyrokinetic and gyrofluid models involved in the paper. The results are summarized in Sec. 4, where also limitations of the present analysis and possible future directions of investigation are discussed.

2. Parent gyrokinetic model and its Hamiltonian structure

In a Cartesian coordinate system (x, y, z) we consider a gyrokinetic model in slab geometry in the δf approximation. The model consists of the following evolution equations

$$\frac{\partial g_i}{\partial t} + \frac{c}{B}\left[J_0\left(\phi - \frac{v}{c}A\right), g_i\right] + v\frac{\partial}{\partial z}\left(g_i + \frac{\mathcal{F}_i}{T_i}eJ_0\left(\phi - \frac{v}{c}A\right)\right) = 0, \tag{1}$$

$$\frac{\partial g_e}{\partial t} + \frac{c}{B}\left[\phi - \frac{v}{c}A, g_e\right] + v\frac{\partial}{\partial z}\left(g_e - \frac{\mathcal{F}_e}{T_e}e\left(\phi - \frac{v}{c}A\right)\right) = 0, \tag{2}$$

of the quasi-neutrality relation

$$\frac{e^2 n_0}{T_i}(\Gamma_0 - 1)\phi + e\int d\mathcal{W}(J_0 g_i - g_e) = 0, \tag{3}$$

and of Ampère's law

$$\nabla_\perp^2 A - 4\pi\frac{e^2}{c^2}n_0\left(\frac{\Gamma_0}{M_i}+\frac{1}{M_e}\right)A + 4\pi\frac{e}{c}\int d\mathcal{W} v(J_0 g_i - g_e) = 0. \tag{4}$$

Eqs. (1) and (2) govern the evolution of $g_i = f_i + \mathcal{F}_i(e/T_i)(v/c)J_0 A$ and $g_e = f_e - \mathcal{F}_e(e/T_e)(v/c)A$, where $f_{i,e}(x,y,z,v,\mu,t)$ indicate the perturbations of the ion and electron equilibrium distribution functions $\mathcal{F}_{i,e}$, respectively, which are defined as

$$\mathcal{F}_i(v,\mu) = n_0\left(\frac{M_i}{2\pi T_i}\right)^{3/2} \mathrm{e}^{-\frac{M_i v^2 + 2\mu B}{2T_i}}, \qquad \mathcal{F}_e(v,\mu) = n_0\left(\frac{M_e}{2\pi T_e}\right)^{3/2} \mathrm{e}^{-\frac{M_e v^2 + 2(M_e/M_i)\mu B}{2T_e}}. \tag{5}$$

We indicate with v the velocity coordinate along the magnetic guide field direction (which we take to be the z direction) and with μ the ion magnetic moment. Ion and electron masses are denoted as M_i and M_e, respectively, whereas $T_{i,e}$ indicate constant temperatures characterizing the ion and electron equilibrium distribution functions. The constants c and e indicate the speed of light and the proton charge, respectively. The flux function $A(x,y,z,t)$ is related to the magnetic field by $\mathbf{B} = \nabla A \times \hat{z} + B\hat{z}$, with B indicating the uniform and constant amplitude of the guide field component. The constant n_0 is the equilibrium density of both electron and ions, whereas $\phi(x,y,z,t)$ denotes the electrostatic potential. The symbol $\nabla_\perp^2$ in Eq. (4) indicates the Laplacian operator restricted to the $xy-$plane. The integrations in Eqs. (3) and (4) are carried out over the element $d\mathcal{W} = 2\pi B d\mu/M_i$. We assume that all fields are periodic along the x, y and z directions and that $f_{i,e}$ decay to zero sufficiently fast as $v \to \infty$ and $\mu \to +\infty$. The operators J_0 and Γ_0 are the standard operators appearing in gyrokinetic theory and correspond to the multiplication, in Fourier space, times the zeroth order Bessel function $J_0(k_\perp\sqrt{2\mu B/M_i\omega_{ci}^2})$ and times the function $I_0(k_\perp^2\rho_i^2)\mathrm{e}^{-k_\perp^2\rho_i^2}$, respectively, where $k_\perp^2 = k_x^2 + k_y^2$ and where we indicated with $\omega_{ci} = eB/M_i c$ the ion cyclotron frequency, and with $\rho_i = \sqrt{T_i/M_i}/\omega_{ci}$ the ion thermal gyroradius. Finally, the canonical bracket appearing in Eqs. (1)-(2) is defined by $[f,g] = \partial_x f \partial_y g - \partial_y f \partial_x g$, for two functions f and g.

In the model (1)-(4) we neglected, as is customary by virtue of the small electron/ion mass ratio, the effects of the gyration radius for the electrons. However, the results present in the paper, can easily be extended to account for them as well.

Showing that the model (1)-(4) possesses a Hamiltonian structure amounts to show that it can be cast in the form

$$\frac{\partial \chi}{\partial t} = \{\chi, H\}, \tag{6}$$

where χ indicates a vector of dynamical variables, corresponding, in our case, to g_i and g_e. $H = H(\chi)$ is a functional, denoted as Hamiltonian, and $\{,\}$ is a Poisson bracket, that is an antisymmetric bilinear operation satisfying the Leibniz identity $\{FG,H\} = F\{G,H\} + G\{F,H\}$, $\forall F,G,H$ and the Jacobi identity

$$\{\{F,G\},H\} + \{\{G,H\},F\} + \{\{H,F\},G\} = 0, \qquad \forall F,G,H. \tag{7}$$

Because the Hamiltonian physically represents the conserved total energy of the system, a good candidate for H for our model is the conserved quantity [8]

$$\begin{aligned} H(g_i,g_e) = \frac{1}{2}\int d^3x d\mathcal{W}\Bigg[&\frac{T_i}{\mathcal{F}_i}g_i^2 + \frac{T_e}{\mathcal{F}_e}g_e^2 + e(J_0 g_i - g_e)L_0^{-1}\int d\mathcal{W}'(J_0 g_i - g_e) \\ &- e\frac{v}{c}(J_0 g_i - g_e)L_1^{-1}\int d\mathcal{W}'(J_0 g_i - g_e)\Bigg]. \end{aligned} \tag{8}$$

The expression (8) assumes the validity of the relations

$$\phi = L_0^{-1} \int d\mathcal{W}(J_0 g_i - g_e), \qquad A = L_1^{-1} \int d\mathcal{W} v (J_0 g_i - g_e), \tag{9}$$

that is the existence of invertible operators $L_{0,1}$ that permit, from Eqs. (3) and (4), to express ϕ and A in terms of g_i and g_e. In Fourier space it is not difficult to see that this inversion can be carried out for each $\mathbf{k}$, apart from the case concerning Eq. (3) and modes with $k_x = k_y = 0$, for which $\Gamma_0 - 1 = 0$. We restrict then to electrostatic potentials such that $\phi_{0,0,k_z} = 0$.

The operators L_0^{-1} and L_1^{-1} also turn out to satisfy the symmetry property $\int d^3x f(\mathbf{x}) L_{0,1}^{-1} g(\mathbf{x}) = \int d^3x g(\mathbf{x}) L_{0,1}^{-1} f(\mathbf{x})$ and the operator J_0 is known (see, e.g. Ref. [5]) to satisfy the relation $\int d^3x d\mathcal{W} f(\mathbf{x}, v, \mu) J_0 g(\mathbf{x}, v, \mu) = \int d^3x d\mathcal{W} g(\mathbf{x}, v, \mu) J_0 f(\mathbf{x}, v, \mu)$. By virtue of these properties, we obtain that the functional derivatives of H with respect to $g_{i,e}$ can be written as

$$\frac{\delta H}{\delta g_i} = \frac{T_i}{\mathcal{F}_i} g_i + e J_0 \left(\phi - \frac{v}{c} A\right), \qquad \frac{\delta H}{\delta g_e} = \frac{T_e}{\mathcal{F}_e} g_e - e \left(\phi - \frac{v}{c} A\right). \tag{10}$$

Direct calculations show that, when using (8) as Hamiltonian, with the help of Eqs. (10) and of integration by parts, the Poisson bracket that yields Eqs. (1)-(2), accounting for Eqs. (3) and (4), is given by

$$\{F, G\} = \sum_{s=i,e} \int d^3x d\mathcal{W} \left(\mp \frac{c}{eB} g_s [F_{g_s}, G_{g_s}] - v \frac{\mathcal{F}_s}{T_s} F_{g_s} \frac{\partial}{\partial z} G_{g_s} \right), \tag{11}$$

where the upper (lower) sign should be taken for ions (electrons). The subscripts on functionals, on the other hand, indicate functional derivatives, so that, for instance $F_{g_s} = \delta F / \delta g_s$.

The operation (11) is the direct sum of two independent Poisson brackets and consequently satisfies the properties of a Poisson bracket. We conclude then that the gyrokinetic model (1)-(4) possesses a Hamiltonian structure. A similar Hamiltonian structure was presented in Ref. [15] for the case of a drift-kinetic model.

3. Derivation of the Hamiltonian gyrofluid model

We define moments of $g_{i,e}$ as $P_{i,e_{jk}} = \int d\mathcal{W} v^j \mu^k g_{i,e}$, where j and k are non-negative integers. Functionals $F(g_i, g_e)$ can be expressed as functionals of the infinite moments $P_{i,e_{jk}}$ by using scalar invariance, so that $F(g_i, g_e) = \bar{F}(\{P_{i,e_{jk}}\}_{j,k \in \mathbb{N}})$ and the functional derivatives consequently transform as $F_{g_{i,e}} = \sum_{j,k \in \mathbb{N}} v^j \mu^k \bar{F}_{i,e_{jk}}$, where we denoted $\bar{F}_{i,e_{jk}} = \delta \bar{F} / \delta P_{i,e_{jk}}$. Inserting this expression into (11) one formally obtains the Poisson bracket in terms of all the moments, which reads

$$\{F, G\} = \sum_{s=i,e} \sum_{m,n,j,k \in \mathbb{N}} \int d^3x \left(\mp \frac{c}{eB} P_{s_{m+j\, n+k}} [F_{s_{mn}}, G_{s_{jk}}] - F_{s_{mn}} \frac{\partial}{\partial z} G_{s_{jk}} \int d\mathcal{W} v^{m+j+1} \mu^{n+k} \frac{\mathcal{F}_s}{T_s} \right), \tag{12}$$

In practice, however, one is interested in deriving a Hamiltonian fluid model, describing the evolution of a finite number of moments, for instance the first $M(N)$ moments for the parallel (perpendicular) direction, for both the ions and the electrons, where M and N are two positive integers. This means that the functionals of interest will be of the kind $F(P_{i_{00}}, P_{i_{10}}, P_{i_{01}} \cdots P_{i_{MN}}, P_{e_{00}}, P_{e_{10}}, P_{e_{01}} \cdots P_{e_{MN}})$ and will no longer depend on the infinite set of moments. The operation (12) between two such functionals, however, is not closed, in the sense that $\{F, G\}$, in general, will not be again a functional of the same kind, but it will depend also on moments of order higher than $M(N)$ in the parallel (perpendicular) direction. In order

Journal of Physics: Conference Series **561** (2014) 012018 doi:10.1088/1742-6596/561/1/012018

to obtain a closed system, the way we follow is that of modifying the Poisson bracket, imposing that such extra higher order moments be functions of the first $M(N)$ moments in the parallel (perpendicular) direction. Imposing such constraint in the bracket, however, can break the properties defining a Poisson bracket. In particular, the Jacobi identity can be violated by this operation. However, in some cases, which we refer to as to Hamiltonian closures, imposing the closure relation in the Poisson bracket, preserves the properties of the Poisson bracket. Hamiltonian fluid models can then be derived adopting Hamiltonian closures.

In particular, we consider here the case $M = 1$ and $N = 0$, that is we restrict to functionals of the first two moments in the parallel direction. The bracket (12) then becomes

$$\{F,G\} = \sum_{s=i,e} \int d^3x \left(\mp \frac{c}{eB} P_{s0}[F_{s0}, G_{s0}] \mp \frac{c}{eB} P_{s1}([F_{s1}, G_{s0}] + [F_{s0}, G_{s1}]) \mp \frac{c}{eB} P_{s2}[F_{s1}, G_{s1}] \right.$$
$$\left. - \frac{n_0}{M_s} \left(F_{s1} \frac{\partial}{\partial z} G_{s0} + F_{s0} \frac{\partial}{\partial z} G_{s1} \right) \right), \tag{13}$$

where, in order to simplify the notation, given that the index for the powers of μ has been fixed equal to zero, we have defined $P_{s_j} = P_{s_{j0}}$, for $j = 0, 1, 2$. It is evident that, as above anticipated, the result of the operation $\{F, G\}$ in (13) depends also on P_{s_2} and consequently the resulting equations of motion cannot lead to a closed fluid model, for any Hamiltonian depending on P_{s_0} and P_{s_1} only. One remedy to this is to look for two functions $\mathcal{P}_s(P_{s_0}, P_{s_1})$, if they exist, such that, when inserting $P_{s_2} = \mathcal{P}_s$ into Eq. (13), the resulting operation (which will now be closed) still satisfies the Jacobi identity and can consequently still provide a Poisson bracket. It turns out that, for brackets of the form (13), it has been shown [11] that the constraint of the Jacobi identity (7) selects, among all the possible closures (not considering, however, those involving integro/differential operators or an explicit dependence on the spatial coordinates), the following one:

$$P_{s_2} = a_s P_{s_0}, \qquad \text{for } s = i, e, \tag{14}$$

where a_s are two arbitrary constants, which dimensionally must be homogeneous to a velocity squared. We remark that the Hamiltonian closure (14) does not depend on first order parallel moments, which is a consequence of having chosen a Maxwellian with zero mean flow as equilibrium distribution function [11].

Once that a valid Poisson bracket, inherited from the original gyrokinetic bracket (11), has been found, in principle, Hamiltonian fluid models can be constructed by choosing the total energy functional (the Hamiltonian) $H(P_{s_0}, P_{s_1})$ that one would like to be conserved, and then using the definition of Hamiltonian system (6) to get the four dynamical model equations, by replacing χ with the four moments P_{s_0} and P_{s_1}.

A relevant example of Hamiltonian closure, which is also consistent with the Hamiltonian of the original gyrokinetic model, is obtained by choosing $a_s = T_s/M_s$. This corresponds to an isothermal closure, based on the equilibrium temperature, for the linearized pressure. The Poisson bracket then reads

$$\{F,G\} = \sum_{s=i,e} \int d^3x \left(\mp \frac{c}{eB} P_{s0}[F_{s0}, G_{s0}] \mp \frac{c}{eB} P_{s1}([F_{s1}, G_{s0}] + [F_{s0}, G_{s1}]) \mp \frac{c}{eB} \frac{T_s}{M_s} P_{s0}[F_{s1}, G_{s1}] \right.$$
$$\left. - \frac{n_0}{M_s} \left(F_{s1} \frac{\partial}{\partial z} G_{s0} + F_{s0} \frac{\partial}{\partial z} G_{s1} \right) \right). \tag{15}$$

Analogously to Ref. [8], one can then make use of the approximate decomposition

$$g_s = \mathcal{F}_s \left(\frac{P_{s0}}{n_0} + \frac{M_s}{T_s} v \frac{P_{s1}}{n_0} \right), \qquad \text{for } s = i, e, \tag{16}$$

which respects the moments definition as well as the isothermal closure relation. Inserting the expression (16) into Eq. (8) yields

$$H(P_{i_0}, P_{i_1}, P_{e_0}, P_{e_1}) = \frac{1}{2}\sum_{s=i,e}\int d^3x\left(\frac{T_s}{n_0}P_{s_0}^2 + M_s\frac{P_{s_1}^2}{n_0}\right) + \frac{e}{2}\int d^3x(\Gamma_1 P_{i_0} - P_{e_0})L_0^{-1}(\Gamma_1 P_{i_0} - P_{e_0})$$
$$- \frac{e}{2c}\int d^3x(\Gamma_1 P_{i_1} - P_{e_1})L_1^{-1}(\Gamma_1 P_{i_1} - P_{e_1}), \tag{17}$$

where we made use of the common closure approximations (see, e.g. Ref. [8]) $\Gamma_1 = \Gamma_0^{1/2}$, $\int d\mathcal{W} J_0 g_i = \Gamma_1 P_{i_0}$ and $\int d\mathcal{W} J_0 v g_i = \Gamma_1 P_{i_1}$. From such closure approximations it follows also that, using Eqs. (3) and (4), the following well-known relations hold:

$$\nabla_\perp^2 A = -\frac{4\pi e}{c}n_0(\Gamma_1 u_i - u_e), \qquad n_0 e\frac{\Gamma_0 - 1}{T_i}\phi = -(\Gamma_1 n_i - n_e), \tag{18}$$

where we introduced the ion and electron density fluctuations $n_{i,e} = \int d\mathcal{W} f_{i,e}$ as well as the parallel fluid velocities $u_{i,e} = \int d\mathcal{W} v f_{i,e}/n_0$. Making use of Eqs. (18), the fluid Hamiltonian can be rewritten, in a physically more transparent way, as

$$H = \frac{1}{2}\sum_{s=i,e}\int d^3x\left(\frac{T_s}{n_0}n_s^2 + M_s n_0 u_s^2\right) - \frac{e^2 n_0}{2T_i}\int d^3x\phi(\Gamma_0 - 1)\phi + \frac{1}{8\pi}\int d^3x|\nabla_\perp A|^2. \tag{19}$$

The first and the second term in Eq. (19) account for the thermal and kinetic energy, whereas the third and the fourth term represent the electrostatic energy due to the polarization term and the magnetic energy, respectively.

Due to the symmetry of the operators $L_{0,1}^{-1}$ and Γ_1 involved in Eq. (17), one obtains the relations

$$\frac{\delta H}{\delta P_{i_0}} = \frac{T_i}{n_0}P_{i_0} + e\Gamma_1\phi, \qquad \frac{\delta H}{\delta P_{i_1}} = \frac{M_i}{n_0}P_{i_1} - \frac{e}{c}\Gamma_1 A, \tag{20}$$

$$\frac{\delta H}{\delta P_{e_0}} = \frac{T_e}{n_0}P_{e_0} - e\phi, \qquad \frac{\delta H}{\delta P_{e_1}} = \frac{M_e}{n_0}P_{e_1} + \frac{e}{c}A. \tag{21}$$

From Eq. (6), using the Poisson bracket (15), the Hamiltonian (17) and with the help of the relations (20)-(21), one can finally obtain the Hamiltonian gyrofluid model

$$\frac{\partial n_i}{\partial t} + \frac{c}{B}[\Gamma_1\phi, n_i] - \frac{n_0}{B}[\Gamma_1 A, u_i] + n_0\frac{\partial u_i}{\partial z} = 0, \tag{22}$$

$$\frac{\partial D}{\partial t} + \frac{c}{B}[\Gamma_1\phi, D] - \frac{T_i}{M_i B}[\Gamma_1 A, n_i] + \frac{T_i}{M_i}\frac{\partial n_i}{\partial z} + \frac{e n_0}{M_i}\frac{\partial \Gamma_1\phi}{\partial z} = 0, \tag{23}$$

$$\frac{\partial n_e}{\partial t} + \frac{c}{B}[\phi, n_e] - \frac{n_0}{B}[A, u_e] + n_0\frac{\partial u_e}{\partial z} = 0, \tag{24}$$

$$\frac{\partial F}{\partial t} + \frac{c}{B}[\phi, F] - \frac{T_e}{M_e B}[A, n_e] + \frac{T_e}{M_e}\frac{\partial n_e}{\partial z} - \frac{e n_0}{M_e}\frac{\partial \phi}{\partial z} = 0, \tag{25}$$

where we introduced the quantities $D = P_{i_1} = n_0(u_i + e\Gamma_1 A/M_i c)$ and $F = P_{e_1} = n_0(u_e - eA/M_e c)$, corresponding to the first order fluid momenta and which represent the parallel canonical momenta of the ion and electron fluid, respectively.

By multiplying Eq. (25) times M_e one sees that, because of terms proportional to the electron inertia, the magnetic flux A is not frozen into any fluid. Indeed, in spite of the Hamiltonian character of the model, the presence of electron inertia violates the frozen-in condition, allowing for magnetic reconnection.

The model (22)-(25) and its Hamiltonian structure were presented in Ref. [14], but in that work the Hamiltonian structure was "guessed" a posteriori, after deriving the model from the more general gyrofluid model of Ref. [6]. The present derivation, on the other hand, originates from the gyrokinetic description, and guarantees, by construction, the Hamiltonian structure of the resulting model.

The above gyrofluid model is characterized by the presence of four Casimirs, that is functionals C, such that $\{C, F\} = 0$, for every functional F. From this definition and from Eq. (6) it follows that $dC/dt = 0$, so that Casimir are invariants of motion and consequently provide information about constraints on the dynamics, also in the nonlinear phase. The four Casimirs for the gyrofluid model can be written as $\int d^3x C_{s\pm}$ for $s = i, e$, where

$$C_{s\pm} = \sqrt{\frac{M_s}{T_s}} P_{s1} \pm P_{s0}. \tag{26}$$

The parent gyrokinetic bracket (11), on the other hand, possesses the two Casimirs $\int d^3x dW g_s$, for $s = i, e$. Note that the Casimirs of the gyrofluid model are "inherited" from those of the gyrokinetic model, in the sense that they can be obtained from those by using the decomposition (16) and replacing v with plus or minus the equilibrium thermal speed $\sqrt{T_{i,e}/M_{i,e}}$.

In the two-dimensional limit $\partial/\partial_z = 0$ the Casimirs of both the gyrokinetic and the gyrofluid model extend to infinite families, given by $\int d^3x dW \mathcal{G}_s(g_s)$ and $\int d^3x \mathcal{C}_{s\pm}(C_{s\pm})$, respectively, for $s = i, e$, where $\mathcal{G}_s$ and $\mathcal{C}_{s\pm}$ are arbitrary functions. The relevance of the Casimirs of the gyrofluid model for two-dimensional reconnection has been shown in Ref. [16].

4. Conclusions

In this paper we have shown that the electromagnetic gyrokinetic model (1)-(4) possesses a noncanonical Hamiltonian structure. From such model we have then derived a gyrofluid model, which, by construction, also possesses a Hamiltonian structure. The resulting model corresponds to a known model [14] for magnetic reconnection. The derivation, however, is new and shows, by taking advantage of the results of Ref. [11], that the constraint of the Jacobi identity selects this model as a Hamiltonian model derived consistently from the Poisson bracket and the Hamiltonian of the parent gyrokinetic model. The Casimir invariants of the gyrokinetic parent model have been shown to be related to those of the final gyrofluid model, which were known to play a role in the cascade toward small scale characterizing collisionless reconnection. Clearly the present results possess evident limitations and they only represent a first step in a long term project aiming at deriving Hamiltonian closures for more realistic models. An obvious direction for improvement is that of including elements of tokamak geometry and physics in the Hamiltonian model. Obvious examples in this context would be toroidal geometry , magnetic, temperature and density equilibrium gradients (although the latter have already been partially treated in Ref. [12]). Perhaps even more fundamental is the problem of considering Hamiltonian closures at higher-order moments. Advanced fluid and gyrofluid models such as those of Refs. [1–8] reduced the gap between kinetic and fluid descriptions by evolving moments up to the heat fluxes, in both the parallel and perpendicular directions. A Hamiltonian derivation of three or four-moment models from kinetic theories is a natural step for the future, although the difficulty in identifying Hamiltonian closures might rapidly and nonlinearly increase when moving to higher order moments.

Acknowledgments

The author acknowledges useful discussions with P.J. Morrison and with the Nonlinear Dynamics team of the Centre de Physique Théorique. Financial support was received from the Agence

Nationale de la Recherche (ANR GYPSI n. 2010 BLAN 941 03) and from the CNRS through the PEPS project GEOPLASMA2.

References

[1] G.W. Hammett, F.W. Perkins, Phys. Rev. Lett., **64**, 3019 (1990).
[2] A. Brizard, Phys. Fluids B, **4**, 1213 (1992).
[3] M.A. Beer, G.W. Hammett, Phys. Plasmas, **3**, 4046 (1996).
[4] P.B. Snyder, G.W. Hammett, W. Dorland, Phys. Plasmas, **4**, 3974 (1997).
[5] W. Dorland, G.W. Hammett, Phys. Fluids B, **5**, 812 (1993).
[6] P.B. Snyder, G.W. Hammett, Phys. Plasmas, **8**, 3199 (2001).
[7] T. Passot, P.L. Sulem, Phys. Plasmas, **11**, 5173 (2004).
[8] B. Scott, Phys. Plasmas, **17**, 102306 (2010).
[9] P.J. Morrison, Rev. Mod. Phys., **70**, 467 (1998).
[10] M. Perin, C. Chandre, P.J. Morrison, E. Tassi, Annals of Physics **348**, 50 (2014).
[11] E. Tassi, J. Phys. A: Math. Theor., **47**, 195501 (2014).
[12] E. Tassi, Eur. Phys. J. D, **68**, 196 (2014).
[13] D. Biskamp, Magnetic Reconnection in Plasmas, Cambridge University Press (2000).
[14] F.L. Waelbroeck, E. Tassi, Commun Nonlinear Sci Numer Simulat, **17**, 2171 (2012).
[15] H.J. de Blank, Proc. of the 28th EPS Conference on Contr. Fusion and Plasmas Phys., Funchal, 18-22 June 2001, ECA, **25A**, 1709 (2001).
[16] L. Comisso, D. Grasso, E. Tassi, F.L. Waelbroeck, Phys. Plasmas, **19**, 042103 (2012).

Journal of Physics: Conference Series **561** (2014) 012019 doi:10.1088/1742-6596/561/1/012019

Comparisons of the nonlinear and the quasilinear model for the bump-on-tail instability with phase decorrelation

S Tholerus, T Hellsten, and T Johnson

KTH Royal Institute of Technology, School of Electrical Engineering, Department of Fusion Plasma Physics, SE-100 44 Stockholm, Sweden

E-mail: simon.tholerus@ee.kth.se

Abstract. The dynamics of discrete global modes in a toroidal plasma interacting with an energetic particle distribution is studied, and in particular when the dynamics of the system using the nonlinear and quasilinear descriptions are macroscopically similar. The dynamics can be described with a nonlinear bump-on-tail model in a two-dimensional phase space of particles. A Monte Carlo framework is developed for this model with an included decorrelation of the wave-particle phase, which is used to model extrinsic stochastisation of the wave-particle interactions. From this description, a quasilinear version of the model is also developed, which is described by a diffusive process in energy space due to the added phase decorrelation. Due to the reduced dimensionality of phase space, the quasilinear description is typically less computationally demanding than the nonlinear description. The purpose of the studies is to find conditions when a quasilinear model sufficiently describes the same phenomena of the wave-plasma interactions as a nonlinear model does. Via numerical and theoretical parameter studies, regimes where the two models overlap macroscopically are found. These regimes exist above a given threshold of the strength of the decorrelation, where coherent phase space structures are destroyed on time scales shorter than characteristic time scales of nonlinear particle motion in phase space close to the wave-particle resonance. Specifically for the quasilinear model, a theoretical value of the time scale of quasilinear flattening is derived and numerically verified.

1. Introduction

Wave-particle interaction plays an important role in plasma physics for heating with waves and for transport caused by microinstabilities. Instabilities appear when the distribution function in energy increases along the characteristic describing the wave-particle interactions. The dynamics is either studied with a quasilinear approximation or with a fully nonlinear model, and comparisons are made to evaluate the applicability of the quasilinear approximation. The nonlinear bump-on-tail model for the modeling of discrete global modes was developed by Berk *et al* [1, 2], and has been extensively studied [1–6] (to name a few references). A parameter quantifying extrinsic stochastisation of the wave-particle interactions has been introduced to this model. Above a certain threshold of this stochastisation parameter, the kinetic equation of the wave-particle interaction can be replaced by a diffusion equation in particle energy, independent of the wave-particle phase. Such a description, with a diffusion coefficient similar to the standard quasilinear diffusion of weakly turbulent plasmas [7, 8], is identified as the quasilinear limit of the nonlinear wave-particle interaction model.

2. Model equations

2.1. Nonlinear Monte Carlo model

Based on an action-angle description of the guiding center Hamiltonian in an axisymmetric toroidal system with slowly varying electromagnetic fields [9], it was shown by Berk, Breizman and collaborators [1, 2] that the Alfvén eigenmode-particle system can be described in a region of phase space locally around the wave-particle resonance using a one-dimensional bump-on-tail model. Following these derivations, using proper variable substitutions, and adding an *ad hoc* collision operator acting on the energetic particles, the wave-particle system can be reduced to the following set of equations:

$$\frac{\partial f}{\partial \tau} + u\frac{\partial f}{\partial \phi} + \mathrm{Re}\left[A\mathrm{e}^{-\mathrm{i}\phi}\right]\frac{\partial f}{\partial u} = \left.\frac{\mathrm{d}f}{\mathrm{d}\tau}\right|_{\mathrm{coll}}, \tag{1}$$

$$\frac{\mathrm{d}A}{\mathrm{d}\tau} = -\int \mathrm{d}\phi\,\mathrm{d}u\, f(\phi,u,\tau)\mathrm{e}^{\mathrm{i}\phi} - \gamma_{\mathrm{d}}A, \tag{2}$$

where $f(\phi, u, \tau)$ is the near-resonance energetic ("bump") distribution, (ϕ, u) is the particle position-energy phase space, τ is a parametrization of the time, $A(\tau)$ is the complex amplitude of the eigenmode ($\arg(A) - \phi$ is the relative wave-particle phase), $\mathrm{d}f/\mathrm{d}\tau|_{\mathrm{coll}}$ is the collision operator acting on the bump distribution, and γ_{d} is an explicit wave damping rate, e.g. due to interactions with a thermal background distribution of particles. Assuming that the amplitude evolves much slower than the phase space evolution of particles and taking $\mathrm{d}f/\mathrm{d}\tau|_{\mathrm{coll}} = 0$, the system approximately becomes a pendulum equation, with particles deeply trapped by the wave field oscillating at a frequency $\omega_{\mathrm{B}} = \sqrt{|A|}$ and a trapped region within $|u| < \sqrt{2(|A| - \mathrm{Im}[A\mathrm{e}^{\mathrm{i}\phi}])}$.

In the absence of sources and sinks, applied by the conditions $\mathrm{d}f/\mathrm{d}\tau|_{\mathrm{coll}} = 0$ and $\gamma_{\mathrm{d}} = 0$, the total energy, expressed as

$$U_{\mathrm{tot}} = \frac{|A|^2}{2} + \int \mathrm{d}\phi\,\mathrm{d}u\, f(\phi,u,\tau)u, \tag{3}$$

is a conserved quantity of the wave-particle system. For this considered case, given an initial low-amplitude perturbation of the eigenmode and a positive derivative of the particle distribution with respect to energy around the wave-particle resonance, the evolution of the amplitude perturbation becomes exponentially growing in time after some initial mixing of the phase space distribution of particles [10]. The *linear growth rate* of the amplitude is given by $\gamma_{\mathrm{L}} = \pi/2 \times \mathrm{d}F(u,0)/\mathrm{d}u|_{u=0}$, where $F(u,\tau) = \int \mathrm{d}\phi\, f(\phi,u,\tau)$. In the nonlinear simulations presented in this paper, a sinusoidal perturbation in ϕ is applied to the initial distribution function, since an initial flat ϕ distribution is in unstable equilibrium, for which the dynamics depend critically on statistical noise in the phase space distribution. An applied perturbation makes the results less dependent on statistical fluctuations.

The quasilinear model is based on the assumption that wave-particle interactions are extrinsically decorrelated, such that coherent interactions only occur on linear time scales. In order to gradually go from the fully nonlinear description to the quasilinear one, a collision operator of the form

$$\left.\frac{\mathrm{d}f}{\mathrm{d}\tau}\right|_{\mathrm{coll}} = D_{\phi}\frac{\partial^2 f}{\partial \phi^2} \tag{4}$$

is introduced, where $D_{\phi} \geq 0$ is a constant, quantifying the strength of phase decorrelation of the particles. This specific form of the collision operator linearly preserves the particle energy distribution, which is typically not the case for a physical stochastisation process. However, its energy conservation property facilitates the comparisons with quasilinear theory a lot, which is explained in Sec. 2.2. Using the Kolmogorov forward equation, the system described by eqs. (1), (2) and (4) can be expressed as a system of stochastic differential equations (SDEs) according to

$$\mathrm{d}\phi_k = u_k\,\mathrm{d}\tau + \sqrt{2D_{\phi}}\,\mathrm{d}W_{\tau,k}, \tag{5a}$$

$$\mathrm{d}u_k = \mathrm{Re}\left[A\mathrm{e}^{-\mathrm{i}\phi_k}\right]\mathrm{d}\tau, \tag{5b}$$

$$\mathrm{d}A = -\left(\sum_k w_k \mathrm{e}^{\mathrm{i}\phi_k} + \gamma_\mathrm{d} A\right)\mathrm{d}\tau, \tag{5c}$$

where the bump distribution $f(\phi, u, \tau)$ is described by a set of discrete entities (markers) with a phase ϕ_k, energy u_k and weight w_k, and $W_{\tau,k}$ are individual independent Wiener processes in τ. An added common weight factor to all particles, such that $w_k \to \alpha w_k$, $\alpha > 0$, can be transformed away using the set of substitutions $A \to \alpha^{2/3}A$, $\tau \to \alpha^{-1/3}\tau$, $(u, \gamma_\mathrm{d}, D_\phi) \to \alpha^{1/3}(u, \gamma_\mathrm{d}, D_\phi)$.

A discrete time approximation is used for the numerical simulations. Assuming that the wave amplitude evolves on time scales much longer than the individual ϕ_k and u_k, A can be treated as an independent variable in eq. (5*b*), splitting the complete system into individual two-dimensional systems of SDEs (one for each particle), and two ODEs for the wave amplitude (the real and the imaginary component). Using an Itō-Taylor numerical scheme with a strong convergence of order 1.5 [11], the discrete stepping of particles in phase space is described by

$$\Delta\phi_k = \sqrt{2D_\phi}\Delta W_k + u_k\Delta\tau + \frac{1}{2}\mathrm{Re}\left[A\mathrm{e}^{-\mathrm{i}\phi_k}\right]\Delta\tau^2, \tag{6a}$$

$$\Delta u_k = \mathrm{Re}\left[A\mathrm{e}^{-\mathrm{i}\phi_k}\left(\Delta\tau - \mathrm{i}\sqrt{2D_\phi}\Delta Z_k + \frac{1}{2}\left[\frac{1}{A}\frac{\mathrm{d}A}{\mathrm{d}\tau} - D_\phi - \mathrm{i}u_k\right]\Delta\tau^2\right)\right], \tag{6b}$$

where ΔW_k and ΔZ_k are Itō integrals of the Wiener processes. These can be sampled as $\Delta W_k = \xi_{1,k}\sqrt{\Delta\tau}$, $\Delta Z_k = (\xi_{1,k} + \xi_{2,k}/\sqrt{3})\Delta\tau^{3/2}/2$, where $\xi_{i,k}$ are independent normally distributed random variables of unit variance and zero mean. For the finite time stepping of the wave amplitude A, the standard fourth order Runge-Kutta method is used.

2.2. *Quasilinear Monte Carlo model*

The Brownian motion of the phase, in eq. (5*a*), induces a decorrelation of the wave-particle interaction in eq. (5*b*). When this phase decorrelation is strong, the evolution of the energy may be approximated by a random walk with a *quasilinear diffusion coefficient*, defined as

$$D_u(u) = \frac{A^2 D_\phi}{2(D_\phi^2 + u^2)}. \tag{7}$$

This diffusion coefficient is similar to the standard quasilinear diffusion of a weakly turbulent plasma [7,8]. Since the quasilinear wave-particle interaction is independent of ϕ and $\arg(A)$, the dimensionality of particle phase space is reduced from two to one, and the amplitude is a real quantity in the quasilinear description. The corresponding kinetic equation of the quasilinear description is given by

$$\frac{\partial F}{\partial\tau} = \frac{\partial}{\partial u}\left(D_u(u)\frac{\partial F}{\partial u}\right) = -\frac{\partial}{\partial u}\left(\frac{\mathrm{d}D_u}{\mathrm{d}u}F\right) + \frac{\partial^2}{\partial u^2}(D_u F). \tag{8}$$

The specific form of the added phase decorrelation preserves the total energy of the system of the nonlinear model, as given by eq. (3). Assuming this is true also for the purely quasilinear model, it is straightforward to derive the quasilinear amplitude equation from eqs. (3) and (8):

$$\frac{\mathrm{d}A}{\mathrm{d}\tau} = -\frac{1}{A}\int \mathrm{d}u\,\frac{\mathrm{d}D_u}{\mathrm{d}u}F(u,\tau) - \gamma_\mathrm{d}A = \frac{1}{A}\int \mathrm{d}u\,D_u(u)\frac{\partial F}{\partial u} - \gamma_\mathrm{d}A. \tag{9}$$

On time scales much shorter than the time scale for quasilinear flattening of the distribution (see eq. (20)) the distribution function evolves slowly, and can be approximated with a constant in time. Inserting this approximation into eq. (9) yields an effective growth rate of the wave amplitude,

$$\gamma_{\text{eff}} = \frac{D_\phi}{2} \int \mathrm{d}u \, \frac{\mathrm{d}F_0}{\mathrm{d}u} \frac{1}{D_\phi^2 + u^2}, \tag{10}$$

that includes effects of a finite width of the energy distribution, which was not considered in the derivation of γ_{L} in the nonlinear description.

The system of equations given by eqs. (8) and (9) can be written as a system of SDEs:

$$\mathrm{d}u_k = -\frac{A^2 D_\phi u_k}{(D_\phi^2 + u_k^2)^2} \mathrm{d}\tau + A\sqrt{\frac{D_\phi}{D_\phi^2 + u_k^2}} \mathrm{d}W_{\tau,k}, \tag{11a}$$

$$\mathrm{d}A = \left(\sum_k \frac{A D_\phi w_k u_k}{(D_\phi^2 + u_k^2)^2} - \gamma_{\text{d}} A \right) \mathrm{d}\tau, \tag{11b}$$

and the 1.5 order strong Itō-Taylor numerical scheme [11] yields the following stepping algorithm in u:

$$\Delta u_k = \frac{A}{2}\sqrt{\frac{D_\phi}{D_\phi^2 + u_k^2}} \Big\{ 2\Delta W_k - \Lambda_k u_k \left(\Delta W_k^2 + \Delta\tau\right) - \Lambda_k^2 \left(D_\phi^2 - 3u_k^2\right) \left(\frac{1}{3}\Delta W_k^3 + \Delta Z_k^+\right)$$
$$+ \Lambda_k^2 u_k^2 \Delta Z_k^- + \Lambda_k^3 u_k \left(7D_\phi^2 - 9u_k^2\right) \Delta\tau^2 + \frac{2}{A}\frac{\mathrm{d}A}{\mathrm{d}\tau} \left(\Delta Z_k^- + \Lambda_k u_k \Delta\tau^2\right) \Big\}, \tag{12}$$

where $\Delta W_k = \xi_{1,k}\sqrt{\Delta\tau}$, $\Delta Z_k^\pm = (\xi_{1,k} \pm \xi_{2,k}/\sqrt{3})\Delta\tau^{3/2}/2$, $\Lambda_k \equiv A\sqrt{D_\phi/(D_\phi^2 + u_k^2)^3}$, and $\xi_{i,k}$ are independent and normally distributed random variables of unit variance.

As apparent from eq. (10), the growth rate of the amplitude in the quasilinear model theoretically coincides with the growth rate in the nonlinear model in the limit $D_\phi \to 0$. However, as D_ϕ approaches zero, the maximum allowed $\Delta\tau$ for convergence of the numerical algorithm is reduced. This limitation for low D_ϕ is not present in the nonlinear model. The first order term in eq. (12) differs from the half order term with a factor $\sim \Lambda_k D_\phi \sqrt{\Delta\tau} \sim A/D_\phi^2 \times \sqrt{D_\phi \Delta\tau}$ at most, which must be $\ll 1$ for good convergence.

2.3. Analytic solutions to the quasilinear model

Solving the system of equations of the quasilinear model, eqs. (8) and (9), is possible in principle by first separating the spatial and temporal parts of the kinetic equation. The eigenfunction solutions of the spatial part are integrals over parabolic cylinder functions. The non-trivial eigenfunctions diverge as $\chi \to +\infty$ or $\chi \to -\infty$, which makes them impractical to use as spatial basis functions for the distribution function. When inserting a general set of eigenfunction solutions into the quasilinear amplitude equation, eq. (9), an infinite dimensional system of equations results, which might also be impractical.

An alternative approach is to use an approximate form of the diffusion coefficient that is valid on limited time scales. A model with a parabolic form of the diffusion coefficient,

$$D_u(u) = \begin{cases} \frac{A^2}{2D_\phi}\left(1 - \frac{u^2}{u_{\text{w}}^2}\right) & : |u| \le u_{\text{w}}, \\ 0 & : |u| > u_{\text{w}}, \end{cases} \tag{13}$$

where $u_{\text{w}} \sim D_\phi$ is the effective width of the diffusion in u, turns out to be practically soluble from an analytical point of view. From now on, the diffusion coefficient in eq. (13) is referred to as parabolic diffusion, whereas the coefficient in eq. (7) is referred to as the Lorentzian diffusion. Inserting the parabolic form of the diffusion coefficient into eq. (8) yields

$$\frac{\partial F}{\partial \tau} = \frac{A^2}{2D_\phi u_\mathrm{w}^2}\frac{\partial}{\partial x}\left(\left(1-x^2\right)\frac{\partial F}{\partial x}\right), \quad |x| \leq 1, \tag{14}$$

where $x \equiv u/u_\mathrm{w}$. It should be noticed that the eigenfunctions of the right hand side are the Legendre polynomials, which form an orthogonal set in $|x| \leq 1$. Decomposing the distribution function in the region $|u| \leq u_\mathrm{w}$ into Legendre polynomials according to

$$F(u_\mathrm{w}x, \tau) = \frac{1}{u_\mathrm{w}}\sum_{n=0}^{\infty} Q_n(\tau)P_n(x), \tag{15}$$

inserting into eqs. (14) and (9), and using the orthogonality property of the Legendre polynomials, it can be shown that each Q_n and the amplitude satisfy the ODEs

$$\frac{\mathrm{d}Q_n}{\mathrm{d}\tau} = -\frac{A^2 Q_n}{2D_\phi u_\mathrm{w}^2}n(n+1), \tag{16}$$

$$\frac{\mathrm{d}A}{\mathrm{d}\tau} = \frac{A}{D_\phi u_\mathrm{w}}\sum_{n=0}^{\infty} Q_n \int_{-1}^{1}\mathrm{d}x\, xP_n(x) = \frac{2Q_1 A}{3D_\phi u_\mathrm{w}}. \tag{17}$$

By solving the closed system of equations given by eq. (16) for $n = 1$ and eq. (17), and solving eq. (16) for general n using the obtained $A(\tau)$, one finds the complete analytical solutions:

$$Q_n(\tau) = Q_n(0)\left(\frac{\eta+\psi}{\eta\exp([\eta+\psi]\tau)+\psi}\right)^{n(n+1)/2}, \tag{18}$$

$$A^2(\tau) = \frac{D_\phi u_\mathrm{w}^2\eta(\eta+\psi)}{\eta+\psi\exp(-[\eta+\psi]\tau)}, \tag{19}$$

where $\eta \equiv A^2(0)/D_\phi u_\mathrm{w}^2$ and $\psi \equiv 4Q_1(0)/3D_\phi u_\mathrm{w}$.

One limitation of the parabolic diffusion model is that it only acts on the distribution function within the region $|u| < u_\mathrm{w}$, whereas the Lorentzian diffusion model acts on the whole u space. On time scales where regions $|u| \gtrsim u_\mathrm{w}$ of the distribution function are affected by diffusion in the Lorentzian model, the numerical and analytical solutions are expected to diverge. One can select u_w such that the initial growth rates of the wave amplitude match in the numerical and the analytical solutions. By doing this, one can obtain similar solutions of the two models on time scales $\tau \lesssim \tau_\mathrm{QL}$, where τ_QL is the characteristic time scale for quasilinear flattening of the distribution function. Therefore, the analytical solutions can be used to obtain an analytical expression for τ_QL.

The time scale of quasilinear flattening can be characterized as the time scale at which $\partial F/\partial u|_{u=0}$ is significantly reduced due to quasilinear diffusion. Only P_n for odd n contribute to the derivative of the distribution function at the resonance. The decay rate of Q_{2n+1} is of the order $2n^2(\eta+\psi)$ according to eq. (18). Hence, the term with the slowest decay that contributes to the derivative is Q_1, which is then expected to be the dominant contribution to $\partial F/\partial u$ at $(u,\tau) = (0, \tau_\mathrm{QL})$. The analytical τ_QL can be defined as the time at which $Q_1(\tau)/Q_1(0)$ essentially differs from unity. Setting this ratio e.g. to $1/\mathrm{e}$ results in

$$\tau_\mathrm{QL} = \frac{1}{\eta+\psi}\left[1+\ln\left(1+\frac{\mathrm{e}-1}{\mathrm{e}}\frac{\psi}{\eta}\right)\right]. \tag{20}$$

Assuming that $\mathrm{d}F_0/\mathrm{d}u \approx \mathrm{d}F_0/\mathrm{d}u|_{u=0}$ on energy scales much larger than D_ϕ, and assuming $\eta \ll \psi$ (equivalent with $A(0) \ll A(\infty) - A(0)$, where $A(\infty)$ is the saturation amplitude of the analytical model) a slightly simpler expression for τ_QL than in eq. (20) can be found, namely $\tau_\mathrm{QL} = \gamma_\mathrm{L}^{-1}[\varsigma + \ln(\sqrt{\gamma_\mathrm{L}D_\phi^3}/A(0))]$, where $\varsigma = \ln(3\pi\sqrt{(\mathrm{e}-1)/8}) \approx 1.47$.

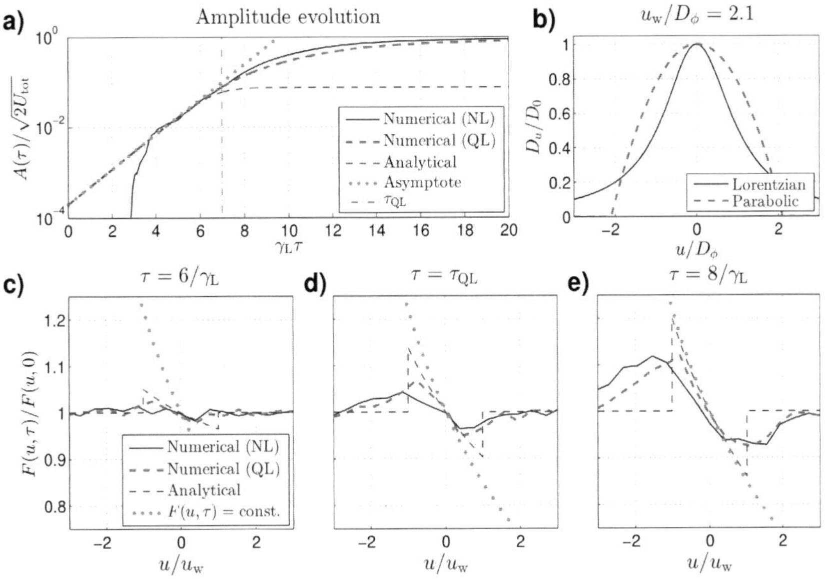

Figure 1. a) Evolution of the wave amplitude using the nonlinear (NL), the quasilinear (QL) and the analytical model. b) Comparison between the Lorentzian and the parabolic diffusion coefficient in the cases presented in (a). $D_0 \equiv A^2/2D_\phi$. c – e) The distribution function around the resonance at three chosen times. The distribution of the NL model is integrated over ϕ.

3. Results

3.1. Quasilinear flattening

Comparisons between the evolutions of the wave amplitude and of the energy distribution functions for simulations of the nonlinear model (NL), the quasilinear (QL) and the analytical model are presented in Fig. 1. An initial triangular bump distribution in energy is used, defined as $F_0(u) = F_0(0)(1 + u/\bar{u})$ for $|u| \leq \bar{u}$, and $F_0(u) = 0$ for $|u| > \bar{u}$, which is chosen to minimize possible effects from higher order derivatives of the energy distribution around the resonance. Using eq. (20) it was found that $\gamma_L \tau_{QL} = 6.98$ for the specific case in Fig. 1. Unlike the quasilinear model, the immediately initial amplitude evolution of the nonlinear model is not exponential. Rather, there is an initial phase mixing state, with a faster growth of the amplitude due to the added sinusoidal perturbation in ϕ-space of the initial distribution function, as discussed in Sec. 2.1. To resolve this discrepancy, the time is shifted for the nonlinear model, such that it matches the wave amplitude of the numerical quasilinear model at $\tau = \tau_{QL}$. Then the distribution functions of the three models can be compared at given times, which is done in Fig. 1.c – 1.e.

The saturation level of the analytical model is much lower than that of the nonlinear and the quasilinear model. This is due to the fact that the wave can only exhaust energy from a localized region $|u| < u_w$ around the resonance using the parabolic diffusion model. Although differences are large between the analytical and numerical solutions during the saturation phase, they approximately agree for times up to the analytical time scale of quasilinear flattening, τ_{QL} (aside from the initial phase mixing state of the nonlinear solution), which can be seen in Fig. 1.a. For $\tau > \tau_{QL}$, the exponential evolution of the wave amplitude gradually ceases for both numerical solutions. This conclusion is consistent with the results presented in Fig. 1.c – 1.e. For $\tau = 6/\gamma_L < \tau_{QL}$, the energy distribution deviates from the initial distribution with a few percent at most. For $\tau \geq \tau_{QL}$, deviations from the initial distribution start to become significant around the resonance, which here corresponds to the process of quasilinear flattening.

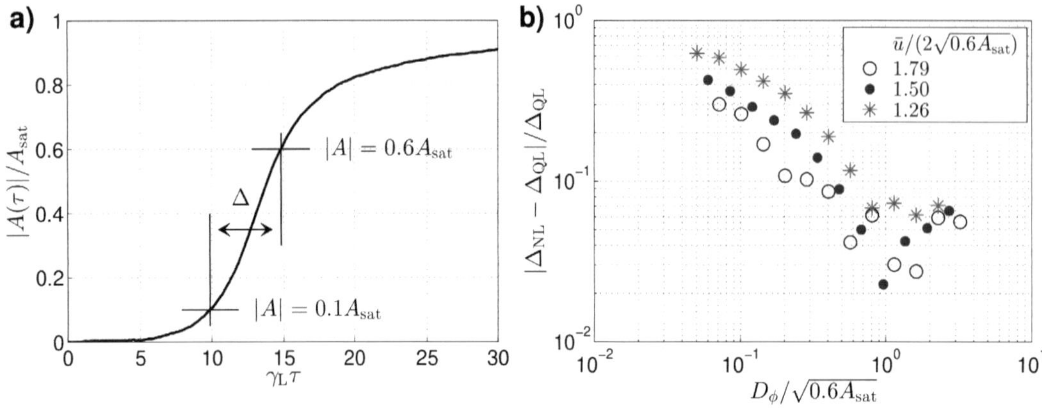

Figure 2. a) The definition of the quantity Δ. $A_{\rm sat}$ is the theoretical value of $|A(\tau)|$ for $\tau \to \infty$. b) The relative difference of Δ between nonlinear simulations and corresponding quasilinear simulations as a function of $D_\phi/\sqrt{|A|}$ at the end of the interval. A triangular initial energy distribution is used, with a full width of $2\bar{u}$, and $\gamma_{\rm d} = 0$.

3.2. Comparisons between the nonlinear and the quasilinear model

In order to conclude in which parameter regimes the numerical nonlinear and the numerical quasilinear model macroscopically agree, one has to find a quantity that primarily depends on the nonlinear dynamics of the system and compare this quantity for a wide set of nonlinear and quasilinear simulations. One nonlinear process is the saturation of the wave amplitude, which can be characterized by a saturation time scale and a value of the saturated amplitude. The latter is however trivial to determine in the presence of phase decorrelation. Turning off the wave damping ($\gamma_{\rm d} = 0$), the saturated amplitude corresponds to a state where the energy difference of the initial particle distribution and a final state of a symmetric energy distribution around the resonance is absorbed by the wave. In the presence of wave damping, the saturated state is simply zero, since the chosen collision operator lacks sources.

The quantity that has been chosen for comparison is the saturation time Δ, here defined as the time between the two states with amplitudes $|A| = 0.1A_{\rm sat}$ and $0.6A_{\rm sat}$ (see Fig. 2.a). In Fig. 2.b, the relative difference between Δ using the nonlinear and the quasilinear numerical models is shown. Effects of the width of the distribution function are studied by performing simulations with different values of $\bar{u}/\sqrt{0.6A_{\rm sat}}$, which is the ratio between the initial full width of the particle distribution in energy and the width of the trapped particle region by the wave field at the end of the Δ-interval. The quantity on the x-axis of Fig. 2.b, $D_\phi/\sqrt{0.6A_{\rm sat}}$, compares the bounce time of particles deeply trapped by the wave field ($\omega_{\rm B}^{-1} = |A|^{-1/2}$, referred to as the nonlinear time scale) with the time scale of phase decorrelation (D_ϕ^{-1}) at the end of the Δ-interval.

As shown in Fig. 2.b, the quasilinear model is able to predict the saturation time scale when the decorrelation time is shorter or similar to the nonlinear time scale ($D_\phi/\sqrt{0.6A_{\rm sat}} \gtrsim 1$). However, when the decorrelation time is long in comparison, e.g. when $D_\phi/\sqrt{0.6A_{\rm sat}} \lesssim 0.1$, the measured relative difference of Δ is larger than 20 %. The decrease of the relative error for increasing decorrelation strength ceases for shorter or similar decorrelation times, which might depend on numerical errors. From Fig. 2.b, one may also observe a better agreement between the models for distribution functions with a wider initial distribution around the resonance relative to the width of the trapped region, i.e. larger $\bar{u}/(2\sqrt{0.6A_{\rm sat}})$. One interpretation is that a large fraction of the complete structure of the distribution function becomes nonlinearly displaced by the wave field when the initial particle distribution is narrow, such that $\delta f/f$ becomes large on short time scales. Another interpretation could be that the discontinuity of the triangular distribution on the positive edge (at $u = \bar{u}$) becomes visible for the wave when the width of the trapped region is similar to $\bar{u}$, which might strongly affect the nonlinear behavior of the system.

4. Conclusions

In this paper, a nonlinear Monte Carlo model and a corresponding quasilinear model for describing the dynamics of discrete global modes interacting with energetic particles in a toroidal plasma in the presence of phase decorrelation have been compared. This study is performed mainly by computing macroscopic quantities in selected parameter regimes using a quasilinear approximation and a fully nonlinear description. There exist parameter regimes where the nonlinear and the quasilinear descriptions approximately coincide macroscopically. These regimes are mainly when the time scale for the destruction of macroscopic phase space structures (due to the added phase decorrelation) are much shorter than the characteristic time scale of phase space evolution of particles around the wave-particle resonance. However, due to the reduced dimensionality of phase space relative to the nonlinear model, there are certain phenomena depending on nonlinear phase space structures that the quasilinear model cannot describe.

Two partly related phenomena common for both the nonlinear and the quasilinear descriptions have been studied for comparison: quasilinear flattening (i.e., local flattening of the energy distribution around the resonance due to quasilinear energy diffusion) and saturation time scales of the wave amplitude in the presence of phase decorrelation. Analytical solutions to a problem similar to the quasilinear description were derived to obtain a theoretical value of the time scale of quasilinear flattening. When compared with numerical simulations using the quasilinear and the nonlinear descriptions, they were found to approximately match the theoretical time scale, both when looking at the deviations from an exponential growth of the wave amplitude and at the flattening of the energy distribution.

The saturation time scale was studied by comparing the time difference (referred to as Δ) between the states where the wave amplitude had reached 10 % and 60 % of the saturated amplitude in the presence of phase decorrelation, using the nonlinear and the quasilinear numerical model. It was found that the value of Δ is similar for the two models when the phase decorrelation is faster than the nonlinear bounce time (from the trapping of the wave field). However, when the decorrelation time scale is reduced, the differences between the quasilinear and the nonlinear model are significant. The energetic distribution of the presented numerical models has a finite width in energy, which affects the macroscopic behavior of the system. Wider initial distributions relative to the trapped particle region in the nonlinear model also give better agreement with the quasilinear model in general. This can be explained by the fact that a large fraction of the complete structure of the distribution function becomes nonlinearly displaced by the wave field when the initial particle distribution is narrow relative to the trapped particle region, such that $\delta f/f$ becomes large on short time scales for these cases.

References

[1] Berk H L, Breizman B N and Pekker M S 1995 *Nucl. Fusion* **35** 1713
[2] Breizman B N, Berk H L, Pekker M S, Porcelli F, Stupakov G V and Wong K L 1997 *Phys. Plasmas* **4** 1559
[3] Berk H L, Breizman B N and Pekker M S 1996 *Phys. Rev. Lett.* **76** 1256
[4] Berk H L, Breizman B N, Candy J, Pekker M S and Petviashvili N V 1999 *Phys. Plasmas* **6** 3102
[5] Lilley M K, Breizman B N and Sharapov S E 2010 *Phys. Plasmas* **17** 092305
[6] Lilley M K and Nyqvist R M 2014 *Phys. Rev. Lett.* **112** 155002
[7] Vedenov A A, Velikhov E P and Sagdeev R Z 1962 *Nucl. Fusion Suppl. Pt. 2* 465
[8] Drummond W E and Pines D 1962 *Nucl. Fusion Suppl. Pt. 3* 1049
[9] Kaufman A N 1972 *Phys. Fluids* **15** 1063
[10] O'Neil T 1965 *Phys. Fluids* **8** 2255
[11] Kloeden P E and Platen E 1995 *Numerical Solution of Stochastic Differential Equations* Second Corrected Printing ed (*Applications of Mathematics* vol 23) (Berlin: Springer-Verlag) ISBN 3540540628

How to apply a turbulent transport model based on a gyrokinetic simulation for the ion temperature gradient mode in helical plasmas

S Toda[1], M Nunami[1], A Ishizawa[1], T -H Watanabe[2] and H Sugama[1]

[1]National Institute for Fusion Science, Oroshi 322-6, Toki, Gifu, 509-5292, Japan
[2]Department of Physics, Nagoya University, Furo-cho, Nagoya, Aichi, 464-8602, Japan

E-mail: toda@nifs.ac.jp

Abstract. How to apply a reduced model for the turbulent ion heat diffusivity [Nunami M *et al.* 2013 Phys. Plasmas, **20** 092307] derived from a gyrokinetic code to a transport simulation is proposed. The reduced model is given by the function of the linear growth rate of the ion temperature gradient (ITG) mode and the decay time of zonal flows. The ion temperature gradient scale length is chosen for the additional modeling to include the linear growth rate of the ITG mode from a gyrokinetic code. The formula for the zonal flow decay time is derived at the given magnetic field configuration. The calculation of an extremely low computational cost in this article reproduces the results of the reduced model for the turbulent ion heat diffusivity within allowable errors.

1. Introduction

Turbulent transport is one of the most critical issues for plasma confinement in magnetic fusion devices. This is because the turbulent transport induces a large amount of the particle and heat loss in toroidal plasmas. Theoretical expressions for the turbulent transport coefficients due to various instabilities were reviewed [1, 2]. Recently, a large number of the gyrokinetic simulations have been done in toroidal plasmas [3]. The results in tokamak [4, 5, 6, 7, 8, 9] and helical [10, 11] plasmas from the gyrokinetic simulations have been studied with the experimental observations. The GKV-X code solving the gyro-kinetic equation has been used to examine the ITG mode and zonal flows in the LHD for studying the turbulent transport in helical plasmas [12], where the ITG mode is assumed to be the main underlying microinstability. The reduced model of $\chi_i \sim \rho_{ti}^2 v_{ti} f(\mathcal{L}, \tilde{\tau}_{ZF})/R$ is taken [13, 14] using the GKV-X code for the high-T_i LHD discharge of the shot number 88343, where f is a function of $\mathcal{L}$ and $\tilde{\tau}_{ZF}$, v_{ti} is the ion thermal velocity, ρ_{ti} is the ion gyroradius and R is the major radius. Here, $\mathcal{L}$ is the mixing length estimate $\tilde{\gamma}_{\tilde{k}_y}/\tilde{k}_y^2$ integrated over the $\tilde{k}_y$ space, where $\tilde{\gamma}_{\tilde{k}_y}$ is the normalized linear growth rate of the ITG mode for the normalized poloidal wavenumber $\tilde{k}_y$ and $\tilde{\tau}_{ZF}$ is the normalized decay time of zonal flows [15, 16]. The nonlinear gyrokinetic simulation results are quantitatively reproduced by the reduced model calculations. However, it is costly to carry out linear calculations of the growth rate by the gyrokinetic simulation at each time step of the dynamical transport code such as TASK3D [17], because the transport analysis of helical plasmas demands a high radial resolution so as to accurately evaluate the radial electric field, which strongly affects the neoclassical transport, and the field configuration. It is needed to speed up the predictive transport simulation, which proposes the experimental scenario to realize the favorable plasma state such as the improved confinement plasma mode, for performing the transport simulation in the wide plasma parameter regime. (In tokamak plasmas, the gyrokinetic simulations at each time step are globally done in the dynamical transport simulation [18, 19].)

In this study, how to apply the reduced model of the turbulent heat diffusivity for the ITG mode derived from the gyrokinetic simulation to the transport code is shown with a low computational cost, while the accuracy of the gyrokinetic simulation is maintained. Modeling of the term $\mathcal{L}$ in the reduced model for the ITG mode is necessary to be involved with a parameter dependence of the plasma instability in the dynamical transport code. The ion temperature gradient scale length $L_{T_i}(= -T_i/(\partial T_i/\partial r))$ is chosen for the parameter to apply $\mathcal{L}$ to avoid the calculations of the linear growth rate by the gyrokinetic code at each time step of the transport simulation. The field configuration is fixed at the initial state in the transport simulation. The linear gyrokinetic analysis is performed using the GKV-X code at this magnetic field configuration before the dynamical transport simulation. The dependence of $\mathcal{L}$ on the ion temperature gradient scale length is examined at the different radial points. We have newly developed the formula for $\mathcal{L}$ in terms of the ion temperature gradient scale length. The decay time of zonal flows depends on the magnetic field configuration and is independent of the the density and temperature profiles. The formula of the zonal flow decay time is needed to be calculated only at the initial state in the transport simulation, because the magnetic field configuration is dynamically fixed in this study. The calculation by substituting these formulae for $\mathcal{L}$ and $\tilde{\tau}_{ZF}$ to the reduced model [13] reproduces the results of the reduced model itself within allowable errors. The computational cost to obtain the value of the turbulent ion heat diffusivity by this modeling at each time step of the transport simulation is much smaller than that of the linear gyrokinetic simulation. This additional modeling for the turbulent ion heat diffusivity is applied to the transport code and enables us to study the simulation results with the experimental results in LHD.

2. The additional modeling of the turbulent ion heat diffusivity

This section addresses the adaption of the reduced transport model to a transport simulation for reducing a computational cost. Firstly, the linear analysis is done using the GKV-X code for the additional modeling of the turbulent ion heat diffusivity, before the dynamical transport simulation. The ITG instability is examined in the high-T_i LHD discharge #88343 [20]. The saturation level of the linear growth rate for the perturbation of the electrostatic potential is calculated. The value of the turbulent ion heat diffusivity $\chi_i^{(1)}/\chi_i^{GB}$ was fitted [13] only by the function $\mathcal{L}\left(\equiv \int(\tilde{\gamma}_{\tilde{k}_y}/\tilde{k}_y^2)d\tilde{k}_y\right)$ as $\chi_i^{(1)}/\chi_i^{GB} = C_0\,(C_T\mathcal{L})^{\delta}$, where χ_i^{GB} is the gyroBohm diffusivity, $\chi_i^{GB} = \rho_{ti}^2 v_{ti}/R$, $\tilde{\gamma} = \gamma/(v_{ti}/R)$ and $\tilde{k}_y = k_y\rho_{ti}$ with $C_0 = 0.11$, $C_T = 9.8 \times 10$ and $\delta = 0.83$. A reduced model for the ITG turbulent heat diffusivity in terms of the functions $\mathcal{L}$ and $\tilde{\tau}_{ZF}(= \tau_{ZF}/(R/v_{ti}))$ was obtained as $\chi_i^{(2)}/\chi_i^{GB} = A_1\mathcal{L}^{\alpha}/(A_2 + \tilde{\tau}_{ZF}/\mathcal{L}^{1/2})$. The numerical coefficients are given by $A_1 = C_1 C_T^{\alpha+1/2} C_Z^{-1}$ and $A_2 = C_2 C_T^{1/2} C_Z^{-1}$, where $\alpha = 0.38$, $C_Z = 0.202$, $C_1 = 6.3\times10^{-2}$ and $C_2 = 1.1\times10^{-2}$. For reducing a computational cost, the additional modeling is needed for $\mathcal{L}$ in terms of the physical parameter which is included in the transport codes. The characteristic length of the ion temperature gradient is considered to be the important parameter for the ITG instability. The parameter L_{T_i} is considered to be more sensitive to the ion heat flux than the electron temperature gradient scale length for the ITG mode examined here. The radial profiles of the density and the electron temperature are fixed in the dynamical transport simulation. Therefore, the parameters, $L_n(= -n/(\partial n/\partial r))$ and $L_{T_e}(= -T_e/(\partial T_e/\partial r))$ do not change at each radial point. As the function of the ion temperature gradient scale length L_{T_i},

Journal of Physics: Conference Series **561** (2014) 012020 doi:10.1088/1742-6596/561/1/012020

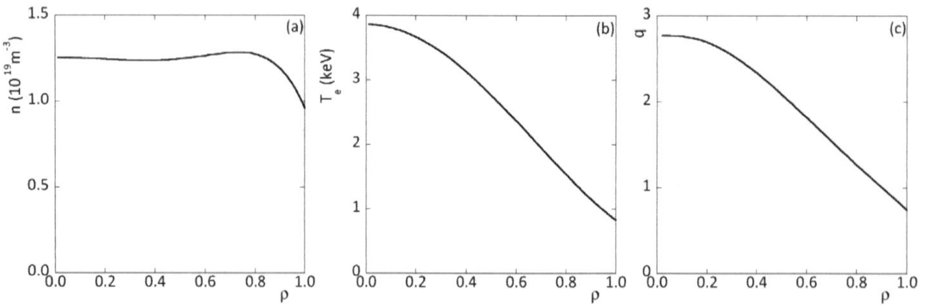

Figure 1. Radial profiles of (a) the density, (b) the electron temperature and (c) the safety factor

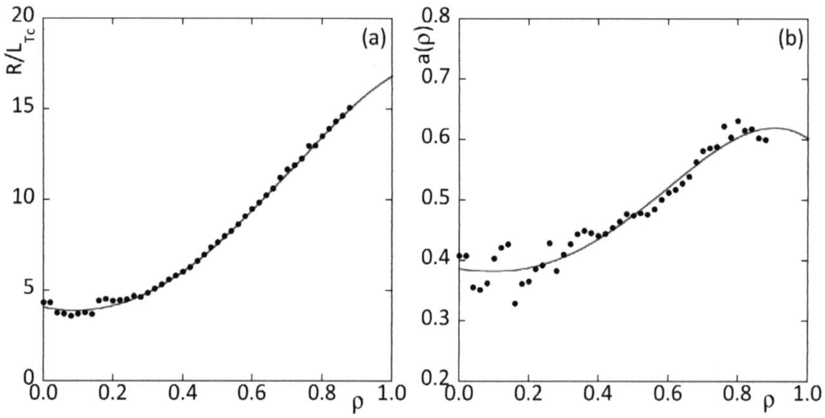

Figure 2. The radial dependence of (a) R/L_{T_c} and (b) $a(\rho)$ is shown with filled circles. The red curve shows the fitting function with respect to the radial position ρ. These are results from the GKV gyrokinetic linear code.

the parameter $\mathcal{L}$ is modeled by

$$\mathcal{L} = a(\rho)\left(\frac{R}{L_{T_i}} - \frac{R}{L_{T_c}}\right), \tag{1}$$

where L_{T_c} is the normalized critical ion temperature gradient for the ITG instability and ρ is the radial axis normalized by the minor radius.

To find the critical ion temperature gradient for the ITG mode, the dependence of $\mathcal{L}$ on R/L_{T_i} is examined [21] with all plasma parameters fixed except the ion temperature gradient at forty five radial points in the radial region $0.00 \le \rho \le 0.88$. When we examine the linear growth rate of the ITG mode, the radial profiles of the density (n), the electron temperature (T_e) and the safety factor (q) are used in figure 1. The density and ion temperature profiles are obtained from the experimental results at $t = 2.233$s in the high-T_i LHD discharge #88343. The T_e profile is set as $T_e = T_i$ in a gyrokinetic simulation. This simulation is performed for the three dimensional equilibrium field configuration with $R = 3.75$m, using the plasma profiles explained here in the VMEC calculation. The slope $a(\rho)$ and the critical ion temperature

Journal of Physics: Conference Series **561** (2014) 012020 doi:10.1088/1742-6596/561/1/012020

gradient L_{T_c} depend on the values of L_{T_e}, L_n and the safety factor, which change due to the radial positions. We calculate the linear fitting function (1) at each radial point and obtain the critical values of R/L_{T_i}, R/L_{T_c}, where $\mathcal{L}$ becomes zero. The critical ion temperature gradient for the ITG mode, (a) R/L_{T_c} and the slope (b) $a(\rho)$ in terms of R/L_{T_i} are obtained in figure 2, when the GKV gyrokinetic linear simulation is performed. When we calculate the value of the ion heat diffusivity in the integrated transport code, the fitting polynomials of R/L_{T_c} and $a(\rho)$ are used as $R/L_{T_c} = 4.0929 - 3.7681\rho + 19.712\rho^2 + 11.087\rho^3 - 14.272\rho^4$ and $a(\rho) = 0.38661 - 0.070919\rho + 0.2571\rho^2 + 0.95949\rho^3 - 0.92978\rho^4$. If the degree of the fitting polynomial increases, the accuracy of the fitting gets better. Figure 3 shows the comparison between the right hand side $(a(\rho)\,(R/L_{T_i} - R/L_{T_c}))$ and the left hand side $(\mathcal{L})$ in (1) with the root mean square of $((a(\rho)\,(R/L_{T_i} - R/L_{T_c}))/\mathcal{L} - 1)$ given by $\sigma = 0.13$, when the fourth degree fitting polynomial is used. The zonal flow decay time $\tilde{\tau}_{ZF}$ [16], which only depends on the magnetic field structure, is examined at forty four points in the radial region $0.02 \leq \rho \leq 0.88$.

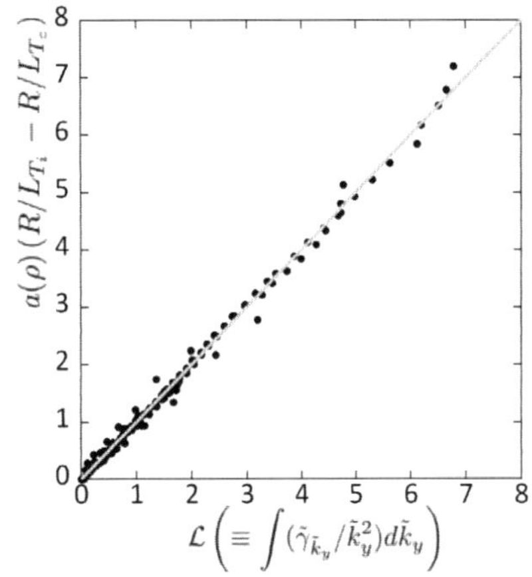

Figure 3. Comparison between the modeled function in terms of the ion temperature gradient scale length and the integral of the linear growth rate divided by the square of the poloidal wavenumber over the $\tilde{k}_y$ space.

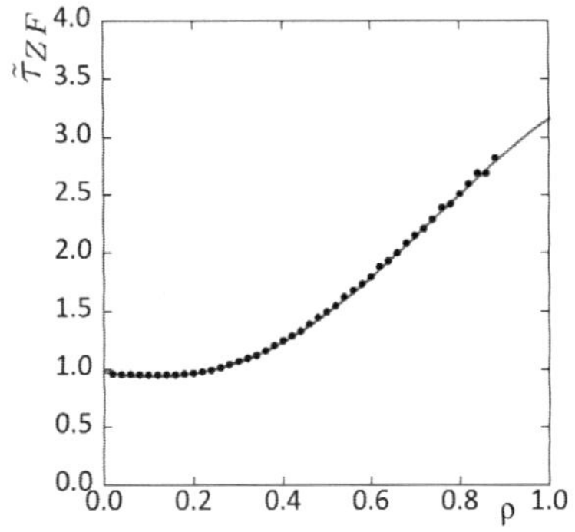

Figure 4. The radial profile of $\tilde{\tau}_{ZF}$ is shown with filled circles. The red curve shows the fitting function for the zonal flow decay time with respect to the radial position ρ.

The radial profile of the zonal flow decay time $\tilde{\tau}_{ZF}$ is shown in figure 4. The fitting function for the zonal flow decay time: $\tilde{\tau}_{ZF}(fit) = 0.98565 - 0.65943\rho + 2.4471\rho^2 + 3.2337\rho^3 - 2.8382\rho^4$ is used throughout the transport simulation, because the field configuration is dynamically fixed. The first modeled turbulent ion heat diffusivity, where the term $a(\rho)(R/L_{T_i} - R/L_{T_c})$ is substituted for $\mathcal{L}$ in $\chi_i^{(1)}/\chi_i^{GB}$, is shown as

$$\frac{\chi_i^{FTS(1)}}{\chi_i^{GB}} = C_0 \left(C_T a(\rho) \left(\frac{R}{L_{T_i}} - \frac{R}{L_{T_c}} \right) \right)^{\delta}, \tag{2}$$

which is the function of only the ion temperature gradient scale length. The second modeled turbulent ion heat diffusivity, where the term $a(\rho)(R/L_{T_i} - R/L_{T_c})$ is substituted for $\mathcal{L}$ and the fitting function $\tilde{\tau}_{ZF}(fit)$ is used for $\tilde{\tau}_{ZF}$ in $\chi_i^{(2)}/\chi_i^{GB}$, is also shown as

$$\frac{\chi_i^{FTS(2)}}{\chi_i^{GB}} = \frac{A_1 \left(a(\rho) \left(\frac{R}{L_{T_i}} - \frac{R}{L_{T_c}} \right) \right)^{\alpha}}{A_2 + \tilde{\tau}_{ZF}(fit) / \left(a(\rho) \left(\frac{R}{L_{T_i}} - \frac{R}{L_{T_c}} \right) \right)^{1/2}}, \tag{3}$$

which is the function of L_{T_i} and the zonal flow decay time. The values for the root mean square of $(\chi_i^{FTS(1)}/\chi_i^{(1)} - 1)$ and $(\chi_i^{FTS(2)}/\chi_i^{(2)} - 1)$ are 0.13 and 0.12, respectively. Therefore, the present modeling (1) reproduces the results of the reduced model, while the accuracy of the linear gyrokinetic simulation is maintained. To obtain the value of the turbulent ion heat diffusivity at each time step in the integrated transport code, the values of (2) and (3) are calculated instead of the reduced model calculation by the linear gyrokinetic simulation. Note that this modeling for $\mathcal{L}$ and the fitting function for the zonal flow decay time is applicable to the magnetic field structure used here and the fixed profiles of the density and the electron temperature.

3. Transport analysis using the additional modeled turbulent diffusivity

In section 2, we have modeled the turbulent ion heat diffusivity, (2) and (3) to reduce a computational cost. The modeled turbulent ion heat diffusivities expressed in terms of the ion temperature gradient scale length for the ITG mode are useful in a transport simulation.

Now, the transport dynamics is examined using the modeled turbulent ion heat diffusivity, when the integrated transport code, e.g., TASK3D [17] is performed. The radial profiles of the density and the electron temperature are fixed, which are shown in figure 1(a) and (b). This is because the radial profiles of the density and the electron temperature are considered to be almost stationary in the experimental results [20]. The T_i profile of the experimental results at $t = 2.233$s in the high-T_i LHD discharge #88343 is used as an initial state for the transport simulation. At the initial state, we set $T_e = T_i$ and use the dynamically fixed q profile in figure 1(c). The dynamics of the radial T_i profile is simulated by solving the diffusion equation as

$$\frac{\partial}{\partial t} \left(\frac{3}{2} n T_i \right) = -\frac{1}{V'} \frac{\partial}{\partial \rho} \left(V' Q_i \right) + P_{hx} + P_{hi}, \tag{4}$$

where V is the plasma volume, $V' = dV/d\rho$, P_{hx} is the heat exchange term and P_{hi} is the absorbed power of ions. The ion heat flux Q_i is set as $Q_i = -\langle |\nabla\rho|^2 \rangle n (\chi_i^{FTS} + \chi_i^{NEO}) \partial T_i / \partial \rho$, where χ_i^{NEO} is the neoclassical diffusion coefficient of ions and <> represents the magnetic surface average. The flux Q_i does not contain the convective part, because the core particle source is negligibly smaller than that in the edge region. The neoclassical diffusion coefficient is derived from DGN/LHD database with the low-β limit ($\beta = 0$) [22]. The profile of the radial electric field E_r is derived from the ambipolar condition at the initial plasma state. A

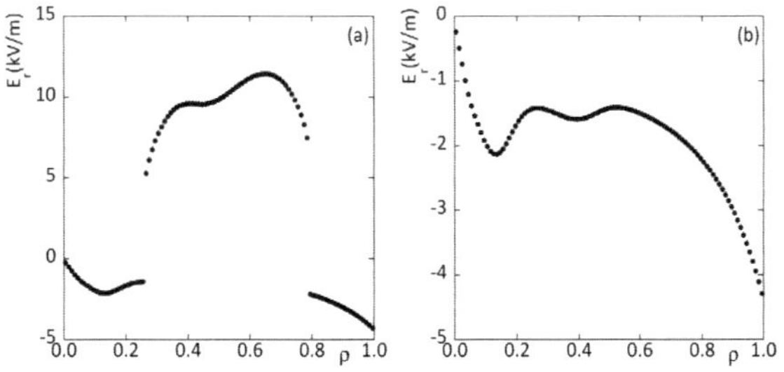

Figure 5. The radial profile of the electric field, which is dynamically fixed, is shown with filled circles. The (a) largest positive or (b) negative radial electric field is chosen from three ambipolar electric fields.

positive E_r is observed in the region $0.8 < \rho < 1.0$ at $t = 2.24$s in the high-T_i LHD discharge of #88343 [20]. Three solutions of the ambipolar radial electric field are found in the radial region $0.265 \leq \rho \leq 0.785$. For the first case, the largest positive radial electric field, which is chosen from three solutions of the ambipolar conditions, is shown in figure 5(a). For the second case, the negative radial electric field is chosen in figure 5(b). The profiles of the radial electric field are dynamically fixed for two cases in this study. This is because the power balance is almost satisfied at the plasma initial state when the reduced model for the turbulent ion heat diffusivity [13] is used and therefore it is predicted that the profile of T_i does not change much in the transport simulation. The turbulent heat diffusivity is calculated using (2) and (3) with the additional modeling (1). The value of the ion temperature is fixed at the initial state in the shaded region $0.785 \leq \rho \leq 1.000$ shown in figure 6 and 7, because the modeling for the turbulent ion diffusivity in section 2 is done in the region $0.00 \leq \rho \leq 0.88$ and in many studies the gyrokinetic analysis does not predict the results which agree with the experimental results especially in the edge region. At the initial state, the radial profile for the absorbed power of ions, P_{hi} is calculated using TASK3D and is dynamically fixed. At first, the positive E_r is chosen from the three ambipolar radial electric fields to calculate the value of χ_i^{NEO}. We show the simulation results for the stationary ion temperature profile at $t = 0.1$s with the solid line in figure 6. In figure 6(a), the simulation result for the ion temperature profile is obtained when we use $\chi_i^{FTS(1)}$ for the turbulent ion heat diffusivity. We show the simulation result of the T_i profile in figure 6(b) when $\chi_i^{FTS(2)}$ is used. The dashed line indicates the radial profile of T_i at $t = 2.233$s in the LHD discharge #88343. In the figure 6 (a) and (b), the simulation results for the radial T_i profile show a good agreement with the experimental results. Specially, in figure 6(b) when the turbulent ion heat diffusivity is given by the function of $\tilde{\tau}_{ZF}$ in addition to L_{T_i}, the better agreement between the experimental and the simulation results is obtained. The stationary profiles of the turbulent and neoclassical diffusivities are also shown in figure 7. The solid and dashed lines represent the turbulent and neoclassical ion diffusivities. In both cases, the ITG mode is destabilized in the radial region $0.135 \leq \rho \leq 1.00$. The turbulent transport is dominant compared with the neoclassical transport in the radial region where the positive electric field is chosen. Table 1 shows the comparison between the ion heat diffusivities calculated by the nonlinear gyrokinetic simulation at the initial state in the transport simulation χ_i^{NL} [13] and from the additional modeling in this study, $\chi_i^{FNS(1)}$ and $\chi_i^{FNS(2)}$. The values of $\chi_i^{FNS(1)}$ and $\chi_i^{FNS(2)}$ are estimated at the stationary state as the result of the transport simulation. The results of the additional modeling in this study (specially, the value of $\chi_i^{FNS(2)}$) quantitatively agree with the results of the nonlinear gyrokinetic simulation in the radial region $0.46 \leq \rho \leq 0.72$,

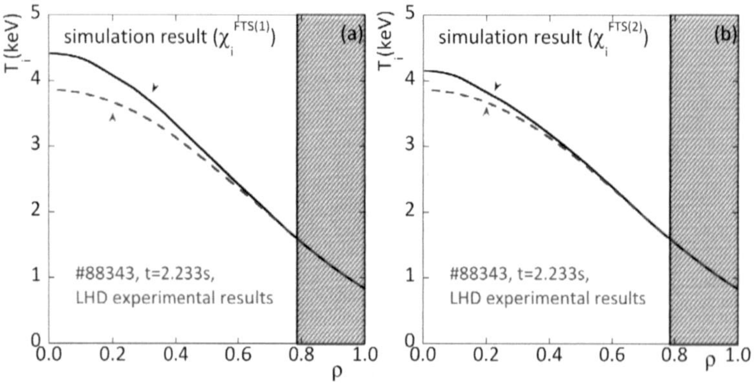

Figure 6. Simulation results for the T_i profile are shown with solid lines using (a) $\chi_i^{FTS(1)}$and (b) $\chi_i^{FTS(2)}$for the turbulent ion diffusivity, when the positive radial electric field is chosen. The dashed lines show the T_i profile at $t = 2.233$s in the LHD discharge #88343.

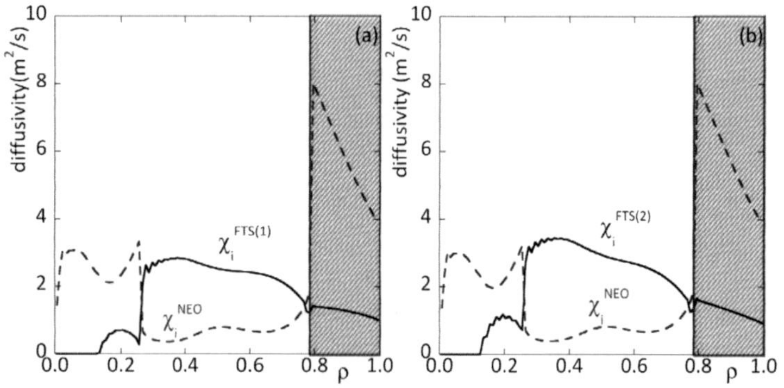

Figure 7. Simulation results for the neoclassical and turbulent diffusion coefficients are shown using (a) $\chi_i^{FTS(1)}$ and (b) $\chi_i^{FTS(2)}$ for the turbulent ion diffusivity, when the positive E_r is chosen. The solid and dashed lines represent the radial profiles of the turbulent and neoclassical ion diffusivities.

even if the stationary profile of T_i slightly changes from the initial state. For modeling $\mathcal{L}$ in terms of the ion temperature gradient scale length, three runs at the different values of L_{T_i} are needed at a radial point before the dynamical transport simulation. It takes about one hour per one time of the program run. If the value of $\mathcal{L}$ is calculated at each time step in the transport simulation, TASK3D, about one hundred thousand times of the program run are necessary at a radial point. Therefore, the transport simulation of an extremely low computational cost can be achieved due to the additional modeling of the turbulent ion heat diffusivity based on the gyrokinetic simulation.

Next, the negative radial electric field is chosen from the three radial electric fields which satisfy the ambipolar condition in figure 5(b). The stationary ion temperature profile at $t = 0.1$s is obtained. When the turbulent heat diffusivities, (2) and (3) are used, the radial profile of the ion temperature does not agree with the experimental results. The ion temperature is found to

Table 1. The ion heat diffusivities by the nonlinear gyrokinetic simulation, χ_i^{NL} and from the additional modeling for transport simulation, $\chi_i^{FNS(1)}$ and $\chi_i^{FNS(2)}$

ρ	$\chi_i^{NL}(\mathrm{m}^2/\mathrm{s})$	$\chi_i^{FTS(1)}(\mathrm{m}^2/\mathrm{s})$	$\chi_i^{FTS(2)}(\mathrm{m}^2/\mathrm{s})$
0.46	3.60	2.64	3.10
0.50	3.30	2.53	2.94
0.57	3.04	2.44	2.77
0.65	2.77	2.33	2.53
0.72	2.40	1.99	2.11
0.79	2.35	1.33	1.51
0.83	1.84	1.38	1.52

become low from the initial T_i profile and the lower ion temperature ($T_i \simeq 3.2$keV at $\rho = 0$) is obtained than the case when the positive radial electric field is chosen. The neoclassical transport is found to be dominant compared with the turbulent transport in the almost whole radial region. The neoclassical ion heat diffusivity in the region $0.265 \leq \rho \leq 0.785$ when the negative E_r is chosen is much larger than those in figure 7. This is because the absolute value of E_r when the positive radial electric field is chosen is larger than that when the negative E_r is chosen. Therefore, the lower ion temperature is obtained due to the enhanced neoclassical transport, when the negative E_r is chosen.

4. Summary

To reduce the simulation cost, how to apply the transport model based on a gyrokinetic simulation to a dynamical transport code is shown within the accuracy of the gyrokinetic simulation. Before the dynamical transport simulation, the GKV-X code is performed for the additional modeling in terms of the ion temperature gradient scale length instead of the linear growth rate term for the ITG mode. The formula of the zonal flow decay time in the reduced model is derived at the given magnetic field structure. The value of the reduced model is reproduced by those of the modeled turbulent ion heat diffusivity with an extremely low computational cost. The simulation results by applying the additional modeled turbulent ion heat diffusivity to the dynamical transport code are shown. In the dynamical transport simulation, the linear gyrokinetic simulation is not needed at all, because of the additional modeling in section 2. This simulation results agree with the experimental results of the ion temperature profile in the high-T_i LHD discharge #88343, when the positive radial electric field is chosen. If the negative radial electric field is chosen, the less agreement of the simulation results with the experimental result of the ion temperature is shown. Therefore, the transport analysis including the neoclassical transport is important in helical plasmas. A critical temperature gradient model has been proposed in [23], which is similar to the models, (2) and (3) in this study. The value of the critical ion temperature gradient for the ITG stability in (2) and (3) is calculated with the GKV linear simulation. So far, we concentrate on the ITG turbulence in helical plasmas. Reduced model of the turbulent heat diffusivity in helical plasmas will be constructed in terms of the linear growth rate of the other modes, such as the trapped electron mode and the electron temperature gradient mode. To apply the reduced model of the turbulent diffusivity to a transport code, the similar method discussed in this study can be used, if we choose the parameters such as L_{T_e} and/or L_n in the transport simulation for the additional modeling. In this study, the promising technique of a low computational cost for adapting the gyrokinetic turbulent transport model to a transport code is proposed within allowable errors.

Note that the form for the turbulent ion heat diffusivity by this additional modeling is valid for the high-T_i LHD discharge #88343. The dependence of the turbulent heat diffusivity on the field configuration and the plasma profiles for the LHD and different devices will be investigated. The effect of the radial electric field on the turbulent diffusivities should be examined for the future study.

Acknowledgments

S. T. would like to thank Dr. M. Sato and Dr. M. Yokoyama for the useful comments when the TASK3D is performed. The authors are grateful to the LHD Experimental Group for making the experimental data available. This work was partly supported by JSPS KAKENHI Grant Number 23561002 and the NIFS Collaboration Research program (Plasma Simulator), NIFS14KNST062.

References

[1] Connor J W and Wilson H R 1994 *Plasma Phys. Control. Fusion* **36** 719
[2] ITER Physics Expert Group on Confinement and Transport *et al* 1999 *Nucl. Fusion* **39** 2175
[3] Garbet X, Idomura Y, Villard L and Watanabe T -H 2010 *Nucl. Fusion* **50** 0433002
[4] Kotschenreuther M, Dorland W, Beer M A and Hammett G W 1995 *Phys. Plasmas* **2** 2381
[5] Bateman G, Kritz A H, Kinsey J E, Redd A J, Weiland J, 1998 *Phys. Plasmas* **5** 1793
[6] Holland C *et al* 2011 *Phys. Plasmas* **18** 056113
[7] Rhodes T L *et al* 2011 *Nucl. Fusion* **51** 063022
[8] Artaud J F *et al* 2010 *Nucl. Fusion* **50** 043001
[9] Senichenkov I Y *et al* 2014 *Plasma Phys.Control. Fusion* **56** 055002
[10] Mynick H E, Pomphrey N and Xanthopoulous P 2010 *Phys. Rev. Lett.* **105** 0950094
[11] Tallents S, *et al* 2014 *Plasma Phys. Control. Fusion* **56** 075024
[12] Nunami M, Watanabe T -H, Sugama H and Tanaka K 2012 *Phys. Plasmas* **19** 042504
[13] Nunami M, Watanabe T -H and Sugama H 2013 *Phys. Plasmas* **20** 092307
[14] Nunami M, Watanabe T -H and Sugama H 2013 *Plasma Fusion Res.* **8** 1203019
[15] Sugama H and Watanabe T -H 2006 *Phys. Plasmas* **13** 012501
[16] Ferrando-Margalet S, Sugama H and Watanabe T -H 2007 *Phys. Plasmas* **14** 122505
[17] Yokoyama M, Wakasa A, Seki R, Sato M, Murakami S, Suzuki C, Nakamura Y, Fukuyama A and LHD Experimental Group 2012 *Plasma Fusion Res.* **7** 2403011
[18] Candy J, Holland C, Waltz R E, Fahey M R and Belli E 2009 *Phys. Plasmas* **16** 060704
[19] Barnes M, Abel I G, Dorland W, Görler T, Hammet G W and Jenko F 2010 *Phys. Plasmas* **17** 056109
[20] Tanaka K, *et al* 2010 *Plasma Fusion Res.* **5** S2053
[21] Nunami M, Watanabe T -H, Sugama H and Tanaka K 2011 *Plasma Fusion Res.* **6** 001
[22] Wakasa A, Murakami S, Itagaki M and Oikawa S 2007 *Jpn. J. Appl. Phys.* **46** 1157
[23] Imbeaux F, Ryter F and Garbet X 2001 *Plasma Phys. Control. Fusion* **43** 1503

Journal of Physics: Conference Series **561** (2014) 012021 doi:10.1088/1742-6596/561/1/012021

Multi-Hierarchy Simulation of Collisionless Driven Reconnection by Real-Space Decomposition

S. Usami[1], **R. Horiuchi**[1,2], **H. Ohtani**[1,2], **and M. Den**[3]

[1] Department of Helical Plasma Research, National Institute for Fusion Science, Toki 509-5292, Japan.

[2] The Graduate University for Advanced Studies (Soken-dai), Toki 509-5292, Japan.

[3] National Institute of Information and Communications Technology, Koganei 184-8795, Japan.

E-mail: usami.shunsuke@nifs.ac.jp

Abstract. The first results on analysis of collisionless driven reconnection with a multi-hierarchy simulation model are reported. In the multi-hierarchy simulation model, real space in a simulation consists of three parts: a magnetohydrodynamics (MHD) domain to deal with macroscopic dynamics, a particle-in-cell (PIC) domain to solve microscopic kinetic physics from the first principle, and an interface domain to interlock the two domains. By means of multi-hierarchy simulations, the influence of macroscopic dynamics on microscopic physics of magnetic reconnection is investigated. Dynamical behaviors of collisionless reconnection in the PIC domain depend strongly on plasma inflows from the MHD domain. It is found that if the width of an MHD inflow increases as $v_w \lesssim v_{A0}$, where v_{A0} is the Alfvén speed at the upstream boundary, magnetic reconnection has only a single X-point, while in cases of $v_w \gtrsim 2.0 v_{A0}$, reconnection with multiple X-points takes place.

1. Introduction

A fusion plasma consists of multiple hierarchies. It contains various space-time scales from macroscopic transports, which cover the entire system, to microscopic processes relating to the dynamics of individual electrons and ions. In simulation studies on fusion plasmas, such different hierarchies are treated by different models. For example, macroscopic equilibrium and instabilities in core plasmas are studied by using magnetohydrodynamics (MHD) or extended MHD models, peripheral plasma transport in the boundary region is investigated by fluid codes for plasmas and transport codes for neutrals, and turbulent transport is studied by gyrokinetic simulations. Moreover, a fusion plasma is a typical complex system which consists of not only multiple hierarchies but also multiple physics. All phenomena described above are in the area of plasma physics, while material sciences including chemistry are needed for plasma-wall interaction and nuclear physics is necessary for fusion.

Studies on each hierarchy or physics are advancing [1], while interactions between hierarchies or physics are poorly understood. For the complete comprehension of fusion plasmas, simulation models which can deal with multiple hierarchies and physics self-consistently and simultaneously are required. In the first step, we have developed a multi-hierarchy model which interlocks an MHD code and a PIC code in the slab geometry. Using this multi-hierarchy model, we studies on collisionless magnetic reconnection, since magnetic reconnection is a typical multi-hierarchy

Journal of Physics: Conference Series **561** (2014) 012021 doi:10.1088/1742-6596/561/1/012021

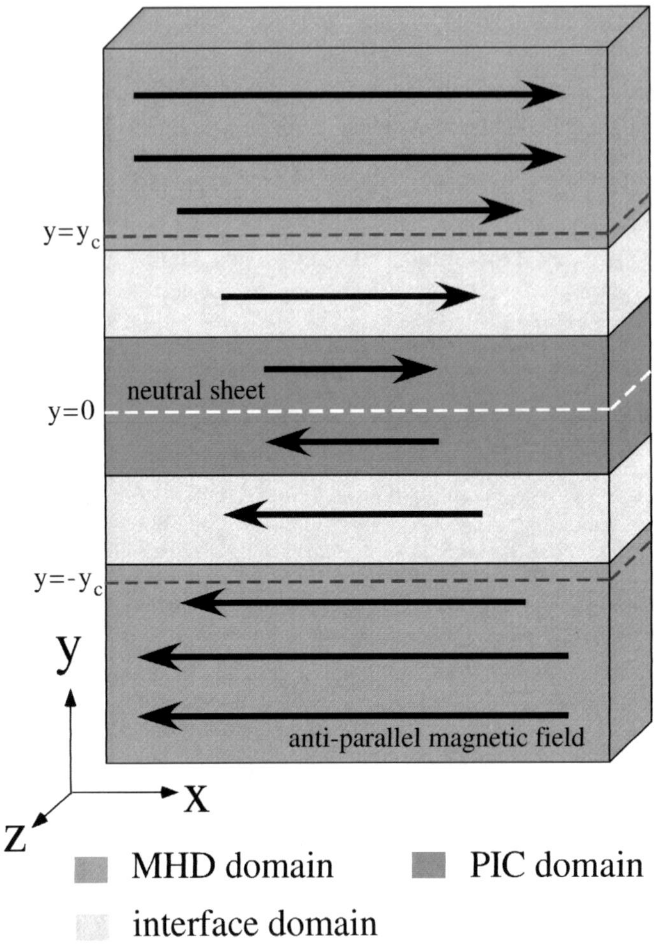

Figure 1. Schematic diagram of a multi-hierarchy model for magnetic reconnection simulation. Real space in a simulation domain is divided into three domains: MHD, PIC, and interface domains.

phenomenon and is believed to be strongly involved with sawtooth oscillations [2, 3, 4, 5] which cause tokamak disruption.

In this paper, with our multi-hierarchy simulations, the influence of global dynamics on microscopic physics of magnetic reconnection is discussed. In Sec. 2, we explain our multi-hierarchy simulation model. It is shown that real-space in a simulation domain is divided and also how and where macro- and micro hierarchies are coupled. In Sec. 3, we introduce results from our multi-hierarchy simulations. It is demonstrated that dynamical behaviors of magnetic reconnection depend sensitively on macroscopic physics. Section 4 gives a summary of this work and discusses on improvement of our multi-hierarchy model for application to sawtooth oscillations.

2. Multi-hierarchy model

2.1. Domain decomposition method

Magnetic reconnection has hierarchical structure in the upstream direction that the characteristic space-time scales change with distance from the neutral sheet [6]. In the vicinity of the neutral sheet, microscopic kinetic physics play important roles, thus models based on the first principle are required. On the other hand, far away from the neutral sheet, plasma phenomena relax to large-scale and slow behaviors so that a one-fluid model can give a good approximation to express dynamics in a region outside the ion inertia length.

Based on this hierarchical structure, our multi-hierarchy model employs the domain decomposition method for real space [7, 8, 9, 10, 11, 12, 13]. Figure 1 shows the schematic diagram of the multi-hierarchy simulation box. Hierarchies are one-dimensionally interlocked in the upstream direction (the direction of inflow injection) along the y-axis. The simulation domain consists of three parts; an MHD domain to express global dynamics, a PIC domain to describe kinetic physics, and an interface domain to couple the MHD and PIC domains. The PIC domain is a microscopic domain, which covers the region where microscopic kinetic effects play crucial roles. We refer to this domain as the PIC domain because the physics in the microscopic domain are solved by a PIC simulation algorithm [14, 15, 16, 17, 18, 19]. On the other hand, the MHD domain is a macroscopic domain. This domain treats the region where MHD conditions are fully satisfied. The dynamics in this domain are expressed by an MHD simulation algorithm [20]. We therefore call this domain MHD domain. Between the PIC and MHD domains, an interface domain with a finite width is inserted in order to prevent noises produced by the PIC algorithm from propagating to the MHD domain directly. Physics in the interface domain is calculated by the PIC and MHD algorithms. Therefore MHD conditions need to be fully satisfied in the interface domain, too. In other words, the interface domain is located in a region where MHD condition is fully satisfied.

In the MHD domain, we adopt non-uniform space grids [12] to extend our model to more realistic wide systems of magnetic reconnection. The grid spacing in the y direction Δy is a function of the space coordinate y. In the regions between $y = -y_c$ and $y = y_c$ illustrated in Fig. 1 (regions covering the PIC domain, the interface domain, and the MHD domain in the vicinity of the interface domain), Δy is the constant minimum size, $0.25(c/\omega_{ce})$, where c is the speed of light and ω_{ce} is the electron gyrofrequency. For $|y| > y_c$, Δy is smoothly larger as $|y|$ increases, and at the upstream boundary, Δy is taken to be $1.0(c/\omega_{ce})$.

2.2. Interlocking algorithm

In this subsection, an interlocking scheme in the interface domain is explained in detail. We first discuss an interlocking scheme for macroscopic quantities. A macroscopic quantity in the interface domain, $Q_{\text{interface}}$, is given by the following hand-shake scheme [7, 10, 12, 13]:

$$Q_{\text{interface}} = FQ_{\text{MHD}} + (1 - F)Q_{\text{PIC}}. \tag{1}$$

Here Q_{MHD} is a macroscopic quantity calculated only by the MHD basic equations and Q_{PIC} is a macroscopic quantity obtained only by the PIC fundamental equations (Newton-Lorentz and Maxwell equations). In the PIC algorithm, macroscopic quantities are obtained by assembling particle velocities and positions statistically. F is a function of the coordinates (x, y, z). In case of one-dimensional hierarchy-interlocking, F is given as a function of only y:

$$F(y) = \frac{1}{2}\left[1 + \cos\left(\pi\frac{y - y_{\text{MHD}}}{y_{\text{PIC}} - y_{\text{MHD}}}\right)\right], \tag{2}$$

where y_{MHD} and y_{PIC} are the boundary positions of the interface domain on the MHD and PIC sides, respectively [8, 9, 10, 11, 12].

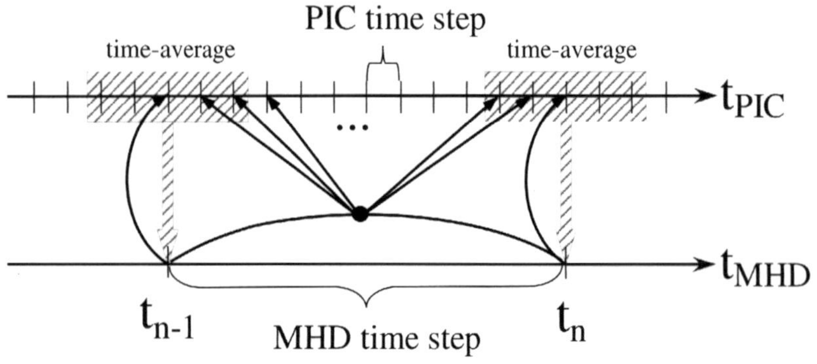

Figure 2. Flow in time for multi-hierarchy simulations. The time steps for the MHD and PIC algorithms are different from each other.

In the interface domain, the PIC algorithm needs microscopic quantities, namely information of individual particles. There however is an immense gap in the number of degrees of freedom between the MHD and PIC algorithms; the MHD algorithm has 8 degree of freedom, and that for the PIC algorithm is almost infinity. In order to bridge this large gap, we assume that shifted Maxwellian velocity distribution is fully satisfied in the interface domain. At every PIC time step, particle velocities and positions in the interface domain are removed and freshly loaded so as to satisfy the profiles with the fluid velocity $\boldsymbol{u}_{\text{interface}}$, mass density $\rho_{\text{interface}}$, and pressure $P_{\text{interface}}$ (the thermal velocity $v_{T,\text{interface}}$ and the number density $n_{\text{interface}}$). Electron particle velocities however are given so that their averaged velocity, namely the electron fluid velocity is $\boldsymbol{u}_{\text{interface}} - \boldsymbol{J}_{\text{interface}}/n_{\text{interface}}$ in order to generate the current density $\boldsymbol{J}_{\text{interface}}$. Let us note that the current density and electric field are not independent physical quantities for the MHD algorithms and they need to be treated specifically. (See Refs. [10, 12] for detail.)

2.3. Flow in simulation time

We introduce how to advance time of our multi-hierarchy simulation in Fig. 2. The MHD and PIC algorithms have different time steps from each other. Large time steps are for the MHD algorithm, and small time steps are for the PIC algorithm. Let us explain how time advances from t_{n-1} to t_n.

- Step 1: Suppose that at $t = t_{n-1}$ physical quantities of the MHD and PIC algorithms are given.
- Step 2: The MHD algorithm sends MHD data at t_{n-1} to the PIC algorithm.
- Step 3: The PIC algorithm refers to MHD data at t_{n-1} received at Step 2 as Q_{MHD} in Eq. (1), advances several time steps, and arrives at $t = t_{n-1} + \delta t$.
- Step 4: PIC information averaged over the period from $t_{n-1} - \delta t$ to $t_{n-1} + \delta t$ are sent to the MHD algorithm.
- Step 5: The MHD algorithm advances one time step and arrives at t_n, referring to PIC data obtained at Step 4 as Q_{PIC} in Eq. (1).
- Step 6: PIC data at $t = t_{n-1} + \delta t$ which were obtained at Step 3 are removed and the PIC algorithm returns to $t = t_n$.
- Step 7: The MHD algorithm sends MHD data at t_{n-1} and MHD data at t_n to the PIC algorithm.
- Step 8: During $t_{n-1} < t < t_n$, the PIC algorithm refers to MHD data interpolated between at t_{n-1} and at t_n received at Step 7 as Q_{MHD} in Eq. (1) and advances to t_n.

Table 1. Normalization constants in the MHD and PIC basic equations.

Quantity	MHD normalization	PIC normalization
length	L_n	c/ω_{ce}
velocity	v_A	c
time	L_n/v_A	$1/\omega_{ce}$
magnetic field	B_n	$m_e^{\rm SP} c\omega_{ce}/q_e^{\rm SP}$
electric field	$v_A B_n$	$m_e^{\rm SP} c\omega_{ce}/q_e^{\rm SP}$
charge	-	$q_e^{\rm SP}$
mass	-	$m_e^{\rm SP}$
number density	-	$(c/\omega_{ce})^{-3}$
mass density	ρ_n	$m_e^{\rm SP}(c/\omega_{ce})^{-3}$
current density	$B_n/(4\pi L_n)$	$q_e^{\rm SP} c(c/\omega_{ce})^{-3}$
pressure	$\rho_n v_A^2$	$m_e^{\rm SP} c^2 (c/\omega_{ce})^{-3}$

We call the above method a multi-time step scheme.

2.4. Unit-transformation

Table 1 represents physical quantities and their normalization constants in the MHD and PIC algorithms. Here, L_n, B_n, and ρ_n are arbitrary length, magnetic field, and mass density, respectively, and v_A is the Alfvén speed defined as $v_A = B_n(4\pi\rho_n)^{-1/2}$, $m_e^{\rm SP}$ is the electron mass, and $q_e^{\rm SP}$ is the electron charge. (The superscript SP represents super-particle.) Normalization constants are completely different from each other and arbitrary constants are included, thus an appropriate unit-transformation between the MHD and PIC unit systems needs to be determined for using the hand-shake scheme (1). In this work, we determine three following relations;

$$B_n = \frac{m_e^{\rm SP} c\omega_{ce}}{q_e^{\rm SP}}, \tag{3}$$

$$L_n = \alpha(c/\omega_{ce}), \tag{4}$$

$$\rho_n = \left(\frac{m_i}{m_e}\right)\left[n_{e0}^{\rm SP}(c/\omega_{ce})^3\right] m_e^{\rm SP}(c/\omega_{ce})^{-3}, \tag{5}$$

The parameter α can be determined freely. In this paper, α is taken to be 1.0 for all simulations. Furthermore, in particle simulations, the ion-to-electron mass ratio m_i/m_e, the ratio of the electron plasma frequency to the electron gyrofrequency ω_{pe}/ω_{ce}, and the normalized electron number density for standard $n_{e0}^{\rm SP}(c/\omega_{ce})^3$ are taken freely by a simulation performer. By virtue of Eqs. (3) - (5), unit-transformation for all other quantities are obtained uniquely [10, 12].

3. Simulation results

In past works, in order to examine physical reliability of the interlocking method, our multi-hierarchy model has been applied to propagation of Alfvén waves [7, 8] and plasma injection from the MHD domain to the PIC domain [10, 12]. Furthermore, we have performed multi-hierarchy simulations of collisionless driven reconnection and demonstrated the physical validity of our model [9, 11, 12].

In this work, we investigate the influence of macroscopic behaviors on microscopic physics of magnetic reconnection. In 2001, W. Pei *et al.* studied collisionless driven reconnection with PIC

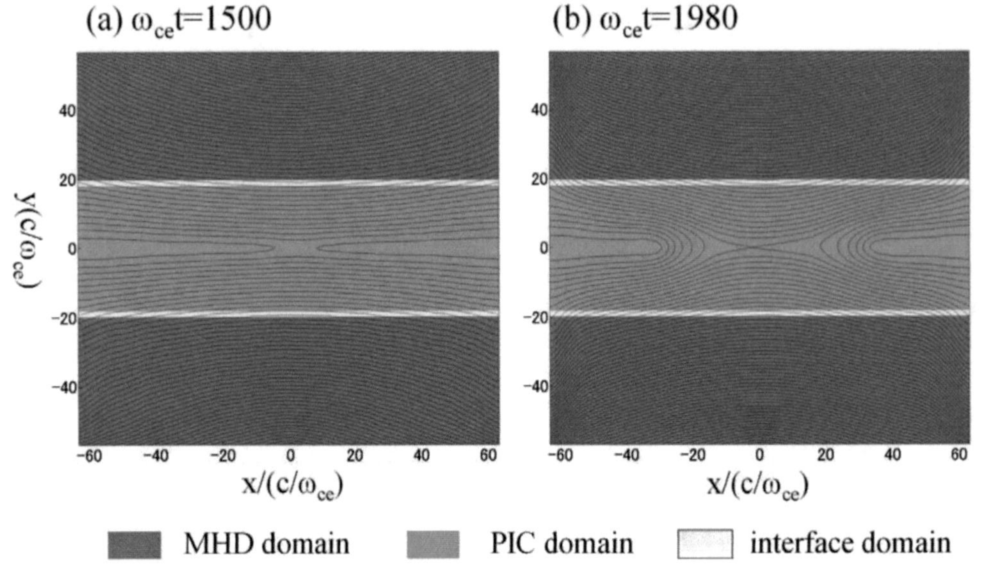

Figure 3. Spatial profiles of the magnetic field lines at (a) $\omega_{ce}t = 1500$ and $\omega_{ce}t = 1980$. The single X-point is located almost at the center in the PIC domain.

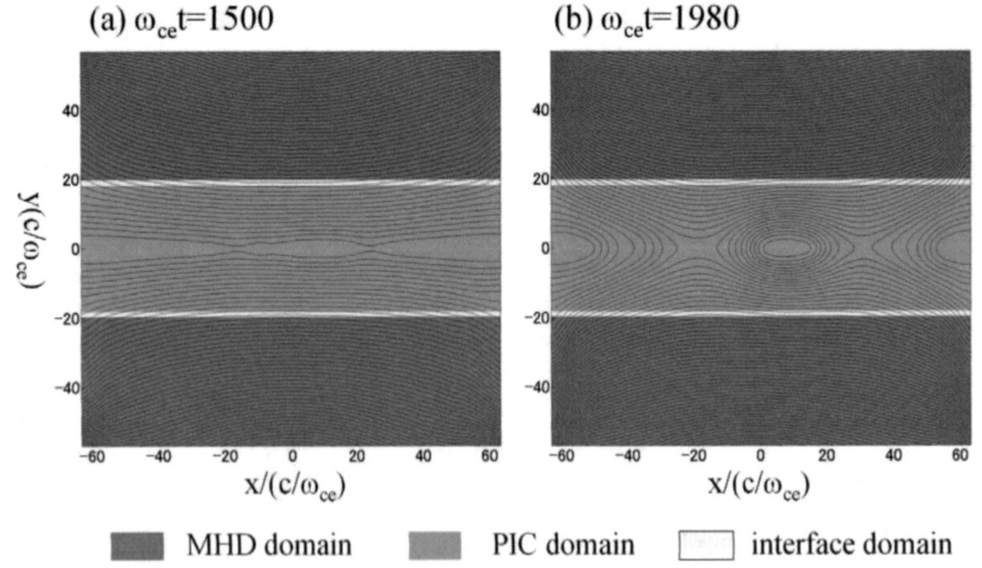

Figure 4. Spatial profiles of the magnetic field lines at (a) $\omega_{ce}t = 1500$ and $\omega_{ce}t = 1980$. In this case, reconnection with multiple X-points occurs.

simulations and showed that dynamical behaviors of reconnection are controlled dominantly by plasma inflow conditions [15, 16]. According to Ref. [16], reconnection evolves to steady state in the case of small non-uniformity of a plasma inflow and intermittent reconnection takes place in the case of large non-uniformity This result implies that the microscopic physics of magnetic reconnection depend strongly on global dynamics, which were simplified as the upstream boundary conditions in their PIC simulations.

In order to clarify the influence of macroscopic dynamics on microscopic physics, we perform simulations with the developed multi-hierarchy model. The simulation domain is shown in Fig. 1. The domain is implemented on a (512 × 303) point grid and the domain size is $128.0(c/\omega_{ce}) \times 114.75(c/\omega_{ce})$. The simulation domain is divided as follows. The MHD domain: $19.875 < |y/(c/\omega_{ce})| < 57.375$; the interface domain: $17.875 < |y/(c/\omega_{ce})| < 19.875$; and the PIC domain: $|y/(c/\omega_{ce})| < 17.875$.

The simulation parameters are taken as follows. The ion-to-electron mass ratio is $m_i/m_e = 100$, and the ratio of the electron plasma frequency to the electron gyrofrequency is $\omega_{pe}/\omega_{ce} = 1.5$. The MHD time step is 0.05 in the MHD unit system and the PIC time step is 0.05 in the PIC unit system. Using Eqs. (3) - (5), we find that 15 PIC time steps correspond to 1 MHD time step. The initial condition is given by a one-dimensional Harris-type equilibrium as $B_x(y) = B_0 \tanh(y/L_y)$ for the magnetic field and $n_p(y) = n_{p0}/\cosh^2(y/L_y)$ for the number density of foreground plasmas, where B_0 and n_{p0} are constant and L_y is the spatial scale. The number density of non-uniform background plasmas is expressed as $n_b(y) = n_{b0}[1 - 1/\cosh^2(y/L_y)]$ [18, 21], where n_{b0} has a constant value $n_{b0}/n_{p0} = 0.25$. The initial particle velocity distribution is a shifted Maxwellian distribution with the averaged velocity equal to the diamagnetic drift velocity. The temperature is spatially constant and those of ions and electrons are the same.

In order to study magnetic reconnection physics under various macroscopic configurations, we carry out simulation runs with different spatial patterns of plasma inflows, which are injected inward from the MHD domain. We impose the driving electric field $E_{zd}(x,t)$ at the upstream boundary of the MHD domain so as to generate plasma inflows. The field $E_{zd}(x,t)$ is set to zero at the initial time and begins to grow first around the center of the upstream boundary ($x = 0$). The inflow region expands with the velocity $v_{\rm w}$. The field $E_{zd}(x,t)$ develops until it reaches $E_0 = -0.04B_0$ (in the PIC unit system) on the entire upstream boundary of the MHD domain. On the other hand, the downstream boundary is free, across which plasmas can go out and in.

Figure 3 shows the spatial profiles of the magnetic field lines at $\omega_{ce}t = 1500$ and at $\omega_{ce}t = 1980$ in case of $v_{\rm w} = 0.6v_{A0}$, where v_{A0} is the Alfvén speed at the upstream boundary. We can see that plasma and magnetic flux are injected from the MHD domain to the PIC domain and magnetic reconnection is driven in the PIC domain. This reconnection phenomenon clearly has a single X-point at the almost center in the PIC domain, and this position is at rest. The system is likely to transit to the steady state. Also, it is confirmed that the pressure tensor term originating from particle meandering motions dominantly sustains reconnection electric field in the vicinity of the reconnection point as shown in Refs. [10, 12].

In Fig. 4, we display the spatial profiles of the magnetic field lines at $\omega_{ce}t = 1500$ and at $\omega_{ce}t = 1980$ in the case of $v_{\rm w} = 4.0v_{A0}$. That is, inflow plasmas propagating inward from the MHD domain more quickly expand into the x direction. It is observed that magnetic reconnection with multiple X-points takes place. A magnetic island is formed, grows, and moves to the downstream boundary. It seems that the system does not relax to the steady state, but intermittent reconnection continues.

Furthermore, we perform simulation runs with various cases of plasma inflows from the MHD domain. Table 2 summarizes the parameter $v_{\rm w}$ for MHD inflows and the corresponding results, i.e. whether single X-point or multiple X-point. When the width of an MHD inflow increases as $v_{\rm w} \lesssim v_{A0}$, we can see the tendency that magnetic reconnection with a single X-point is driven and the system seems to relax to the steady state. On the other hand, in cases of $v_{\rm w} \gtrsim 2.0v_{A0}$, a magnetic island is produced, i.e. intermittent reconnection with multiple X-points occurs.

Table 2. Simulation parameters and corresponding results.

v_w	Tendency
0.25 v_{A0}	single X-point
0.35 v_{A0}	multiple X-point
0.50 v_{A0}	single X-point
0.60 v_{A0}	single X-point
2.0 v_{A0}	multiple X-point
3.0 v_{A0}	multiple X-point
4.0 v_{A0}	multiple X-point

This tendency can be explained as follows. The expanding speed of an MHD inflow controls the aspect ratio of the current sheet. When the width of an MHD inflow increases slowly, the current sheet pushed by the inflow is short. The region where reconnection points could emerge is limited to a small central region. Thus, reconnection with only single X-point trends to take place. On the other hand, in cases in which an MHD inflow expands fast, the current sheet pushed by the inflow, i.e., the region where reconnection points could emerge becomes long. Actually for $2.0v_{A0} < v_w < 4.0v_{A0}$, reconnection with two X-points was driven.

Here let us refer to the phase diagram for magnetic reconnection in Fig. 5 of Ref. [22]. The phase diagram graphically organizes knowledge of magnetic reconnection from results in experiments and observations. We focus on the regime where the Lundquist number is extremely high, since we consider collisionless reconnection. According to this diagram, whether single X-line or multiple X-line reconnection occurs depends on $\lambda = L/d_i$, where L is the plasma size (or the length of the current sheet) and d_i is the ion inertia length. That is, single X-line reconnection occurs for $\lambda < \lambda_c$, multiple X-line reconnection takes place for $\lambda > \lambda_c$. It can be said that slow expanding of MHD inflows in our multi-hierarchy simulations means that the length of the current sheet pushed by MHD inflows is small and effectively corresponds to small values of L, i.e., λ in the phase diagram and vice versa. Our results are consistent with the phase diagram of magnetic reconnection from a qualitative viewpoint at least.

4. Summary and discussion

We have reported the first results on analysis of collisionless driven reconnection with a multi-hierarchy simulation model. Our multi-hierarchy simulation model is based on the domain decomposition method. Real space in a simulation domain is divided into three parts: an MHD domain with non-uniform space grids to macroscopic dynamics away from reconnection points, a PIC domain to solve microscopic kinetic physics in the vicinity of the neutral sheet, and an interface domain to commute the MHD and PIC information so as to interlock the two domains. In the interface domain, macroscopic quantities are given by the hand-shake scheme, while shifted Maxwellian velocity distribution is assumed for determining information of individual particles.

By using the multi-hierarchy model, we have investigated how macroscopic dynamics affect microscopic physics of magnetic reconnection. Magnetic reconnection in the PIC domain is driven by plasma inflows from the MHD domain, and dynamical behaviors of reconnection depend sensitively on MHD inflows. We have found that when MHD inflows expand as $v_w \lesssim v_{A0}$, magnetic reconnection with a single X-point is driven and the system is likely to relax to the steady state. In cases in which the expanding speed of an MHD inflow is $v_w \gtrsim 2.0v_{A0}$, reconnection dynamics are intermittent and multiple X-points appear. In the near future, we would like to investigate the influence of MHD inflows on reconnection physics by using larger-scale simulations, in which especially the size in the x direction is larger (the current sheet is longer) for long periods of time. We predict that magnetic reconnection with more than two X-points will be driven by certain conditions in MHD dynamics.

In addition, physics of magnetic reconnection in the presence of a guide magnetic field need to be investigated, since a strong guide field exists in fusion devices such as tokamak. Studies on the influence of a guide field on magnetic reconnection physics also are advancing all over the world. For spontaneous reconnection, internal instabilities such as tearing mode or ballooning mode in the presence of a guide field have been analyzed [23, 24]. For driven reconnection, with PIC simulations, R. Horiuchi *et al.* have found the pressure tensor term is dominant at the reconnection point, even when a guide field is strong [19]. In the next step, we will study hierarchy-interactions for collisionless driven reconnection in the presence of a guide field, first in the slab geometry.

Furthermore, we plan to improve our multi-hierarchy simulation and apply the improved model to sawtooth oscillations. When sawtooth oscillation occurs, stored plasma energy in the core is released to the periphery region. For fusion, sawtooth oscillation needs to be under control. In spite of a long history of researches, the detailed physical mechanism of sawtooth oscillations is not completely clarified. We believe a fluid model alone can not explain the mechanism of sawtooth oscillations, and microscopic kinetic processes play crucial roles in some regions, in particular in the vicinity of the reconnection points. On the other hand, simulations based on the first principle such as PIC models need immense computer resources, and the entire region of a fusion device can not be simulated even by using the "K" computer provided by the RIKEN Advanced Institute for Computational Science, which is one of the supercomputers with the highest performance in the world. As a multi-hierarchy model for studies on sawtooth oscillations, we would employ the idea that an extended MHD code including non-ideal terms such as finite Larmor effects and diamagnetic effects, and a PIC code are coupled in torus geometry. In this improved multi-hierarchy model, the major part, i.e., almost the entire poloidal surface in tokamak would be treated by an extended MHD code , and small parts (for instance, regions in the vicinity of the $q = 1$ surface) would be solved by a PIC code.

Acknowledgments

This simulation work was performed by employing the Plasma Simulator at the National Institute for Fusion Science. This work was partially supported by a Grant-in-Aid for Scientific Research from the Japan Society for the Promotion of Science (Grant No. 23340182, No. 24740374), the Research Cooperation Program on Hierarchy and Holism in Natural Sciences at the National Institutes of Natural Sciences, and the General Coordinated Research at the National Institute for Fusion Science (NIFS13KNTS020, NIFS14KNSS048).

References

[1] Horiuchi R and NSRP members 2011 *Plasma Fusion Res.* **6** 2101055
[2] Kadomtsev B B 1975 *Sov. J. Plasma Phys.* **1** 389
[3] Park H K, Lühmann Jr. N C, Donne J H, Glassen I G J, Domier C W, Mazzucato E, Munsat T, van de Pol M J, Xia Z, and TEXTOR Team 2006 *Phys. Rev. Lett.* **96** 195003
[4] Park H K, Donne J H, Lühmann Jr. N C, Glassen I G J, Domier C W, Mazzucato E, Munsat T, van de Pol M J, Xia Z, and TEXTOR Team 2006 *Phys. Rev. Lett.* **96** 195004
[5] Nicolas T, Sabot R, Garbet X, Lutjens H, Luciani J F, Sirinelli A, Decker J, Merle A, and JET-EFDA Contributors 2013 *Plasma Fusion Res.* **8** 2402131
[6] Ishizawa A and Horiuchi R 2005 *Phys. Rev. Lett.* **95** 045003
[7] Usami S, Ohtani H, Horiuchi R, and Den M 2008 *Comm. Comput. Phys.* **4** 537
[8] Horiuchi R, Usami S, Ohtani H and Den M 2009 *J. Plasma Fusion Res.* SERIES **8** 184
[9] Usami S, Ohtani H, Horiuchi R, and Den M 2012 *Comm. Comput. Phys.* **11** 1006
[10] Usami S, Ohtani H, Horiuchi R, and Den M 2009 *Plasma Fusion Res.* **4** 049

[11] Horiuchi R, Usami S, Ohtani H, and Moritaka T 2010 *Plasma Fusion Res.* **5** S2006
[12] Usami S, Horiuchi R, Ohtani H, and Den M 2013 *Phys. Plasmas* **20** 061208
[13] Sugiyama T and Kusano K 2007 *J. Comput. Phys.* **227** 1340
[14] Moritaka T, and Horiuchi R 2008 *Phys. Plasmas* **15** 092114
[15] Pei W, Horiuchi R, and Sato T 2001 *Phys. Plasmas* **8** 3251
[16] Pei W, Horiuchi R, and Sato T 2001 *Phys. Rev. Lett.* **87** 235003
[17] Li B and Horiuchi R 2008 *Phys. Rev. Lett.* **101** 215001
[18] Ohtani H and Horiuchi R 2009 *Plasma Fusion Res.* **4** 024
[19] Horiuchi R, Usami S, and Ohtani H 2014 *Plasma Fusion Res.* **9** 1401092
[20] Horiuchi R and Sato T 1989 *Phys. Fluids* B **1** 581
[21] Shinohara I, Suzuki H, Fujimoto M, and Hoshino M 2001 *Phys. Rev. Lett.* **87** 095001
[22] Ji H and Daughton W 2011 *Phys. Plasmas* **18** 111207
[23] Lembérge B and Pellet R 1982 *Phys. Fluids* **25** 1995
[24] Pritchett P L 2013 *Phys. Plasmas* **20** 080703

Turbulence and zonal flow structures in the core and L-mode pedestal of tokamak plasmas

L Villard[1], B F McMillan[2], O Sauter[1], F Hariri[1], J Dominski[1], G Merlo[1], S Brunner[1] and T M Tran[1]

[1] Centre de Recherches en Physique des Plasmas, Ecole Polytechnique Fédérale de Lausanne, Switzerland

[2] Centre for Fusion, Space and Astrophysics, Department of Physics, University of Warwick, United Kingdom

E-mail: laurent.villard@epfl.ch

Abstract. Zonal flows (ZF) play a crucial role in regulating ion temperature gradient (ITG) turbulence. In previous global gyrokinetic simulations [1] using the ORB5 code with the adiabatic electron model, it was observed that long-lived ZF structures, leading to a corrugated transport and temperature gradient pattern, could develop in shaped tokamak plasmas much more than in circular shaped plasmas, resulting in reduced transport. These studies are extended to a hybrid electron model in which trapped electrons are kinetic while passing electrons are assumed to have a Boltzmann response for a case dominated by ITG modes. These confirm the results of the fully adiabatic electron model. Simulations done in "gradient-driven" mode, with a Krook-like relaxation towards a given profile, result in non-realistic corrugated heat source/sink profiles. However, after switching off completely the heat source/sink, it is shown that the ZF and transport corrugation remains. Thus the heat source corrugation is merely a consequence, not a cause, of the zonal structures and related radial transport pattern. Considering then core profiles with constant logarithmic gradients and pedestal profiles with linear gradients for L-mode plasmas, as in Ref.[2], we analyze how ITG transport and zonal structures react by independently varying the logarithmic gradients in the core and the linear gradients in the pedestal, using the adiabatic electron model. Results show the presence of large radial zones straddling the core-pedestal transition region. Avalanche-like events propagate over the radial zone at constant speed and repeat with a well defined frequency somewhat below the local geodesic acoustic mode (GAM) frequency. These avalanches are observed on the $E \times B$ ZFs, effective heat diffusivity and heat flux, thus a change of gradient in the core affects transport in the pedestal and vice versa. In spite of these non-local effects, attempt is made to characterize transport, and in particular its stiffness, quasi-locally. Global simulation results show that with increased input power the logarithmic gradient in the core is only slightly increased while the linear gradient in the pedestal is substantially enhanced.

1. Introduction

Thanks to the development of gyrokinetic theory and advanced numerical techniques adapted to high performance computing platforms, the fusion plasma research community has tools to study turbulent transport from first principles [3] with an increasing level of detail and realism. In this paper, we focus on non-local transport effects mediated by zonal flows (ZF) in situations where the dominant primary micro-instability is the ion temperature gradient mode (ITG).

The role of ZFs in regulating ITG turbulence was established some time ago [4, 5]. Let us recall that the ZFs (in this paper we shall deal with $E \times B$ ZFs) are non-linearly generated by turbulence. The ZFs exist linearly in two branches, a zero frequency one, which was shown to be undamped in the collisionless limit [6] and a branch with finite frequency due to geodesic curvature, the geodesic acoustic mode (GAM) [7]. Nonlinearly, the ZFs exhibit another type of dynamics characterized by avalanches and bursts [8]. These avalanches result in intermittent, radially propagating bursts that carry a fraction of the turbulent heat, particle and momentum fluxes.

In global gyrokinetic flux-driven [9] and gradient-driven [1] simulations, it was shown that these avalanches are related to non-local, non-diffusive spatial behaviour, resulting in a corrugated self-organized flow structure with radially modulated temperature gradients and heat transport. It was furthermore shown that the geometry of the magnetic background configuration plays an essential role in this dynamics: plasma shaping, in particular elongation, increases both the steady residual ZF and the GAM damping, resulting in stronger, more persistent ZF structures. Since steady ZFs are more effective than oscillatory ZFs in suppressing turbulence [10], ITG turbulent ion heat transport was shown to be reduced in ITER-shaped plasmas as compared to circular cross-section plasmas [1]. These simulations were performed assuming an adiabatic electron response. Here, we examine what happens to the shaping effect when a more realistic model for electrons is considered, namely for which trapped electrons are kinetic and only passing electrons are adiabatic. Since these results were obtained with "gradient-driven" simulations with a relaxation term to a prescribed equilibrium, the question remains as to whether the observed corrugations were not an artifact of the particular source term used. This question is also addressed in this paper.

All the above-mentioned simulations were performed focusing on tokamak core parameters. For such parameters, it is generally found that turbulent transport is characterized by a threshold logarithmic gradient of temperature above which heat transport increases sharply. In other words, it is found (both in numerical simulations and in experiments) that a large increase in input power results in only a modest increase in logarithmic temperature gradient: this phenomenon has been named "profile stiffness". This explains, partly at least, the degradation of energy confinement time with increasing input power observed in virtually all magnetic fusion devices. Recent experimental investigations [2] on the TCV tokamak have carefully examined this question with an unprecedentel level of detail. In particular, it appears that experimental profiles of temperature and density can be parametrized by a "core" region (excluding the inside of the sawtooth inversion radius) with nearly constant logarithmic gradients and a "pedestal" region (even though these were L-mode plasmas) characterized by constant gradients. It was also found that increasing the input power results in profiles with almost no increase in the core logarithmic gradient whereas the pedestal linear gradients were substantially increased, hence the conclusion that core profiles are stiff while pedestal profiles are non-stiff. The question arises whether such profiles are compatible with the computed transport from global gyrokinetic simulations. As a first step in this direction, we address this question with a set of gradient-driven simulations in an ITG turbulent situation, using the adiabatic electron model. Spatially non-local effects will be evidenced: varying the core logarithmic gradient affects transport in the pedestal and vice-versa. The dynamics of ZFs, exhibiting avalanches that propagate through the pedestal and core regions, will be investigated, and provide a possible explanation for these nonlocal effects. The Krook-like source term used in gradient-driven simulations acts as an effective source / sink of power throughout the whole plasma. In order to obtain realistic power profiles, i.e. with a source localized in the core and a sink localized in the pedestal, specific combinations of core and pedestal gradient parameters are searched. For such profiles, it will be found that stiffness in the core is much higher than in the pedestal.

The remainder of the paper is structured as follows. In Section 2 the global gyrokinetic model is introduced and parameters are described. In Section 3, the core turbulent heat transport using kinetic trapped electron model is computed for an ITER-shaped plasma and compared to a circular plasma of otherwise same profiles. In Section 4, the role of the source term in the ZF and transport self-organization is examined and it is shown that the source corrugation is merely a consequence, not a cause, of the observed ZF, ion heat transport and ion temperature gradient profile corrugation. In Section 5, global ITG simulations are performed with initial profiles characterized by constant logarithmic gradients in the core ($\rho/a \leq\sim 0.8$) and constant (linear) gradients in the pedestal ($\sim 0.8 < \rho/a \leq 1.0$) and the question of core vs pedestal stiffness is examined. Discussion and conclusions are presented in the last section.

2. Global gyrokinetic model and parameters

The basis of our investigations is the gyrokinetic theory in which conservation properties are ensured, at the appropriate ordering level [11]-[13]. Numerical simulations are carried out with the global gyrokinetic code ORB5 [14] based on a Lagrangian particle-in-cell (PIC) approach. The equations for the distribution function of each plasma species σ reads:

$$\frac{\partial f_\sigma}{\partial t} + \dot{\mathbf{R}} \cdot \frac{\partial f_\sigma}{\partial \mathbf{R}} + \dot{p_\parallel} \frac{\partial f_\sigma}{\partial p_\parallel} = \sum_{\sigma'} C(f_\sigma, f_{\sigma'}) + S(f_\sigma) , \tag{1}$$

where $p_\parallel = m_\sigma U + (e/c) J_0 A_\parallel$ is the canonical parallel momentum, J_0 is the gyroaverage operator, U is the parallel velocity, C is the Coulomb collision operator and S is a source term (see below). In this paper, collisions and electromagnetic effects will be neglected ($C = 0$, $A_\parallel = 0$). The system of equations is complemented by the Euler-Lagrange characteristics equations and the quasi-neutrality equation with linearized polarization density in the long wavelength approximation, $k_\perp \rho_s < 1$. Different approximations will be used for the perturbed electron density. The simplest model assumes purely electrostatic perturbations and an adiabatic electron response along the equilibrium field lines $\delta n_e = (en_{0e}/T_e)\left(\phi - \bar{\phi}\right)$ where $\bar{\phi}$ is the flux surface-averaged perturbed potential. Defining toroidal magnetic coordinates (s, θ_*, φ) and noting J the Jacobian we have $\bar{\phi} = \int \phi J \mathrm{d}\theta_* \mathrm{d}\varphi / \int J \mathrm{d}\theta_* \mathrm{d}\varphi$. Trapped electron mode response can be taken into account in a "hybrid" model, in which only passing electrons are assumed to respond adiabatically $\delta n_e^{\mathrm{passing}} = \bar{\alpha_p}(en_{0e}/T_e)\left(\phi - \bar{\phi}\right)$ where α_p is the fraction of passing electrons, while the trapped electrons are assumed drift-kinetic. More details can be found in Refs.[15, 16]

The source term S in Eq.(1) has three purposes. First, it is used as a heat source, so that the temperature gradients are maintained above their marginal stability point over long times. Second, it is used in order to obtain a quasi-steady state, in which not only the fluxes and gradients, but also the entropy, keep a steady time-averaged value [17]. Third, it is used as a noise control scheme, in order to maintain the signal to noise ratio steady (again, in a time-averaged sense) over indefinitely long times [18]. The source term takes the form of a modified Krook operator, $S = -\gamma_K \delta f + S_{\mathrm{corr}}$, in which $\delta f = f - f_0$, f_0 is a time-independent prescribed function of phase space coordinates, γ_K is chosen small enough not to damp too much the physical eigenmodes (typically smaller than $1/10^{\mathrm{th}}$ of the most unstable linear growth rate) and S_{corr} is such that the zonal flow phase space structures are conserved [18]:

$$\overline{\int \mathrm{d}\mathbf{v} S \left(\frac{v_\parallel}{B} - \left[\frac{v_\parallel}{B}\right]_b\right)} = 0 \tag{2}$$

where the overbar means flux surface-average, and the square bracket with subscript b means bounce-average. Moreover, S_{corr} is designed so as to conserve particle density and parallel momentum. There is the option, in addition, to conserve kinetic energy as well, in which case the source term acts merely as a noise control operator; if this option is not used, the source term acts also as a heat source/sink that tend to restore the temperature profiles towards prescribed ones.

Two different profile definitions are used. First, as logaritmic gradient of the poloidal flux ψ:

$$\frac{1}{T}\frac{\mathrm{d}T}{\mathrm{d}\tilde{\psi}} = -\frac{\kappa_T}{2}\Big(\tanh(s_1/2\rho_*) - \tanh(s_2/2\rho_*)\Big) \times\Big(1 - 1/\cosh^2(s_1/0.05) - 1/\cosh^2(s_2/0.05)\Big) \quad (3)$$

with $s_1 = s - 0.025$, $s_2 = s - 0.975$, $s = \sqrt{\tilde{\psi}}$, $\tilde{\psi} = \psi/\psi_a$, ψ_a is the poloidal flux at the plasma boundary. It produces profiles with $\mathrm{d}\ln T/\mathrm{d}\tilde{\psi} \approx \mathrm{const} = -\kappa_T$ over most of the plasma radius ($s \in [0.2, 0.8]$) and going smoothly to zero outside this region. Second, as functions of $\rho_V = \sqrt{V(\psi)/V(\psi_a)}$, with different functional forms in the core and the pedestal:

$$T(\rho_V) = \begin{cases} \min\left(T_0, T_{\mathrm{ped}}\exp\left(-\kappa_T(\rho_V - \rho_{V,\mathrm{ped}})\right)\right) & \rho \le \rho_{V,\mathrm{ped}} \\ T_1(1 - \mu_T(\rho_V - \rho_{V,\mathrm{edge}})) & \rho_{V,\mathrm{ped}} < \rho \le \rho_{V,\mathrm{edge}} \end{cases} \quad (4)$$

where T_0, T_1, $\rho_{V,\mathrm{ped}}$, $\rho_{V,\mathrm{edge}}$, κ_T and μ_T are given input parameters and $T_{\mathrm{ped}} = T_1 + \mu_T(\rho_{V,\mathrm{edge}} - \rho_{V,\mathrm{ped}})$. Density profiles are defined in a similar way, with parameters n_0, n_1, κ_n and μ_n.

Boundary conditions are the unicity condition for ϕ at the magnetic axis ($s = 0$) and $\phi = 0$ at the plasma boundary ($s = 1$), which sets (unrealistically) turbulent field fluctuations to zero there. On the other hand, we do not impose $\delta f = 0$ at the boundary, but instead put back the particles that eventually leave the plasma in a way that conserves particle energy, magnetic moment and toroidal canonical momentum. These somewhat artificial boundary conditions are a weak point of our current approach, especially when considering turbulent transport in the pedestal. More realistic boundary conditions will be considered in future works.

The equilibrium magnetic configuration is taken from a numerical solution of the axisymmetric ideal MHD equilibrium Grad-Shafranov equation obtained with the CHEASE code [19]. Two cases will be considered: an ITER-shaped configuration (details in Ref.[1]) and a TCV reconstructed equilibrium from experiment [2].

We define an "effective" heat diffusivity as $\chi_{\mathrm{eff}} = -\overline{\mathbf{q}_{\mathrm{H}} \cdot \nabla\rho}/(n\mathrm{d}T/\mathrm{d}\rho\overline{|\nabla\rho|^2})$, where $\mathbf{q}_{\mathrm{H}}$ is the heat flux, ρ is a "radial" variable which can be any arbitrarily chosen function of the magnetic flux and the overbar denotes flux-surface average. We define the gyro-Bohm unit $\chi_{GB0} = \rho_{s0}^2 c_{s0}/a$, with a the minor radius on the equatorial plane, ρ_{s0} the ion sound Larmor radius and c_{s0} the ion sound speed with the electron temperature taken at a reference magnetic surface and the magnetic field on axis. The finite size parameter is defined as $\rho* = \rho_{s0}/a$ and an effective size parameter [20] as $\rho*^{\mathrm{eff}} = \rho_{s0}/w$, with w the radial width of the linearly unstable region. The gyro-Bohm power unit is $P_{\mathrm{GB0}} = n_0 m_i \rho_{s0}^2 c_{s0}^3$.

3. ITG transport in shaped plasmas with the hybrid electron model

We consider an ITER-shaped ideal MHD equilibrium [1] with initial profiles defined by Eq.(3), $T_e = T_i$, with varying κ_T, κ_n, keeping $\eta_i = \kappa_T/\kappa_n$ constant at 3.15, and three different system sizes $1/\rho* = 90, 180, 360$. We use the hybrid electron model. Even though trapped electron response is taken into account, for these parameters the dominant micro-instabilities are ITGs. In Ref.[1] a similar study, except for the use of the adiabatic electron model, showed that plasma shaping was favourable for confinement: ion heat transport was much reduced as compared to a circular cross-section plasma with otherwise similar parameters. The nonlinear critical temperature gradient was found higher and stiffness lower.

The present study with the hybrid electron model largely confirms the previous findings. As expected, for a given value of R/L_T the ITG-driven transport is higher when the non-adiabatic

Journal of Physics: Conference Series **561** (2014) 012022 doi:10.1088/1742-6596/561/1/012022

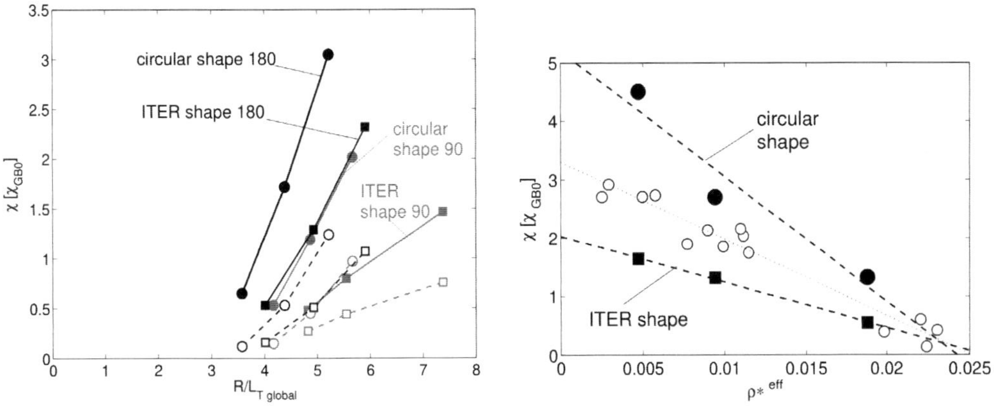

Figure 1. Left: ion (solid lines) and electron (dashed lines) heat diffusivity for an ideal MHD ITER-shaped equilibrium (squares) and a circular configuration (circles) computed with the ORB5 code using the hybrid electron model. The numbers indicate the values of $1/\rho_*$. Right: ion heat diffusivity versus effective normalized size ρ_*^{eff} for the hybrid electron model (filled symbols), circular (circles) and ITER-shaped (squares), $R/L_T = 5$. The open cirles are for the circular, Cyclone case, $R/L_T = 6.9$, with adiabatic electrons, data from Ref.[20] computed with ORB5 and GENE codes. The dashed lines are linear data fits.

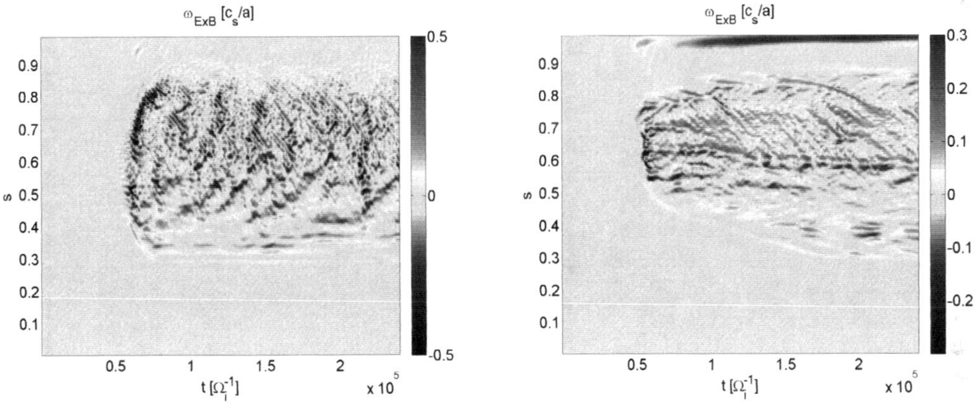

Figure 2. Contours of ZF shearing rate vs radius and time for a circular (left) and an ITER-shaped ideal MHD equilibrium (right), with the hybrid electron model and $1/\rho* = 360$.

trapped electron response is taken into account. Other than that, a systematic comparison with a circular plasma leads to the same qualitative conclusions. Figure 1 (left) shows the effective ion (solid lines) and electron (dashed) heat diffusivity of as a function of R/L_T, for ITER-shaped (squares) and circular (circles) plasmas and two values of $1/\rho* = 90, 180$. For all cases, even though this is basically an ITG situation, there is a finite electron heat transport of about 30%-50% of the ion heat transport. The decrease in both ion and electron heat transport with plasma shaping is evident.

Finite size (finite $\rho*$) effects, computed at a value of $R/L_T = 5$, are shown in Fig.1(b).

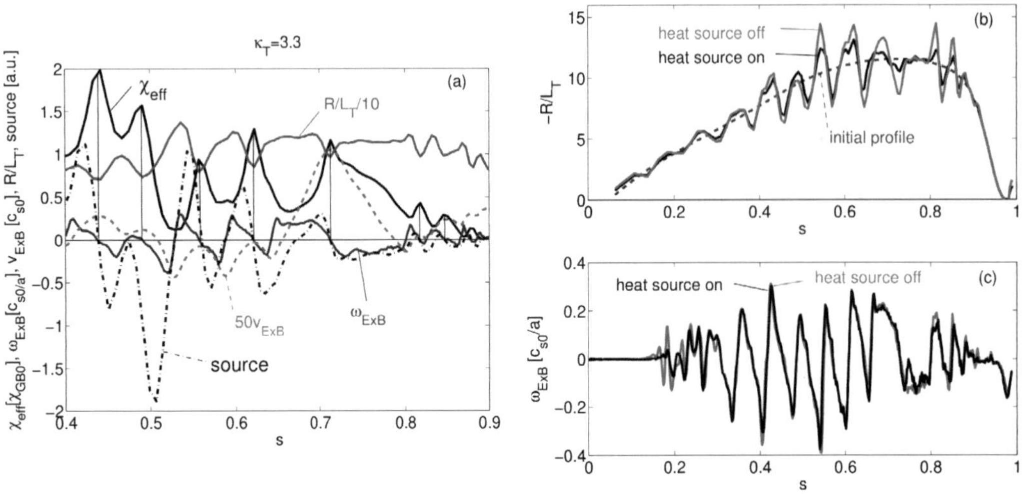

Figure 3. Time-averaged radial profiles of effective ion heat diffusivity $\chi_{\rm eff}[\chi_{\rm GB0}]$, ZF shearing rate $\omega_{E\times B}[c_{s0}/a]$, ZF velocity $v_{E\times B}[c_{s0}]$, logarithmic temperature gradient R/L_T, and heat source/sink [a.u.] (a). The thin vertical lines are drawn to underline the correspondance between local maxima of $\chi_{\rm eff}$, minima of R/L_T, maxima of $v_{E\times B}$ and zeros of $\omega_{E\times B}$. ITER MHD configuration, $1/\rho* = 720$, adiabatic electrons, with modified Krook as heat source and noise control. Time-averaged radial profiles of R/L_T (b) and $\omega_{E\times B}$ (c), with heat source on (black) and off (red). Gyro-Bohm units are defined at mid radius, $\rho_V = 0.5$.

The lower transport for the ITER-shaped plasma is evident at all scales. The results of the Cyclone case (circular, adiabatic electrons, $R/L_T = 6.9$) are also reported on the same graph with open symbols (data from Ref.[20]). For the three cases shown (circular adiabatic electrons, circular hybrid electrons, ITER-shaped hybrid electrons) the strength of the finite $\rho*$ effect is of comparable magnitude. Linear fits through the data (dashed lines) seem to indicate complete turbulence suppression for $\rho*^{\rm eff} \approx 0.025$ [although it is probaly a coincidence that this threshold value is the same for all three cases].

The reduced transport with plasma shaping was attributed to plasma self-organization mediated by ZFs, with more steady ZF in the shaped case as compared to the circular case [1]. The same qualitative behaviour is found with the hybrid electron model: Fig.2 shows contours of the $E \times B$ ZF shearing rate versus radius and time. In both cases avalanches and GAMS are excited, however there is a definitely higher fluctuating component (avalanches) in the circular plasma (left) than in in the ITER-shaped plasma (right).

4. On the effect of sources in gradient-driven simulations

In the previous section, as well as in Ref.[1], simulations have been carried out in "gradient-driven" mode, i.e. with a Krook-like source term that tends to restore the system towards specified initial profiles. These simulations showed a self-organization of the plasma characterized by ZF structures, radially modulated gradients and transport with a corrugated pattern. The temperatures and corresponding gradients can deviate from their original profiles but this results in a heat source/sink that depends on the system evolution. An example is shown in a global ITG simulation of ITER with adiabatic electrons, Fig.3 (a): the time-averaged power heat source shows radial modulations which are clearly correlated to the self-organized corrugations

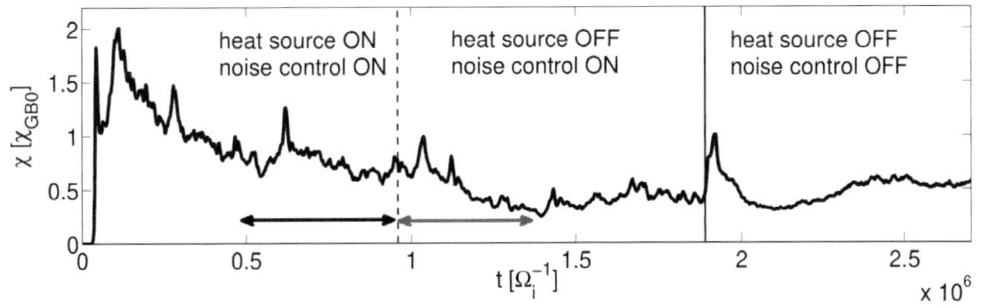

Figure 4. Effective ion heat diffusivity versus time, radially averaged over $.3 < s < .9$. ITER-shaped equilibrium, $\rho* = 720$, adabatic electron model. Gyro-Bohm units are defined at mid radius, $\rho_V = 0.5$. Horizontal arrows mark the time-averaging intervals used in Fig.3.

of ZFs and transport. Hence, it may be suspected that the self-organization is an artefact of the particular "gradient-driven" source model used. We note that the heat source is positive to the left of local maxima of χ (which are also local minima of R/L_T), and negative (i.e. a sink) to the right of these radial positions. This can be explained as follows: corrugations of heat transport lead to local flattening of the temperature profile. On the left side of these flatter regions, the temperature will be lower than prescribed and the Krook term will tend to restore it back, hence a positive heat source. On the right side of these flatter regions, the temperature will be higher than prescribed and the Krook term will tend to restore it back, hence a negative heat source (i.e. a sink).

We have continued the simulation of Fig.3 (a), still with a modified Krook operator but enforcing kinetic energy conservation, effectively switching off the heat source/sink and keeping it as a noise control tool. Our results indicate that, after heat source switch-off, the radial corrugations of the ZFs stay on practically unchanged while the gradient corrugations are even enhanced, Fig.3(b)(c). The effective heat diffusivity corrugation is maintained with a slow gradual decrease over time of the radial averaged χ (Fig.4).

To eliminate the possibility that self-organization would nevertheless result from the Krook-like source term, we have continued the simulation further, switching off the Krook term completely. Again, the ZF, heat transport and gradient corrugation subsists. [However, the absence of noise control is felt, with a signal to noise ratio decreasing over time, preventing us from obtaining physically sound results beyond $t = 2 \times 10^6[\Omega_i^{-1}]$.

This shows without ambiguity that the simulated self-organized structures relating heat transport, Zonal Flows and temperature gradients are not a result of the particular "gradient-driven" source term used. Rather, the observed radial modulation of the heat source is merely a consequence of the temperature profile corrugation, which itself results from the turbulent transport modulation by ZFs. If the heat source corrugation has any effect, it is eventually to reduce the temperature gradient corrugation.

On the other hand, ZFs are essential: in simulations where ZFs are artifically suppressed, the self-organized structures disappear altogether. Due to the self-organization, the local heat diffusivity is not simply a function of the local driving gradient: in Fig.3 we note that χ_{eff} has local *minima* where R/L_T has local *maxima*, and reciprocally. The appearance of these "mini transport barriers" is, again, a consequence of the nonlinear interaction of ZFs with turbulence.

Journal of Physics: Conference Series **561** (2014) 012022 doi:10.1088/1742-6596/561/1/012022

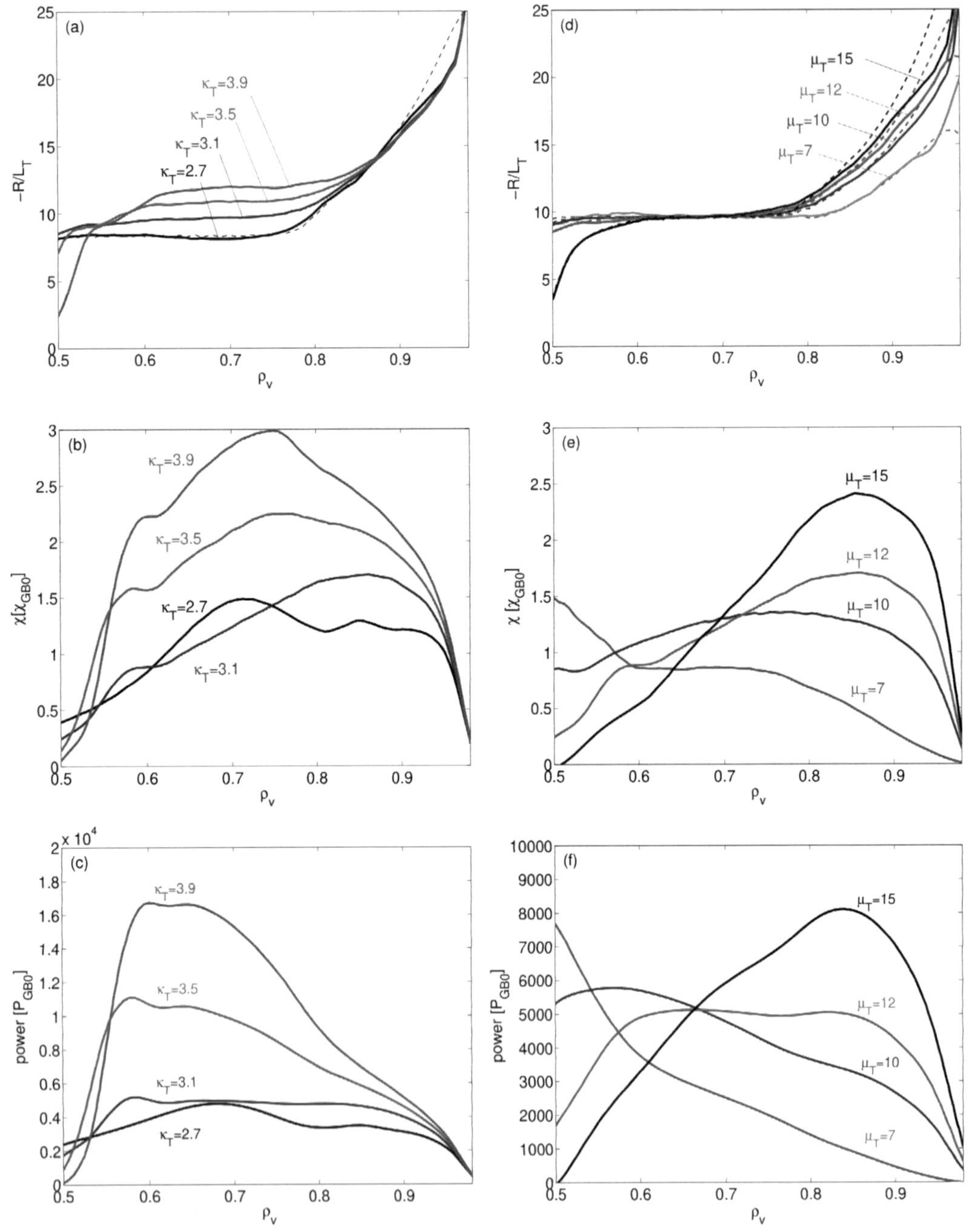

Figure 5. Left column (a)-(c): scan in κ_T at constant $\mu_T = 12$. Right column (d)-(f): scan in μ_T at constant $\kappa_T = 3.1$. Radial profiles of R/L_T (a)(d), effective heat diffusivity χ (b)(e) and power through the magnetic surfaces (c)(f), in gyro-Bohm units defined at $\rho_V = 1$.

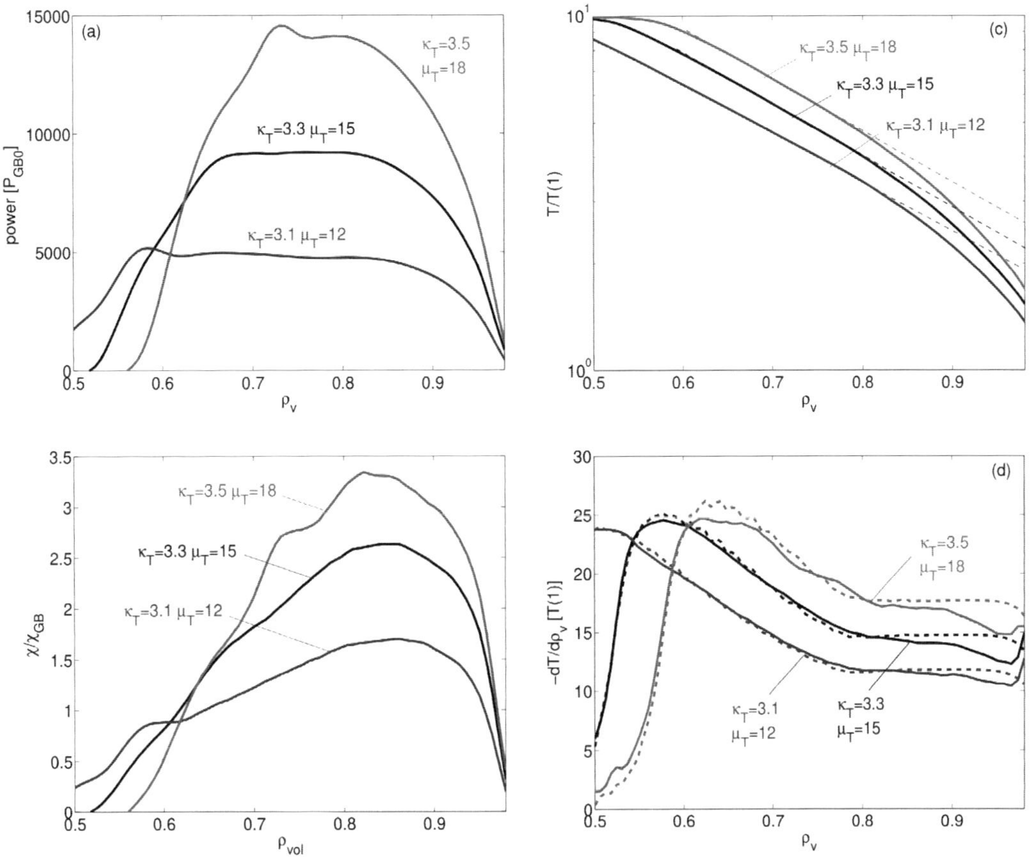

Figure 6. Power through magnetic surfaces $P_{\mathrm{GB0}}]$ (a), heat diffusivity χ $\chi_{\mathrm{GB0}}]$ (b), temperature $T/T(1)$ (c) and $-(1/T(1))dT/d\rho_V$ (d) radial profiles, for (κ_T, μ_T) combinations that give power profiles as flat as possible, i.e. with a corresponding heat source localized well into the core and a heat sink localized in the pedestal near the edge. Gyro-Bohm units are defined at $\rho_V = 1$.

5. Core and pedestal ITG turbulent transport

We now turn to the question of edge and pedestal transport stiffness. Careful investigations [2] conducted on the TCV tokamak in series of experiments have led to the conclusion that, even in L-mode plasmas, a "pedestal" region can be defined, typically covering the outer 20% of the plasma radius, in which temperature and density profiles gradients are constant, whereas in the core region (but outside the sawtooth inversion radius) *logarithmic* gradients are constant. Moreover, power scan studies have revealed that the core logarithmic gradients remain approximately constant whereas pedestal linear gradients srongly depend on the input power. The present study examines whether these findings can be reproduced, at least qualitatively, with global gyrokinetic simulations. For the sake of simplicity, however, the adiabatic electron model will be used and an ITG situation is examined, whereas the experiments were mostly in the TEM regime, so a quantitavive match cannot be expected.

We consider an ideal MHD equilibrium reconstructed from TCV shot #43516 and specify initial profiles as in Eq.(4), with $\rho_{V,\mathrm{ped}} = 0.8$, $\rho_{V,\mathrm{edge}} = 1.0$, $T_0 = 10$, $T_1 = 1$, $n_0 = 5$, $n_1 = 1$,

Journal of Physics: Conference Series **561** (2014) 012022 doi:10.1088/1742-6596/561/1/012022

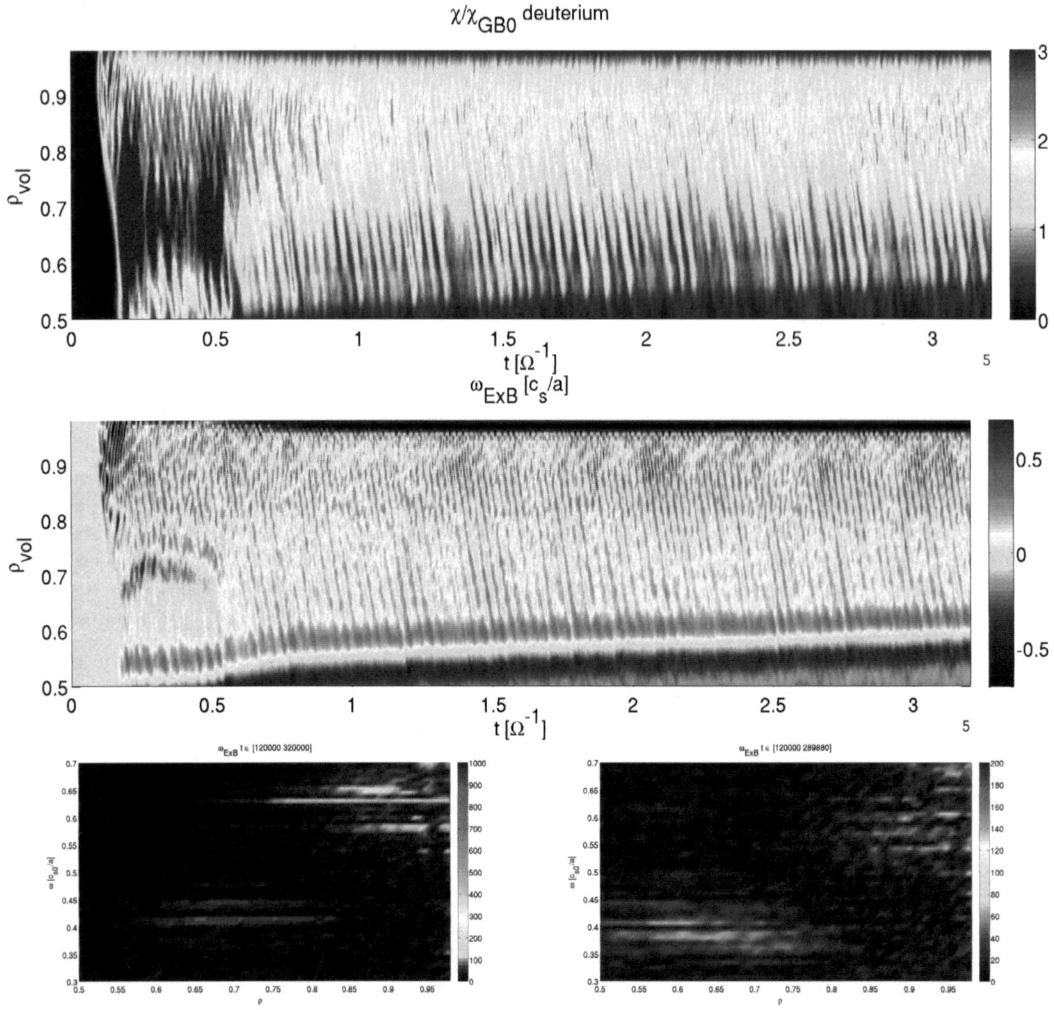

Figure 7. Top: contours of the effective ion heat diffusivity vs radius and time, for $\kappa_T = 3.1$, $\mu_T = 12$. Middle: contours of the $E \times B$ ZF shearing rate, for $\kappa_T = 3.1$, $\mu_T = 12$. Bottom: frequency spectrogram of the $E \times B$ ZF shearing rate vs radius, for $\kappa_T = 3.1$, $\mu_T = 12$ (left) and for $\kappa_T = 3.1$, $\mu_T = 10$ (right).

$\kappa_n = 2.3$ and $\mu_n = 5$. We set $T_e = T_i$ and vary both κ_T and μ_T. The reference surface for ORB5 profile normalization and definition of gyro-Bohm units is taken at $\rho_V = 1$ which, for TCV parameters, leads to $1/\rho* = 245$ [Note that with a reference surface at mid-radius, the $1/\rho*$ parameter is around 100.]. Simulations are performed with 256×10^6 particles, a 3D grid $N_s \times N_{\theta*} \times N_\varphi = 256 \times 512 \times 256$, and a field-aligned Fourier filter with $\Delta m = 5$, which is found appropriate to include all physically-relevant modes with sufficient signal/noise ratio. Note that the field-aligned Fourier filter keeps only modes $m \in [nq(s) - \Delta m \quad nq(s) + \Delta m]$ such that unphysical high $k_{\|}$ modes, which anyway do not satisfy gyrrokinetic ordering, are eliminated [16]. The modified Krook operator, with enforced conservations of density, parallel momentum and ZF residual phase space structure, Eq.(2), is used as a noise control and heat source.

Simulations are run for a long enough time so that the system has reached a quasi-steady state, in the sense that a time-independent power balance is satisfied: we verify that the power heat source integrated over the volume inside any given magnetic surface is equal to the power heat flow through that surface. Typically, we run the simulations for $t = 3.2 \times 10^5 [\Omega_i^{-1}] = 1300[c_{s0}/a]$, with c_{s0} evaluated with the temperature at the plasma boundary. With a reference temperature at mid-radius, this is $t \approx 3200[c_s(0.5)/a]$. In the following, time-averaged quantities are defined over the interval $1.2 \times 10^5 \leq t \leq 3.2 \times 10^5$ $[\Omega_i^{-1}]$.

Fig.5 (left column) shows time-averaged radial profiles obtained with constant pedestal gradient, $\mu_T = 12$, and varying the core logarithmic gradient κ_T from 2.7 to 3.9. Increasing κ_T by a factor 1.26, from 3.1 to 3.9, leads to a heat diffusivity increasing by a factor of about 2, Fig.5(b) and a heat power increasing by a factor of about 3, Fig.5(c). This is the well-known stiffness in the core. We note, however, that the heat diffusivity increases in the pedestal region ($\rho_V > 0.8$) as well, even though the local gradients do not change there. This is a spatially non-local effect: turbulence in the core affects the pedestal region. The case $\kappa_T = 2.7$ is more puzzling, since it leads to comparable core heat diffusivity as for $\kappa_T = 3.1$, whereas diffusivity is lower in the pedestal even though the gradients are unchanged there. This is another manifestation of a spatially non-local effect: turbulence in the pedestal spreads to the core region, lowering the turbulence level in the pedestal and increasing it in the core.

The right column of Fig.5 shows the same quantities as in the left column, however for a fixed value of the core logarithmic gradient $\kappa_T = 3.1$ and varying the pedestal gradient μ_T from 7 to 15. Again, spatially non-local effects are found: transport in the core ($\rho_V < 0.8$) is affected by the change in gradients in the pedestal ($\rho_V > 0.8$), even though the core gradients do not change. The behaviour of input power is remarkably non-monotonic with μ_T: for $\mu_T = 12$, we find that the power profile is mostly flat for $\rho_V \in [0.6, 0.9]$, which corresponds thus to a (positive) heat source localized well inside ($\rho_V < 0.6$) and a heat sink localized in the outer pedestal region ($\rho_V > 0.9$). For μ_T values *both* lower and higher than that, the total input power is larger and the heat source/sink profile is much less realistic. The case ($\kappa_T = 3.1, \mu_T = 12$) thus appears close to what one would obtain in flux-driven simulations, though this needs to be confirmed in future works.

A double scan in (κ_T, μ_T) parameters was then carried out, aiming at obtaining parameter combinations that result in power profiles as flat as possible, i.e. over the widest possible radial interval. The aim is to find a power profile consistent with heat sources well localized in the core and heat sinks close to the edge, as one would expect similar results in truly flux-driven simulations (which are left for future studies). Such cases are shown in Fig.6. Interestingly, the heat diffusivity does not exhibit apparent non-local effects, in the sense that, at every location, χ increases where the temperature gradient increases, and the transport might be interpreted quasi-locally. Concerning stiffness, increasing the power by a factor of about 3 corresponds to an increase in κ_T by only 11% (from 3.1 to 3.5), and an increase in μ_T by 50% (from 12 to 18). Thus the empirical conclusion that "core profiles are stiff and pedestal profiles are non-stiff" appears justified. We note, in passing, the striking similarity of temperature profiles (Fig.6, top right) with those measured experimentally (Fig.3(b) of Ref.[2]), characterized by a clear deviation from constant logarithmic profiles in the pedestal (dotted lines). On the bottom right we show the temperature gradient profile, $-dT/d\rho_V$, with initial profiles as dotted lines and time-averaged quasi-steady profiles as solid lines. For all cases the different radial dependencies in the core and the pedestal are clear. Interestingly, increasing the power leads to an almost uniform increase of the gradients at all radii: if stiffness would be measured by the variation of power vs *linear* gradients (instead of logarithmic), there would be no substantial difference between core and pedestal stiffness.

In spite of the apparent locality of transport in Fig.6, we know from the study in Fig.5 that spatially non-local effects are in fact present. Considering the case ($\kappa_T = 3.1, \mu_T = 12$) as a typical example, we show in Fig.7 that non-local effects show up as avalanches which propagate over a large fraction of the plasma radius that straddles the pedestal and core regions. Most of them propagate inwards and some outwards - though the heat flux associated with avalanches is always outward. Their velocity of propagation is constant, despite the fact that the background temperature varies by a factor of up to 8 over the radial range where they propagate, and their frequency, while close to the local GAM frequency at the edge, is much below the local GAM frequency in the core. Moreover, avalanches are not reflected, and the observed pattern is not that of a global linear eigenmode, even though there is a high radial coherence and a rather well defined frequency constant over radius, $\omega = 0.63[c_{s0}/a]$, see Fig.7 bottom left. In physical units and for TCV parameters this is a frequency of $f = 28.2$kHz, which corresponds very well to the experimentally measured frequency, see Fig.3(a) of Ref.[21]. In our parameter scans, we found that avalanches do not always repeat with a well-defined frequency. It seems that there is a minimal value of μ_T necessary for this to happen. In the μ_T scan of Fig.5 (right column), we find that long-range propagating avalanches are always present, but they have a well-defined frequency for $\mu_T = 12, 15$ and not for $\mu_T = 7, 10$. We show in Fig.7 bottom right the spectrogram for $\mu_T = 10$: no radial coherent mode is present in this case.

6. Conclusions

We have evidenced several spatially non-local turbulent transport effects, for all of which the ZF nonlinear interaction with turbulence appears to play a major role. These are responsible for establishing radial structures modulating the turbulence. First, we have confirmed that these structures are also present, at least in an ITG-dominated situation, when considering a hybrid electron model in which trapped electrons are kinetic and only passing electrons are assumed adiabatic. Moreover, geometrical effects are also present: plasma shaping was shown to lead to more steady ZFs and lower ion and electron heat transport (Figs.1 and 2).

We have then verified that the self-organized structures observed in our gradient-driven simulations were not an artifact of the implied sources and sinks, which also exhibit corrugations. The corrugated structures persist even when heat sources/sinks are removed (Fig.3). The fundamental phenomenon at the basis of radial structure formation, in which avalanches can be present, (self-organization) is the ZF interaction with turbulence. These structures lead to a radial modulation of heat tranport, which then leads to the corrugation of temperature profiles: local *minima* of χ coincide with local *maxima* of temperature gradient, forming a series of "mini transport barriers".

Finally, our studies of profiles with constant logarithmic gradients in the core and constant linear gradients in the pedestal have evidenced that (1) this type of profile appears compatible with global turbulent simulations: sets of parameters were found leading to realistic power deposition profiles (Fig.6). (2) Changing the gradients in the core is affecting turbulence in the pedestal, and vice-versa (Fig.5). These spatially non-local phenomena appear to be related to avalanches mediated by ZFs (Fig.7). (3) In spite of these obvious non-local effects, it seems that transport could be interpreted locally (Fig.6)

Although these results so far do not contradict experimental findings, more work is necessary before claiming that gyrokinetic theory and global numerical simulations are validated. Some more realistic boundary conditions, in particular, should be implemented: currently our model assumes $\phi = 0$ at the edge of the computational domain, leading to an unrealistic vanishing of perturbations there. Flux-driven simulations, with specified input power profile and sinks should be carried out to investigate more realistic profile evolutions. These confinement timescale simulations [22] are very demanding. Also, the study of core and pedestal gradients should be pursued with a more complete physical model and include cases dominated by trapped electron mode (TEM) turbulence. In this context, studies of the core and edge stiffness in the TEM regime including fully kinetic electrons, examining the effect of plasma triangularity, have recently being carried out [23], with results indicating a weaker stiffness near the edge than in the core, similarly to what was found in the present work that dealt with ITG turbulence.

Acknowledgements
This work was partly supported by Eurofusion grant EF-ENR-CRPP-02 and by the Swiss National Science Foundation. Computations were performed on the DAINT and ROSA platforms at CSCS in Switzerland and on the HELIOS platform at IFERC-CSC in Japan.

References
[1] Villard L, et al. 2013 *Plasma Phys. Control. Fusion* **55** 074017
[2] Sauter O, et al. 2014 *Phys. Plasmas* **21** 055906
[3] Garbet X., Idomura Y., Villard L. and Watanabe T.H. 2010 *Nucl. Fusion* **50** 043002
[4] Lin Z., Hahm T.S., Lee W.W., Tang W.M. and White R.B. 1998 *Science* **281** 1835
[5] Dimits A.M. *et al* 2000 *Phys. Plasmas* **7** 969
[6] Rosenbluth M.N. and Hinton F.L. 1998 *Phys. Rev. Lett.* **80** 724
[7] Winsor N. *et al* 1968 *Phys. Fluids 11* 2448
[8] McMillan B.F. *et al* 2011 *Phys. Plasmas* **18** 112503
[9] Dif-Pradalier G. *et al* 2010 *Phys. Rev. E* **82** 025401
[10] Hahm T.S. *et al* 1999 *Phys. Plasmas* **6** 922
[11] Scott B. and Smirnov J. 2010 *Phys. Plasmas* **17** 112302
[12] Hahm T.S. 1988 *Phys. Fluids* **31** 2670
[13] Brizard A.J. and Hahm T.S. 2007 *Rev. Mod. Phys.* **79** 421
[14] Tran T.M. *et al* 1999 *Proc. of the Joint Varenna-Lausanne International Workshop, 1998, Società Italiana di Fisica, Bologna, Italy* 45-58; Jolliet S. *et al* 2007 *Comput. Phys.Commun.* **177** 409
[15] Vernay T. *et al* 2013 *Plasma Phys. Control. Fusion* **55** to appear
[16] Jolliet S. 2009 *"Gyrokinetic particle-in-cell global simulations of ion-temperature-gradient and collisionless-trapped-electron-mode turbulence in tokamaks", PhD Thesis No.4326, Ecole Polytechnique Fédérale de Lausanne*
[17] Jolliet S., *et al* 2009 *Phys. Plasmas* **16** 052307
[18] McMillan B.F. *et al* 2008 *Phys. Plasmas* **15** 052308
[19] Lütjens H., Bondeson A. and Sauter O. 2006 *Comput. Phys. Commun.* **97** 219
[20] McMillan B.F. *et al* 2010 *Phys. Rev. Lett* **105** 155001
[21] de Meijere C.A. *et al* 2014 *Plasma Phys. Control. Fusion* **56** 072001
[22] Idomura I 2014 *Phys. Plasmas* **21** 022517
[23] Merlo G 2014 *Plasma Phys. Control Fusion*, submitted. (This conference).

Institute of Physics
Dirac House, Temple Back
Bristol BS1 6BE UK

CURRAN ASSOCIATES INC
proceedings.com